TRANSACTIONS OF
THE MATERIALS RESEARCH SOCIETY OF JAPAN

TRANSACTIONS OF THE MATERIALS RESEARCH SOCIETY OF JAPAN

TRANSACTIONS OF THE MATERIALS RESEARCH SOCIETY OF JAPAN

Edited by

SHIGEYUKI SŌMIYA

MASAO DOYAMA

MASAKI HASEGAWA

YOSHITAKA AGATA

ELSEVIER APPLIED SCIENCE
LONDON and NEW YORK

ELSEVIER SCIENCE PUBLISHERS LTD
Crown House, Linton Road, Barking, Essex IG11 8JU, England

Sole distributor in the USA and Canada
ELSEVIER SCIENCE PUBLISHING CO., INC.
655 Avenue of the Americas, New York, NY 10010, USA

WITH 42 TABLES AND 221 ILLUSTRATIONS

© 1990 ELSEVIER SCIENCE PUBLISHERS LTD

Softcover reprint of the hardcover 1st edition 1990

British Library Cataloguing in Publication Data

Transactions (Materials Research Society of Japan)
Transactions.—Vol. 1 (1990)—
1. Materials
I. Materials Research Society of Japan
620.11

ISBN-13:978-94-010-6842-0   e-ISBN-13:978-94-009-0789-8
DOI:10.1007/978-94-009-0789-8

Library of Congress Cataloging-in-Publication Data

Transactions of the Materials Research Society of Japan
  edited by Shigeyuki Sōmiya, Masao Doyama, Masaki Hasegawa,
Yoshitaka Agata
      p.   cm.
  Includes bibliographical references.

  1. Materials—Congresses. I. Sōmiya, Shigeyuki. II. Dōyama.
Masao, 1927– . III. Hasegawa, Masaki. IV. Agata, Yoshitaka
TA401.3.A33   1990
620.1′1—dc20   90—49397

To Professor Emeritus Yunoshin Imai

for his encouragement in the establishment and promotion of mutual international understanding and the exchange of ideas.

Preface

The Materials Research Society of Japan (MRS-Japan), formerly the Advanced Materials Science and Engineering Society (AMSES), was established on 16 March 1989 in Tokyo, Japan. AMSES was established following the International Conference on Advanced Materials, held from 30 May to 3 June 1988 in Tokyo (*MRS Bulletin*, October and November 1988). This meeting was similar to the MRS meeting held in Boston, USA, and consisted of 21 symposia, which were published as proceedings in 14 volumes. The number of participants was over 1600.

The first President of AMSES, Professor Masao Doyama, gave the following address:

As advanced technology develops toward its highest goals, a small improvement in existing materials is not enough to meet the demands. The deadlock of advanced technology often brings the invention of new materials.

Human civilization has grown along with materials. The Stone Age, the Bronze Age, and the Iron Age represent the materials most used in those times. Since the beginning of the 20th century, the plastic age, the semiconductor age, the new ceramics age, and the composite materials age have been identified, but no single material dominates.

In addition to the traditional classification of materials (the warp) such as metals, semiconductors, ceramics and organic materials, materials have to be studied by the woof. After World War II, metallurgy hit a deadlock. To overcome the deadlock, metallurgy changed to materials science, absorbing the knowledge of physics, chemistry, chemical engineering, electrical engineering, mechanical engineering, civil engineering, etc., and collaborating with ceramics, semiconductors, and organic materials.

This movement was not successful in Japan because at that time the production of iron and steel in Japan was increasing very rapidly, and the country could not spare a sidelong glance at other fields. Now the iron and steel industry in Japan has reached maturity, and the value of materials science is being rediscovered in Japan. In organic materials, the properties of single molecules reflect those of the entire product. Synthesis is emphasized. Fracture can be treated by itself or in comparison among metals, semiconductors, ceramics and organic, rheology and amorphous materials. The complexity of the problem cannot be adequately addressed by traditional societies.

Process, properties, structure, and environment are the four elements of materials. Without good processes, good materials cannot be made. In the future, the development of materials must be made from a broad perspective and must be useful for mankind.

The Materials Research Society, begun in the United States, has grown rapidly with the strategy of materials science and flexibility. The MRS International Conference on Advanced Materials held in Tokyo, Japan drew 1600 participants and proved a great success. The Advanced Materials Science and Engineering Society of Japan was established by the kind invitation of MRS President R.P.H. Chang through the collaboration of Professor Sōmiya and Dr K. Inoue. This Society is an international society holding a strong connection with MRS.

On the occasion of the foundling of the Advanced Materials Science and Engineering Society, we respectfully request your guidance and warm support.

MRS Bulletin **15**(6) (1989) 29

The International Materials Research Committee was established in September 1989 and AMSES was recognized as one of the founding societies of this Committee world-wide and also as the only society in Japan in November 1989. Since then, many societies related to materials have changed their names and used the abbreviation MRS. The Executive Editors therefore thought it preferable to change the name from AMSES to MRS-Japan, the Materials Research Society of Japan.

It is easier to say MRS-Japan than AMSES, especially in Japanese, and the name MRS is more widespread than AMSES. It was therefore decided to change the name before the publication of these transactions of the Materials Research Society of Japan.

This volume includes the following symposia:

1. Short course by Dr. M.V. Swain
    (8 November 1988, Tokyo)
    Chair: S. Sōmiya

2. Lecture Meeting on Advanced Materials 'Zirconia Ceramics'
    (6 December 1988, Tokyo)
    Chair: S. Sōmiya

3. Lecture Meeting on Advanced Materials
    'Crystal Growth'
    (5 September 1989, Tokyo)
    Chair: S. Sōmiya

4. Lecture Meeting on Hydrothermal Reactions
    (16 November 1989, Tokyo)
    Chair: S. Sōmiya

5. Annual Symposium on Advanced Materials
    (14–15 December 1989, Kanagawa)
    Chair: M. Doyama, S. Sōmiya and M. Hasegawa

Many papers arrived after their deadline or were not written in English. Some will appear in the next volume of the Transactions of the Materials Research Society of Japan.

The papers in this volume reflect the phases of advanced materials. After looking at these papers, we are able to understand the knowledge and level of science and technology in the field of advanced materials. Understanding is the first step in promoting R & D.

We are convinced that this volume will promote understanding and the development of materials science and engineering, as well as promoting international communication among the scientists, engineers, researchers, students, etc, who are involved with materials.

Executive Editors
Shigeyuki Sōmiya (Chief)
Masao Doyama
Masaki Hasegawa
Yoshitaka Agata

Acknowledgements

These Transactions of the Materials Research Society of Japan include the following Symposia. The Editor in Chief expresses his appreciation to the Symposium chairs, without whose efforts the symposia could not have been organised.

1. Short Course by M.V. Swain
    Shigeyuki Sōmiya

2. Zirconia Ceramics
    Shigeyuki Sōmiya

3. Zircon
    Shigeyuki Sōmiya

4. Hydrothermal Reactions
    Shigeyuki Sōmiya

5. Advanced Materials
    Masao Doyama
    Shigeyuki Sōmiya
    Masaki Hasegawa

Support for travel and living expenses for researchers from Australia, China, Korea, Taiwan and Sri Lanka was provided by the Sōmiya Foundation (Tentative), and from the USA and Europe by contributions to Teikyo University from the following companies:

Chichibu Cement Co. Ltd
JEOL Co.
Nippon Soda Co. Ltd
Onoda Cement Co. Ltd
Toshiba Ceramics Co. Ltd

to all of whom I am most grateful.

The secretarial work of the Society was carried out by Mr Yoshitaka Agata, KSP Co., and his colleagues. I wish to say thank you to them, to Rigaka Co. for supporting accommodation expenses, and to all symposium chairs, session chairs, participants and seminar editors.

Shigeyuki Sōmiya
Editor-in-Chief
Transactions of the
Materials Research
Society of Japan

Contents

Advanced Materials

(*Senior Editors*: M. DOYAMA, S. SŌMIYA and M. HASEGAWA

Zirconia and Zircon Ceramics
(*Senior Editor*: S. SŌMIYA)

xvi

R A R E   M E T A L S
< blue skies ahead>

HIDEO KANEKO
Director of the Society of Non-Traditional Technology
1-2-8 Toranomon Minato-ku Tokyo, 105 Japan

A B S T R A C T

This contribution discuss the changing nature of the field related to the
development of novel metals, especially rare earth metals and their
intermetallic compounds. I believe that this will be able to transmit some
of the enthusiasm of materials scientists for the development of advanced
materials.

I N T R O D U C T I O N

This contribution is a general discourse or overview of research and
development of advanced materials, discussing the changing nature of the
field and the development of novel rare earth alloys.
Materials science is entering a new and exciting period and the next
century will bear witness to completely new types of materials. Rare
metals, especially rare earth metals and their alloys are example of these
materials.
Challenges are going on not only in relation to the never ending demand
for superior materials, but also for new and sophisticated materials for
microelectronic,   magnetic,   opto-electronic   and   superconducting
applications. Besides these new materials, you may notice some quite new
manufacturing processes or uses of these materials.
However, this contribution will describe only a few of the many advances
made recently in materials technology, and mention some of the challenges
ahead. It is not intended to be comprehensive, nor to introduce detail,
but rather to capture the spirit of today's renaissance in the field of
materials. Therefore materials experts will find that a number of
potentially important materials have been omitted. This is partly due to

the limitation of pages. But I hope that the materials discussed here are those that particularly excite you and seem most likely to become prominent in the near future.

FUNDAMENTALS OF R & D

The prediction mentioned above in the chapter of introduction are not the results into of gazing a crystal ball, but are based on an appraisal of current research in leading laboratories around the world. However, in discussing the likely materials or processes, it is not enough to consider current research only. In a dynamic world of changing political frontiers, shrinking distances, widening information net-works and growing environmental concern, attitudes to the production and uses of materials will also radically change. This transformation has come about quite suddenly. As you know, it is only a few-years since you were considered to be a very new and specialized field with few applications. What has caused this situation? At least two factors seem to have been involved.
First is an increased need for higher performance materials. The market demand has resulted from increasingly vigorous international competition. This has been the response of materials scientists and engineers to challenges to their creativity.
A second factor is our recently enhanced capability for producing inovative materials with specified properties. For instance, it is possible to build up a structure literally atom layer by atom layer using molecular beam epitaxy. In addition, none-equilibrium concentrations of elements can be introduced by ion bombardment and laser annealing.
Another example is rapid solidification technology. By this method amorphous or fine crystalline materials can be produced. The great advantage of this approach is that the metals has a very high alloy content which can not be obtained by traditional technologies.

CROSS – DISCIPLINARY IN R & D

An important change of the situation would relate to some significant breakthrough in understanding of research and development. Formerly, the inter-disciplinary approach has been evaluated as a fundamental idea, but now a days, this idea is becoming old fashioned, and a new idea the - cross-disciplinary approach taking its place.
The old concept inter-disciplinary means that different fields of specialists co-operated with each other in recognizing the boundary of each field. But what seems to have happened is the cross-disciplinary research development now being conducted in industry. Cross-disciplinary is a new concept, meaning the overlapping of different fields, eliminating their border lines; that is to say a borderless idea. Earlier distinctions between basic and applied, metals and polymer, since and engineering are breaking down and are being replaced by the cross-disciplinary teams forced on an emerging technology.
It is evident that today's highly competitive environment is bringing science and practice closer together then ever before. Consequently, in today's laboratories, it is not unusual to see a theoretical physicist and

3

a technical engineer gathered together in front of a blackboard, discussing production or performance of some new materials. Fig.1 shows the relationship between inter-disciplinary and cross-disciplinary.

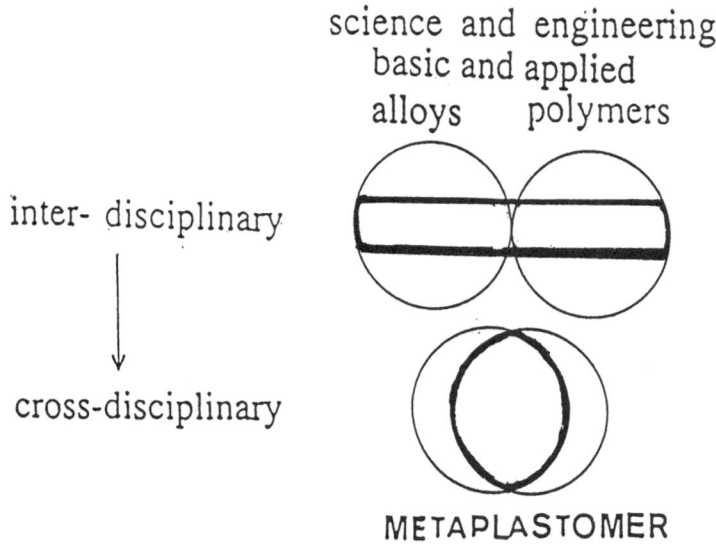

science and engineering
basic and applied
alloys   polymers

inter- disciplinary

cross-disciplinary

METAPLASTOMER

Fig.1   Inter-disciplinary and cross-disciplinary

self-strengthened polymer

{poly-benzimidazole   70 pct   modulus
{rigid rod polybenzthiazole 30 pct   100 GPa

polyazomethine
   density 1.23 g · cc^{-1}
   melts at 520 K
   modulus 160 GPa
   strength 850 MPa
extruded at 500 K to produce tubes

Fig. 2   An example of metaplastomer

4

As a consequence of this new idea. Japanese laboratories and universities are being built in abroad and some industrial plants of foreign countries are being built in Japan. These are good examples of the cross-disciplinary or borderless idea.
A typical example of cross-disciplinary of metals and polymers is metaplastomer , one of which is shown in Fig.2. This new material has the highest specific strength of any kind of material, as can be seen in Fig.3.

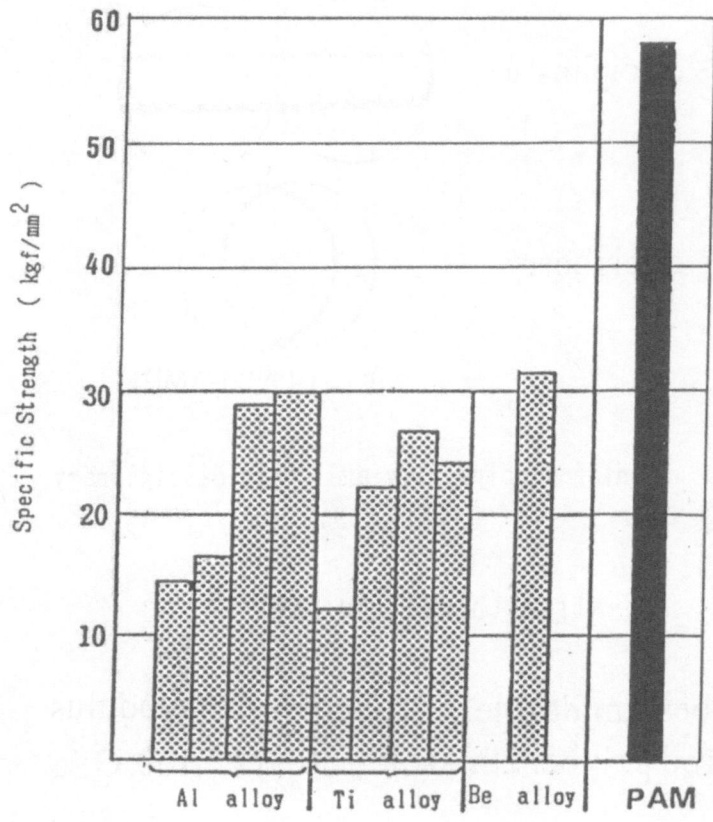

Fig.3   Specific strength of light materials

4 f −M E T A L   B A S E D   M A G N E T I C   M A T E R I A L S

The 3d metal based magnetic materials can now be said to have been older materials, because they have the electron structure shown in Table.1, and therefore are being replaced by newer magnetic materials.

Table 1   Electron structure and magnetic moment of 3d metals

Magnetic Electron

Conduction

Electron

Electron structure										
		1s	2s	2p	3s	3p	3d	4s		
22	Ti	2	2	6	2	6	2	2		
23	V	2	2	6	2	6	3	2		
24	Cr	2	2	6	2	6	5	1		
25	Mn	2	2	6	2	6	5	2		
26	Fe	2	2	6	2	6	6	2		
27	Co	2	2	6	2	6	7	2	} ferro
28	Ni	2	2	6	2	6	8	2	
29	Cu	2	2	6	2	6	10	1	
30	Zn	2	2	6	2	6	10-	2	

Magnetic Moment   (μB/atom)

	Fe	Co	Ni
atom	6	6	5
crystal	2.2	1.7	0.6

what are these newer magnetic materials? It can be said that they are 4f metal-based materials. Then, what are these 4f metals? For what reason were these new materials developed. Here we have an explanation, which initially refers to the reasons and finally to the practical materials available. Though the close relationship between the newer magnetic materials and the periodic table will immediately be revealed.

The 4f metals are generally known as the rare-earth metals, and in a narrow sense, they are the fifteen elements from La, the No.57 in the periodic table, to Lu, the No.71. However, from the practical viewpoint , Sc and Y, which belong to the Ⅲ a group, are usually included in the 4f metals. Then, why can these 4f metals develop as newer magnetic materials? Let's determine the reasons by studying the periodic table.

This can be done by referring to Table.2 to examine the electron structures of the 4f metals, as has been done for the 3d metals. The structures show that the number of electrons is gradually increased on the outer sides of the 4f level with vacancies left on it. Thus, with the rare-earth metals, the 4f-level electrons serve to provide the magnetic properties. The structures given in Table.2 can be represented by the following general expression:

$$\text{(Xe electron shell)} + 4f^n 5s^2 5p^6 5d^1 6s^2$$

When atoms with such an electron structure constitute a crystal, the electrons on the outer two levels, 5d and 6s, will be emitted as free electrons. Consequently, the rare-earth metals are generally trivalent.
The 4f-level electrons, which are responsible for the magnetic properties, are protected by the outer two orbits, 5s and 6p. Therefore, when a crystal is formed, the magnetic properties of the atom are maintained without any change; this is an outstanding feature of the 4f metals. The previously described 3d metals have a disadvantage in that the 3d level is fully exposed to the outside field when a crystal is formed. Fig.4 compares the 4f atom with the 3d by means of model illustrations.

Table 2   Electron structure of 4f metals

		K	L	M	4s	4p	4d	4f	5s	5p	5d	5f	5g	6s	6p	6d	6f	6g	6h
55	Cs	2	8	18	2	6	10		2	6				1					
56	Ba	2	8	18	2	6	10		2	6				2					
57	La	2	8	18	2	6	10	0	2	6	1			2					
58	Ce	2	8	18	2	6	10	1	2	6	1			2					
59	Pr	2	8	18	2	6	10	3	2	6	0			2					
60	Nd	2	8	18	2	6	10	4	2	6	0			2					
61	Pm	2	8	18	2	6	10	5	2	6	0			2					
62	Sm	2	8	18	2	6	10	6	2	6	0			2					
63	Eu	2	8	18	2	6	10	7	2	6	0			2					
64	Gd	2	8	18	2	6	10	7	2	6	1			2					
65	Tb	2	8	18	2	6	10	9	2	6	0			2					
66	Dy	2	8	18	2	6	10	(10)	2	6	(0)			2					
67	Ho	2	8	18	2	6	10	(11)	2	6	(0)			2					
68	Er	2	8	18	2	6	10	(12)	2	6	(0)			2					
69	Tm	2	8	18	2	6	10	13	2	6	0			2					
70	Yb	2	8	18	2	6	10	14	2	6	0			2					
71	Lu	2	8	18	2	6	10	14	2	6	1			2					

Magnetic Electron

Conduction Electron

Unpaired Electron

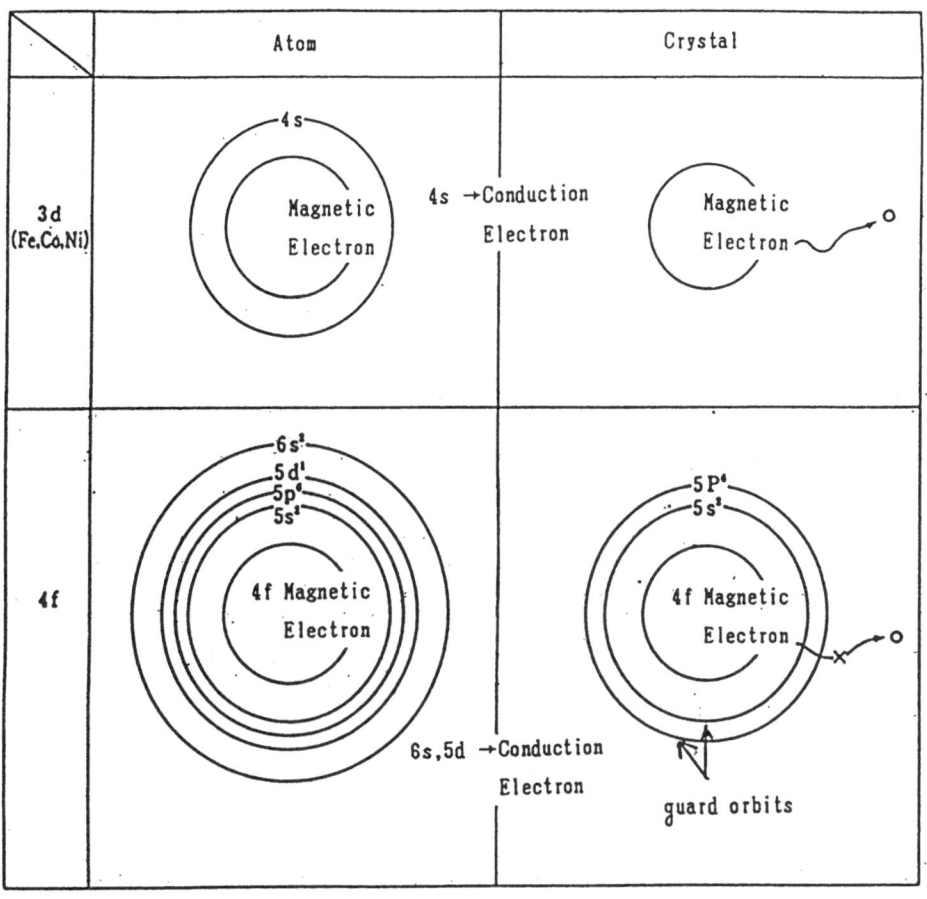

Fig.4   Model illustration of 3d and 4f atoms

Now, let's discuss the magnetic properties of the atom of an rare-earth
metal. As stated above, the 4f-level electrons are protected by the outer
two orbits, 5s and p, and thus, stable as if they are in a castle equipped
with outer and inner moats. Consequently, the 4f-level electrons generate
a magnetic moment through the orbital motion (L) and spin (S). The total
magnetic moment (J) for an atom is the vector sum of (L) and (S) as
expressed by the following equation:

$$J = L = S$$

        where, when n<7, the negative sign must be taken, while , when
n>7, the positive sign must be used, and "n" is the number of electrons.

Table 3   Magnetic moment of 4f atoms

and 4f metals

	Atom	Crystal
La		
Ce	2.5	2.5
Pr	3.5	3.4
Nd	3.6	3.3
Sm	0.8	1.7
Eu	7.9	8.4
Gd	7.9	7.9
Tb	9.7	9.7
Dy	10.6	10.6
Ho	10.6	10.9
Er	9.5	9.9
Tm	7.5	7.6

Table.3 gives comparison with those for the crystals of rare-earth metals. As can be seen from this table, the value of the magnetic moment for the rare-earth metal crystal is equal to the value of the magnetic moment for the atom, and the values of magnetic moment themselves are very large.

Now, let's go back to Table.1 for comparison. It will be found that the atom of the 3d metals provide large values of magnetic moment, however, with the crystals of the 3d metals, the vales of magnetic moment are very small. Thus, as the base metal of magnetic material, the rare-earth metals are the 3d metal based magnetic materials for practical use are being developed one after another to play a leading role in tommorow's industries.

The above description emphasizes only the advantages of the rare-earth metals, however, any advantage is always accompanied by some disadvantage. Recognizing the drawbacks will serve to extend the advantages. Thus, let's discuss the Curle or magnetic transformation points for the rare-earth metals.

Fig.5 shows magnetic transformation points and the effective magnetic moments. From the figures, the following drawbacks can be found. First, the transformation points are all low; that for Gd. which is the highest, being as low as near room temperature. This is a drawback as the practical material. Secondary, from the viewpoint of practical use, there is no good balance between the magnetic transformation point and the magnetic moment. This is another shortcoming.

Thus, the rare-earth metals offer both advantages and disadvantages. What is the measure to be taken to cover the disadvantages and extend the advantages? Solving these problems will provide the key to the development of practical materials, and the solution will be stated in the next section.

One way to extend the advantages and eliminate the disadvantages of the 4f metals is to combine 4f metals (R) with 3d metals (T). The R and T metals can form a wide variety of intermetallic compounds, as follows: (1) (4)

$$R_3T, \ R_4T_3, \ RT, \ RT_2, \ RT_3, \ R_2T_7, \ R_6T_{23}, \ RT_5, \ R_2T_{17}, \ and \ RT_{12}$$

The magnetic moments acting between the R and R, R and T, T and T metals provide various characteristics. In other words, the exchange interactions occurring between the R and R, R and T, T and T metals in a compound vary in behavior, depending upon the combination. The magnetic moments acting between the R and R, R and T, T and T metals offer various behavior; parallel and anti-parallel, or ferro, ferri, and anti-ferro. Consequently, the materials produced offer a variety of magnetic properties which results in availability of various magnetic materials for use in a variety of applications. Thus, the 4f metal-based functional materials for use in electromagnetic applications are replacing the 3d metal-based ones. It must be emphasized that such replacement can be derived from the periodic law.

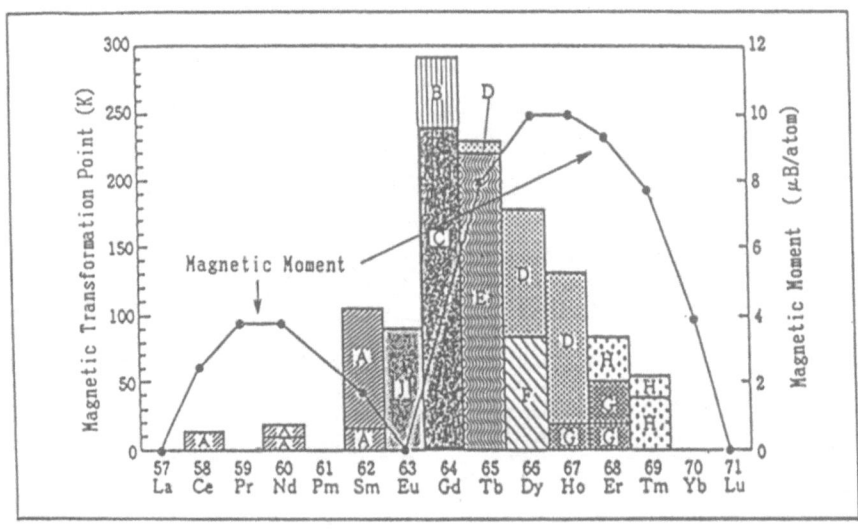

Fig.5   Magnetic transformation point and magnetic moment of 4f metals

f - d   A L L O Y S   A N D   H E A V Y   F E R M I O N

One of the new and exciting theories of f-d alloys is the theory of heavy electron and valance fluctuation. The special types of behavior of electronic and magnetic properties of f-d compounds are the reflection of the characteristic features of the f electronic states. A clear-cut example of heavy electron behavior is found in compounds containing f-element. Among these, Ce, Yb and U are the most common elements forming such materials. It is these same elements which are known to be capable of exhibiting the Kondo effect in dilute alloys and the high temperature properties of heavy electron systems appear to be those of a collection of independent Kondo elements. These compounds are called the dense Kondo alloy system. The rare-earth compounds including the nearly trivalent Ce or Yb ions frequently exhibit the dense Kondo behavior. In the magnetic part of the electric resistance, the logarithmic temperature dependence reveals in the high temperature range above the Kondo temperature Tk, where the 4f electrons behave as the isolated ions. At low temperature below Tk, the resistance behaves proportional to T^2. The specific heat coefficient of dense Kondo compounds are extremely enhanced, which indicate the high states of density at Fermie surface. Therefore, various aspects of the magneto-acoustic effects of valance fluctuation compounds with the unstable 4f electronic state should be studied. And the de Haas-van Alphen effects due to the cyclotron motion of the conduction electron, focusing on the properties of the heavy electron, should be also studied.

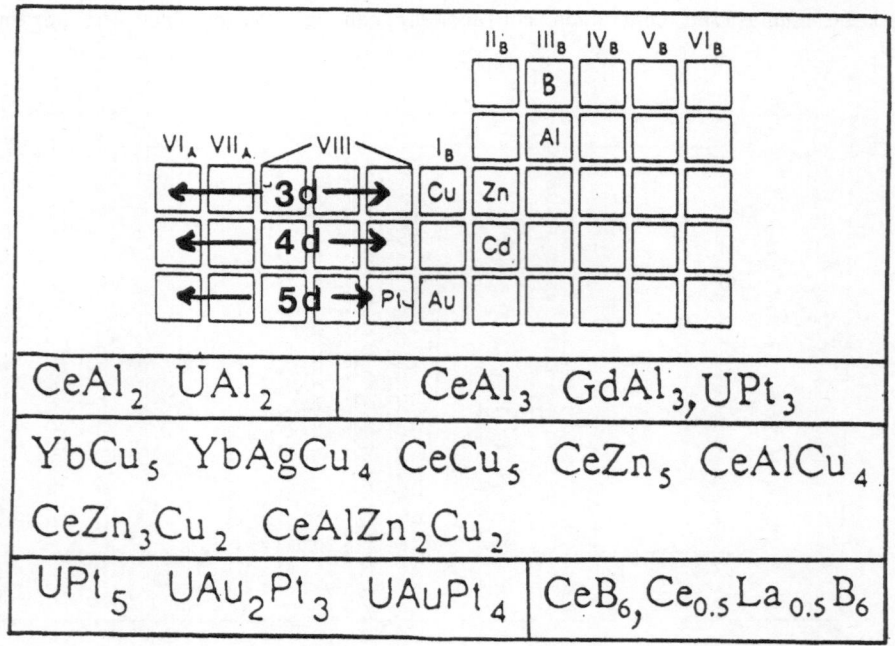

Fig.6   Dense Kondo alloys

In Fig.6 I show where binary heavy electron compounds are found in the periodic table. Elements near and to the right of the border of d elements from such compound with f elements. Compilation of these heavy materials suggest that certain crystal structure are better suited to the forming of a heavy electron ground state. (2)

$CeAl_2$ and UAl_2 are examples of C15 cubic Laves phase sructure. The $CaCu_5$ structure is a simple hexagonal structure. $CeCu_5$ and $CeZn_5$ are examples of this structure. $CeZn_3Cu_2$ is also found in it. What is interesting is that substituing Al to from $CeAlCu_4$ and $CeAlZn_2Cu_2$ leads to very large specific heat coefficient. In a number of other materials, heavy electron behavior has been found. The examples of these materials are shown in Fig.2. (3) (4)

C O N C L U S I O N

I trust that I have been able to transmit some of my personal enthusiasm about my field, and some sense of the incredible number and diversity of challenging opportunities it now presents. Many if not most of the technologies will depend, for the foreseeable future, on rare metals. I believe there are blue skies ahead for rare-earth metals.

R E F E R E N C E S

1. Ning Y.T., The prediction and synthesis of some new intermetallic compounds. J.Less-Common Metals, 1989, 147, 167-173.
2. Fisk Z., Heavy electrons, new materials. J.Mag.Mag.Mat., 1988, 76/77, 637-641
3. Goto T., Magneto-Acoustic effects on heavy electron. J.Mag.Mag.Mat., 1988, 76/77, 305-311.
4. Schultz L., Coercivity in $ThMn_{12}$ type magnets. J.Appl.Phys., 1988, 64, 5711-5713.

Development of the Electron Microscope in Japan

Kazuo Ito
JEOL Co. Ltd.
3-2-1, Musashino, Akishima,
Tokyo, Japan

Abstract:

This article is about how Japan has attained today's top position in electron microscope manufacturing. My observation is that it is simply because of

(1) cooperative, and at the same time, competitive, alliances among universities and industries for R&D, and
(2) from industry's side, continuous and determined efforts to reach the theoretical limit, disregarding the short term efficiency.

Since the first electron microscope, showing higher resolution than the optical microscope, was constructed by E. Ruska and B. von Borries at Siemens more than a half century ago, and due to its high resolution and analytical capabilities, the electron microscope has become the essential research tool for the biological and material sciences.

During the same time, as is well known, Japan became its largest manufacturer country, and perhaps less well known, also its largest user country in the world.

According to our estimation, in 1988 alone, more than 2200 units of the electron microscope were produced in the world, excluding Communist countries: about 500 units of transmission microscope and about 1700 units of scanning microscope. In this estimation, small micro-scopes such as simple scanning and the biggest ones such as 1000kV microscopes, are both counted as units. Of

these numbers, about 70-80% were manufactured, and 40%
installed, respectively, in Japan.

Why and how could Japan reach such a position?
Though there are many products in which Japan occupies
the leading position in manufacturing today, the high
share of the world's electron microscope market is
quite unique compared to other sophisticated scientific
instruments.

In this paper I would like to analyze briefly the
background of what happened in this half century, by
using mostly firsthand facts, due to my privilege of
being one of the founders of JEOL Co. Ltd.

1) Japan was an early starter, though not an inventor.

According to the record, as soon as the report of E.
Ruska and B. von Borries[1] arrived in Japan, it had a
strong impact on the Japanese scientific community. I
was too young at that time, and therefore, I was not a
member of that circle. However, from my point of view,
it is unbelievable that such an ambiguous photograph of
coli bacterium created such enthusiasm in Japan, which at
that time was embittered and poverty-stricken by the war.

The 37th subcommittee was created quickly the next
year, 1939, in the Japan Society for the Promotion of
Science. This subcommittee consisted of 15 members from
universities, national research institutes and research
centers of big industries such as Hitachi and Toshiba,
etc. The research project of this subcommittee was to

study and construct the electron microscope and to
promote its application in three years with a budget of
80,000 Yen. Taking into consideration the inflation
rate, this is roughly equivalent to one million of
today's dollars. The effect of this encouragement was
really astonishing. When the 2nd World War ended in
1945, nearly 20 electron microscopes were working
throughout Japan. The level of these instruments and
their applications were far behind compared to Germany's
and the U.S.'s. However, the basic knowledge was there
which was necessary to appreciate the abundance of
information which was pouring in by much world literature.

It is very interesting to see that the example of a
cooperative R&D program existed even at that time.

When the original JEOL company was created in 1946
for the sole purpose of producing the electron microscope, we did not know of the existence of this official
organization. However, JEOL was certainly supported
during this most difficult time by the continuing
enthusiastic atmosphere of the Japanese community.

2) <u>Though we started from an inferior position, we have
   continued to reach toward the highest theoretical
   limit of resolution, even when we could not find any
   real application for it.</u>

When the war ended in 1945, I believe that Siemens'
microscope had already reached a resolution of 10-20Å. I
was able to witness this fact with my own eyes when I was

fortunate enough to attend the 3rd International Conference of Electron Microscopy in London in 1954. The first introductory report was presented by B. von Borries, who was one of the inventors of the electron microscope, but, unfortunately, he died a few years later of a head tumor. Fig. 1 and Fig. 2[2] are the copy of his first presentation. It is interesting that 12 manufacturers existed in the world at that time, including 4 Japanese, one of which was ours, Japan Electron Optics. As a matter of fact, the name of JEOL comes from our original name, Japan Electron Optics Laboratory Co., Ltd.

The point I would like to mention is that only Siemens' Elmiskop showed a resolution of about 10Å, and no others. And this was clearly demonstrated by the crystal lattice photograph of phthalocyanine taken and published by Menter[3] which was shortly taken with Siemens Elmiskop 1. This was the opening of the new era of the high resolution microscope and I would like to follow the development of the resolution up to today, using this same phthalocyanine. The molecular structure of the phthalocyanine is shown in Fig. 3. In this special case, the center position is Cu and sixteen marginal CH bonds are replaced with C-CL bonds. When this material is vacuum condensed on the cleavage surface of sylvin single crystal, the epitaxial growth gives rise to thin crystallites just as shown in Fig. 4, which are suitable for electron microscope observation. The first photograph, Fig. 5, shows the above mentioned materials, which is similar to that published by Menter, though we took it several years later. This reveals the line structure of 13Å separation, but no details concerning the atom configuration. Next Fig. 6 does show the diamond-shaped molecular image though still does not reveal the atom structure.

Fig. 7 begins to show the atom structure and Fig. 8 clearly illustrates the position of Cu at the center and 16 CL atoms. Photographs from Fig. 5 to Fig. 8 were obtained by N. Uyeda, T. Kobayashi, and E. Suito[4], together with JEOL's people. The time span from Fig. 5 to Fig. 8 was nearly 30 years, and the resolution of the electron microscope which took Fig. 8 was 1.4Å. This means that the improvement of the resolution from 10Å - 1.4Å needed 30 years of continuous effort.

Unfortunately, we don't see the same kind of effort from the front runners' group of 1954. This is somehow quite understandable and reasonable because of the difficulty of making thin film material and because of the wider use of the electron microscope for biological applications, higher resolution than 10Å was considered unnecessary for many years. But when the time finally came for direct atom observation, which should be the ultimate goal for the electron microscope, the most ready instrument was Japanese.

I heard that when Sir Hillary was asked why he climbed Mt. Everest, his answer was because it was there. The same philosophy could also apply to the world of science.

A few examples of the Atom image are a single gold crystal in 110 orientation containing a stacking fault shown in Fig. 9 and YBa Cu high Tc Super Conductor containing the local defect shown in Fig. 10.

3) The concept of the analytical electron microscope was established much earlier in Japan than anywhere else.

Because of the influence from the electron diffraction group which had been traditionally very powerful in Japan even at the early stage, the electron microscope was considered to be an extension of the diffraction instrument, as the optics theory implied. Many researchers joined from the electron diffraction field. For them the information obtained with the electron microscope was complementary to diffraction's, and this was the prime reason for their interest in electron microscopy. So naturally, all techniques utilized in the diffraction field were to be transplanted into the electron microscope.

It was not easy for the manufacturers, because the demand for the specimen stage was quite different in both cases and these analytical capabilities in many cases contradicted the conditions needed to obtain high resolution.

However, we had to accept the market need and just had to try. Now we say very often that, here in Japan, we have to accept the principle of "the Customer is god." It seems that even in the electron microscope this philosophy pushed us to modify the instrument accordingly, and inevitably this was much more complicated mechanically, however, it eventually resulted in final success. It should be noted that this happened even under the conditions of:

   a) biological market, which did not require any complications besides high resolution and low contamination, has always been much larger that the material market;

   b) thinning bulk material, good enough for the transmission electron microscope was not possible until a later day.

Fig. 11 shows the advertisement of our microscope made in the French Journal in 1954. It said besides high resolution, the capabilities of transmission as well as of reflection, together with diffraction and in-situ specimen heating up to 1,000°C. Compared to Fig. 12, which was the Siemens' advertisement of the same time, both were already very different instruments.

Normally the electron microscope means the transmission microscope. In that case then, as I said repeatedly, the bulk material could be studied only by replica and not the material itself, evidently the diffraction of which was not obtainable. However, in the book another type such as reflection method was also mentioned, though not very commonly utilized. I thought perhaps the combination of this method with routinely used reflection diffraction might be very useful for material research and started to market the first commercial instrument, as shown earlier in Fig. 11. Fig. 13 shows an example of the result obtained with this instrument[5]. We could have a surface image (a) with diffraction (c) which looks more interesting than the replica image (b) of the same sample. However, this was not successful at that time because the reflection method required another difficult specimen preparation. Practically, the angle of incidence has to be very small, less than 8 degrees, meaning that the surface of the studied sample must be flat by atom scale to obtain useful results, which was out of reach at that time.

Long after, this method was beautifully used for the study of a single crystal of silicon by K. Yagi[6], an example of which is shown in Fig. 14. The specimen is Si(111)-7x7. One step corresponds to one atom layer of 3.1Å and a screw dislocation is pointed to by an arrow. It is quite true that new ideas become really useful only at a much later time.

Fig. 15 is a series of photographs of heating experiments taken by N. Takahashi et al.[7]. The specimen is the evaporated film of 50-50 Al-Cu alloy of a bout 500Å thickness. Together with diffraction, we could follow the phase transformation of this material, though we could not fully analyze the image contrast. These were the trials to accommodate electron microscope and electron diffraction in one instrument. The real further step for the electron microscope to be an analytical instrument was realized by the successful combination of scanning capability in the transmission electron microscope.

The modern electron microscope has a very strong objective lens, so that the specimen is deep inside the magnetic field, and the upper half of the objective lens before the specimen acts as a strong condenser lens, as illustrated in Fig. 16. By combining with the condenser lens sitting above the objective lens, one can control the size of the illumination from a minimum of 10Å to the whole area of the specimen. With the addition of a scanning coil vertical to the beam axis, one can scan the finest beam across the specimen. Now the only remaining problem for realizing the scanning image is how to collect the secondary electrons created from the sample. Fortunately, the energy of the secondary electron is so small that they are confined along the axis spiralling upward with 100% efficiency due to the strong magnetic force which can be collected by the photomulitplier tube by easily adding the proper potential. This device was patented by us[8] and we started to commercialize the analytical electron microscope, the schematic diagram of which is shown in Fig. 17. Around the specimen, sitting at the center, there are detectors for the secondary electron image (SEI), backscattered image (BEI), two detectors for the energy dispersive X-ray spectrometers, high angle EDS and ultra thin window EDS, and at the bottom, a fluorescent screen for the transmission electron microscope image (TEM image) and several types of diffraction, such as selected area diffraction (SAD), micro beam diffraction (MBD) and convergent beam diffraction (CBD), and scanning transmission electron microscope (STEM) and electron energy loss spectrometer (EELS). TV monitor is also applicable. Some applications are shown in the following figures. Fig. 18 is three photographs of copper oxide powder; (1) secondary electron image, (2) transmission image and (3) replica with transmission. Fig. 19 is the EDS image together with the TEM. The above is the Fe distribution along the line scan and the lower shows the Fe distribution image. Fig. 20 is the EELS image of Boron Nitride.

All these capabilities together with computer image processing, are the modern electron microscope, which we have created step by step in these 30 years.

4) Close cooperation network has been established with
   worldwide top scientists.

In order to develop the instrument by the philosophy of "Market is the God," we have to know the market. In this respect we have had constant good relations with many electron microscopists throughout the world, to whom I express my sincere appreciation. Some I have referred to in this paper, and though I cannot mention all of them, just to show that we are operating inter- nationally, I would like to mention some who are non- Japanese and who personally gave me useful advice and opportunities in the new world: Prof. B. von Borries, Prof. J. Trillat, Prof. G. Dupouy, Sir. P.B. Hirsh, Prof. J.M. Cowley, and Prof. G. Thomas.

5) Conclusion

In these past 50 years, the electron microscope has been metamorphosed to the very complete analytical instrument. The first instrument we manufactured is shown in Fig. 21 and one of the latest, 1,000 kV electron microscope, in Fig. 22. The difference is really unbelievable. Now, how about the future. Though we have nearly reached the limit of resolution, about 1.0$\overset{o}{\mathrm{A}}$, there are still many demands being presented for the application of electron beam, either as an analytical tool or as a fine manufacturing machine. I personally believe that there will be no time yet to be able to rest on past accomplishments.

Reference

1. B. v. Borries, E. Ruska:
    Wiss. Veroff. Siemens-Werk., "Vorlanfige Mitteilungen
    uber Fortschritte im Bau und in der Leistungen des
    Ubermikroskopes", 17-1, 99 (1938).

2. B. v. Borries: The physical situation and the perform-
    ance of high-resolving microscopy using fast
    corpuscles. Proc. International Conf. of Electron
    Microscope, (London) 1954, 4-25

3. J.W. Menter: The direct study by electron microscopy
    of crystal lattices and their imperfections:
    Proc. Roy. Soc. A 236, 119-135 (1956)

4. N. Uyeda, T. Kobayashi, E. Suito, Y. Harada and
    M. Watanabe:
    Direct observation of phthalocyanine molecules in
    epitaxial films. Proc. International Conf. of Electron
    Microscopy Grenoble (1970) Vol. 1, 23-24

5. K. Ito, T. Ito and M. Watanabe:
    An improved reflection type electron microscope. J.
    Electron microscopy Japan, 2, 10-14 (1954)

6. K. Yagi: Observation of surface with ultra clean
    vacuum electron microscope (in Japanese): Electron
    microscopy, 16, 128-135 (1981)

7. N. Takahashi and K. Mihama:
    Etude par microdiffraction et microscopie electroniques
    et la transformation des alliages Al-Cu due au
    chauffage dans le vide. Acta Metallurgica, Vol. 5,
    March, 159-168, (1957)

8. H. Koike and K. Uyeno: Electron microscope: Japanese
    Patent 983332 applied on April 18, 1970.

Fig. 1 and Fig 2   From the introductory paper of B. von Borries at the International Conference in London, 1954.

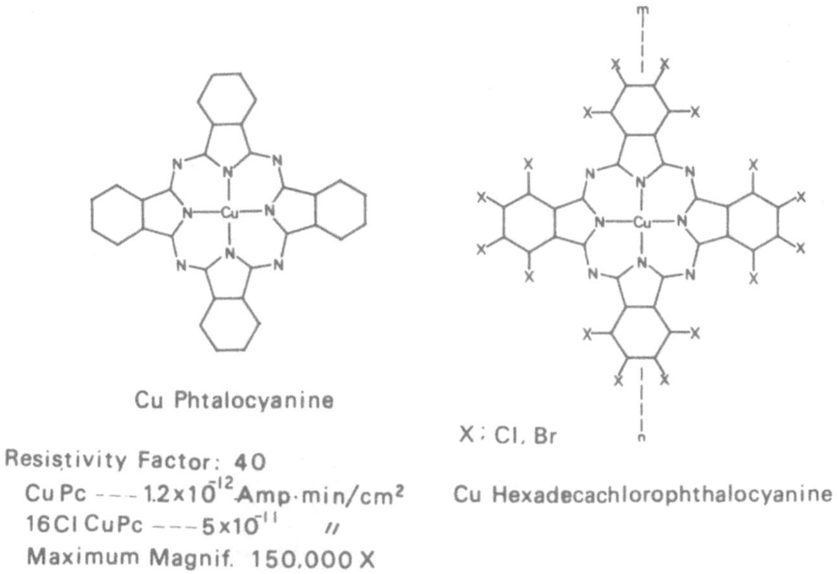

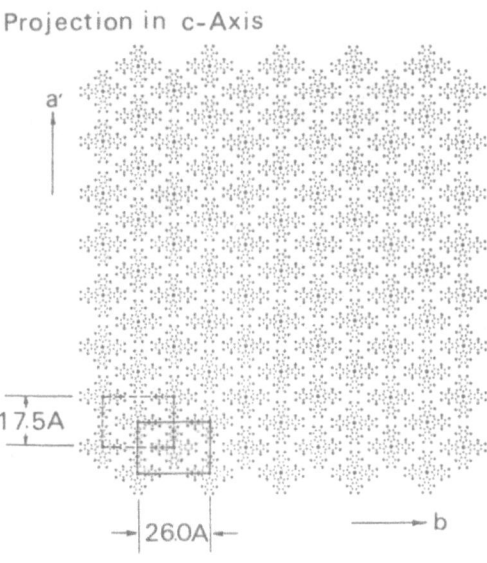

Cu Phtalocyanine

Resistivity Factor: 40
Cu Pc ----1.2x10^{12}Amp·min/cm^2
16Cl CuPc ----5x10^{11} //
Maximum Magnif. 150,000 X

X: Cl, Br

Cu Hexadecachlorophthalocyanine

Fig. 3   Molecular structure of Cu-Phthalocyanine.   From
Fig. 3 to Fig. 8, courtesy of N. Ueda.

Projection in c-Axis

a'

17.5A

26.0A

b

Fig. 4   Crystallized phthalocyanine observed in the
electron microscope.

Fig. 5   Cu-phthalocyanine in the 1950s

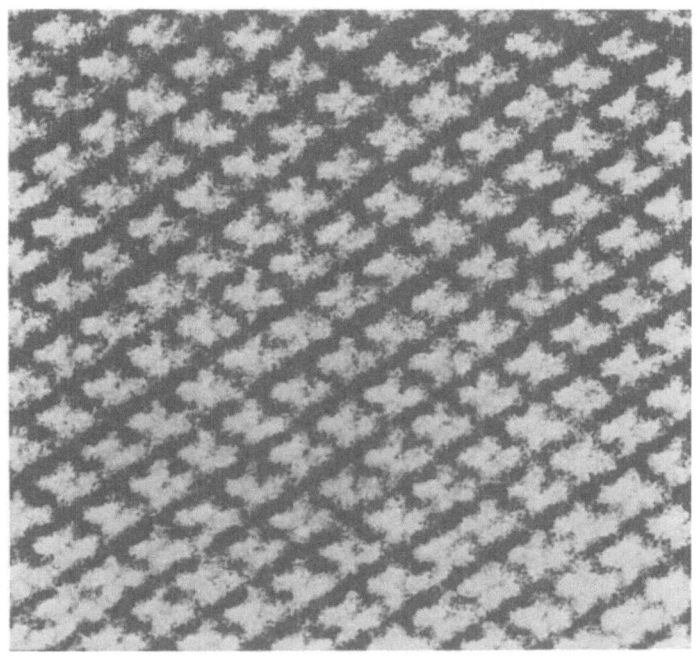

Fig. 6   Cu-phthalocyanine in the 1960s

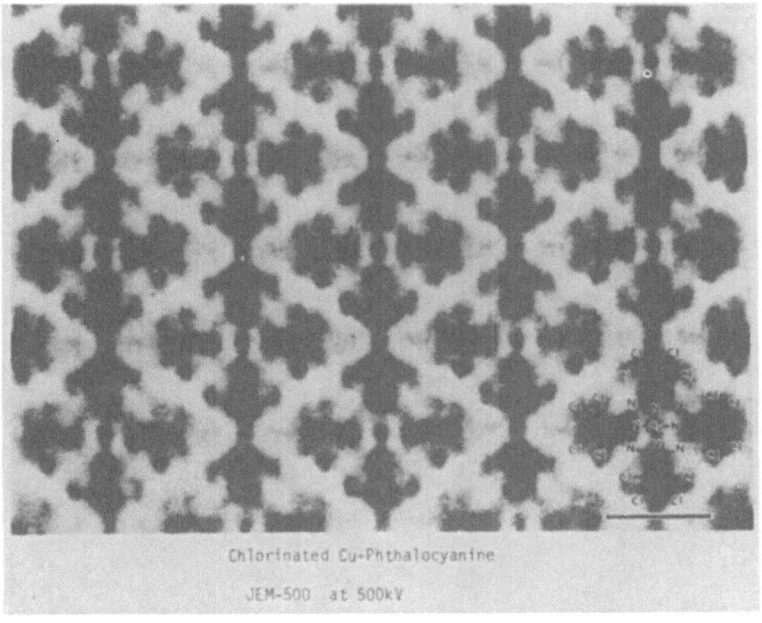

Fig. 7   Cu-phthalocyanine in the 1970s

Fig. 8 Cu-phthalocyanine in the 1980s

Fig. 9 Single gold crystal containing stacking fault,
courtesy of H. Hashimoto.

Fig. 10   YBaCu high Tc Super conductor, courtesy of K.
          Hiraga.

Fig. 11   Advertisement of JEOL in 1954

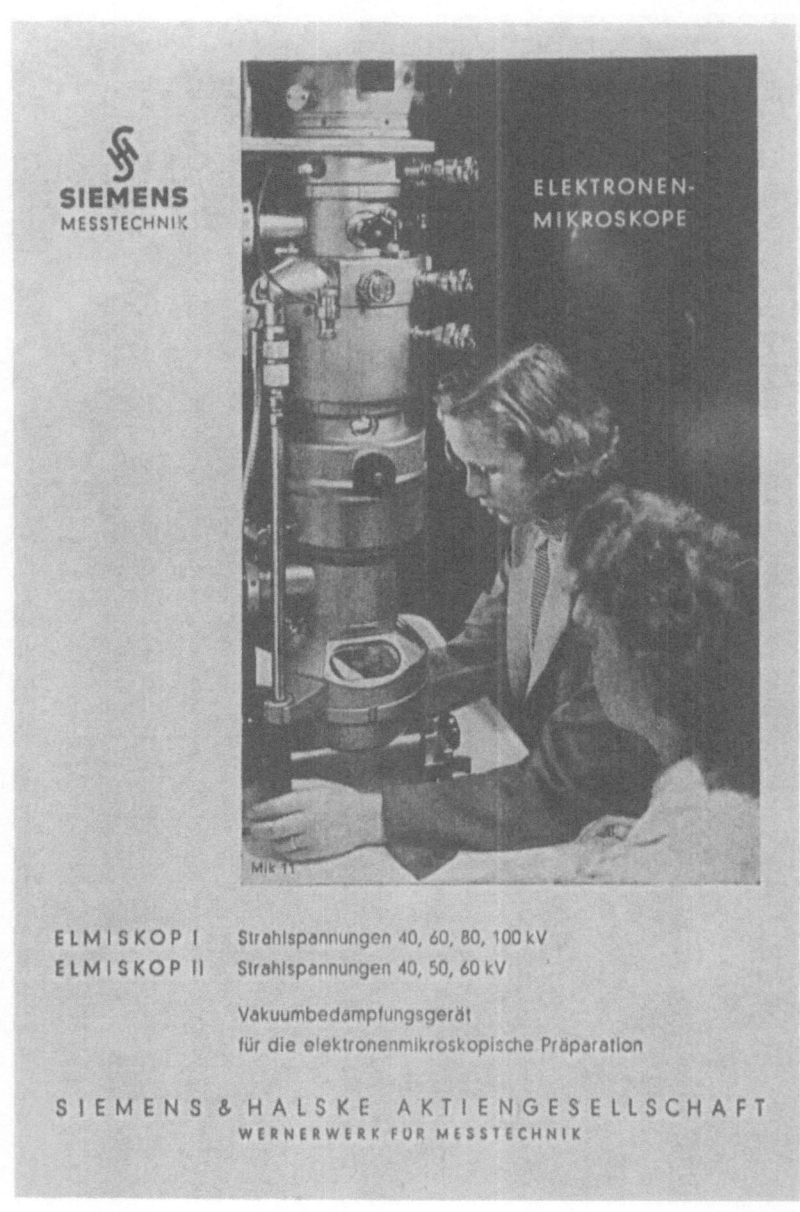

Fig. 12 Advertisement of Siemens in 1955

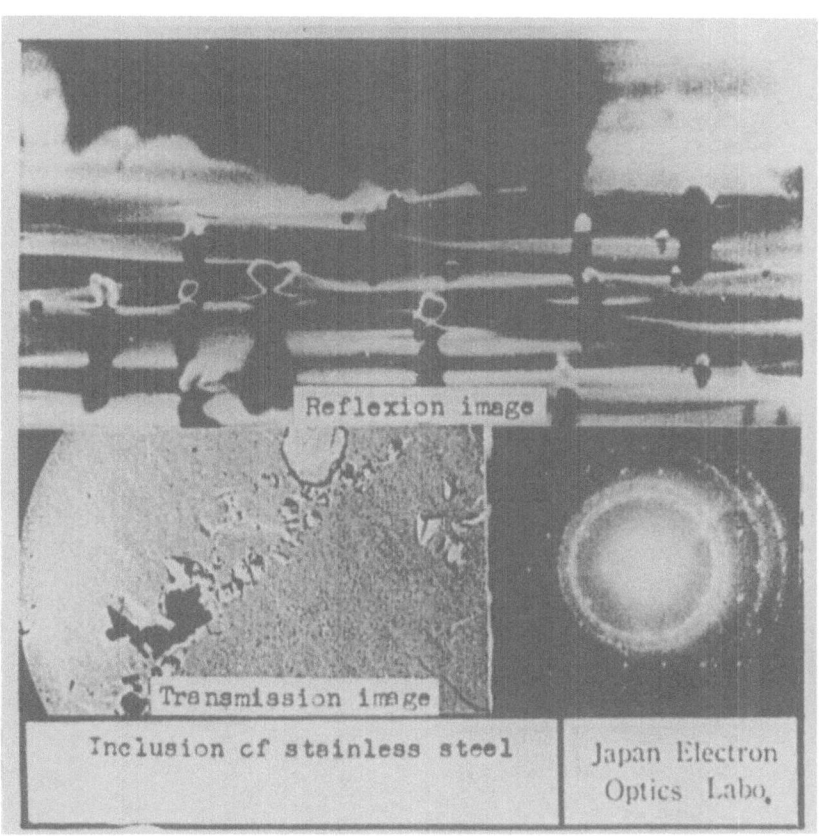

Fig. 13 Reflection microscopy compared to replica method.

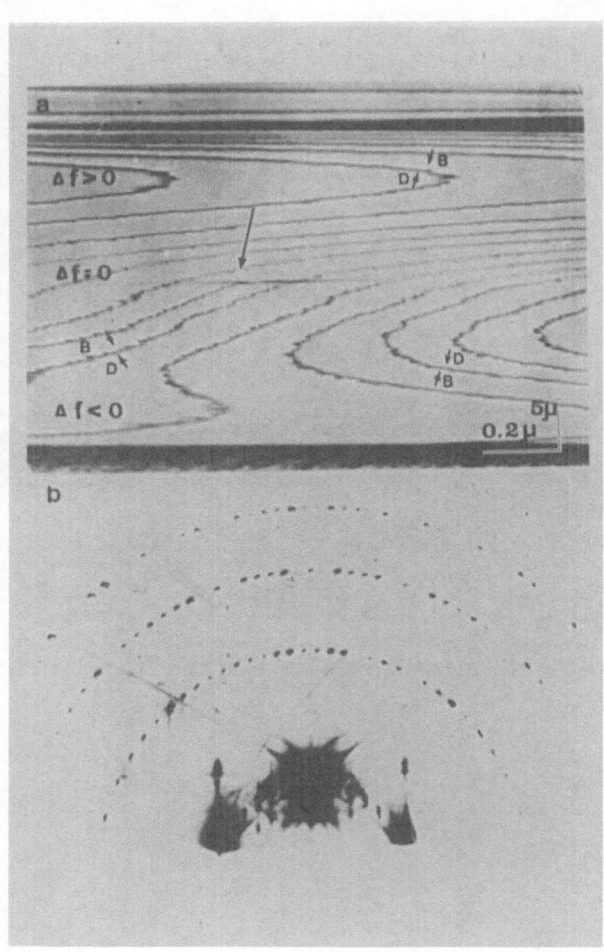

Fig. 14 Surface of Si observed by reflection method,
courtesy of K. Yagi.

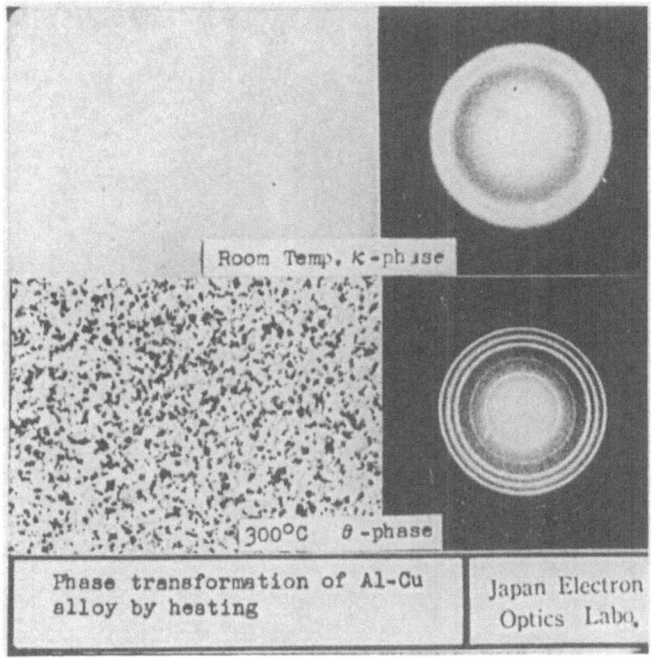

(a)

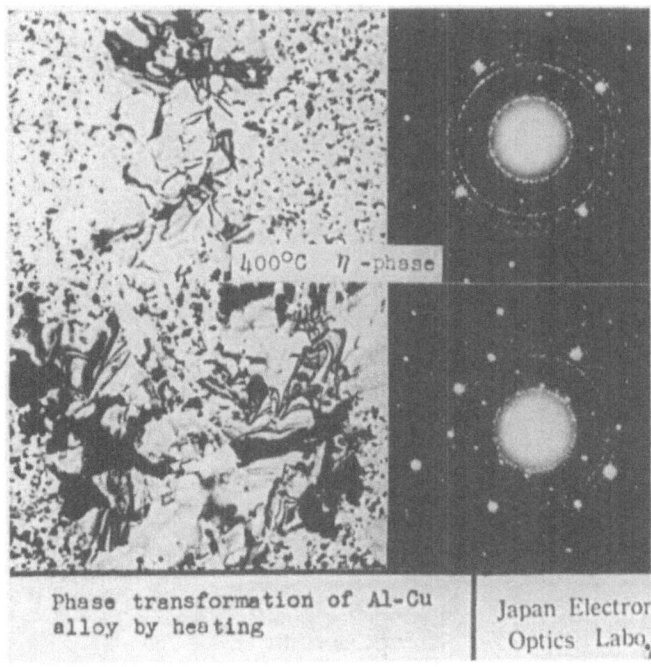

(b)

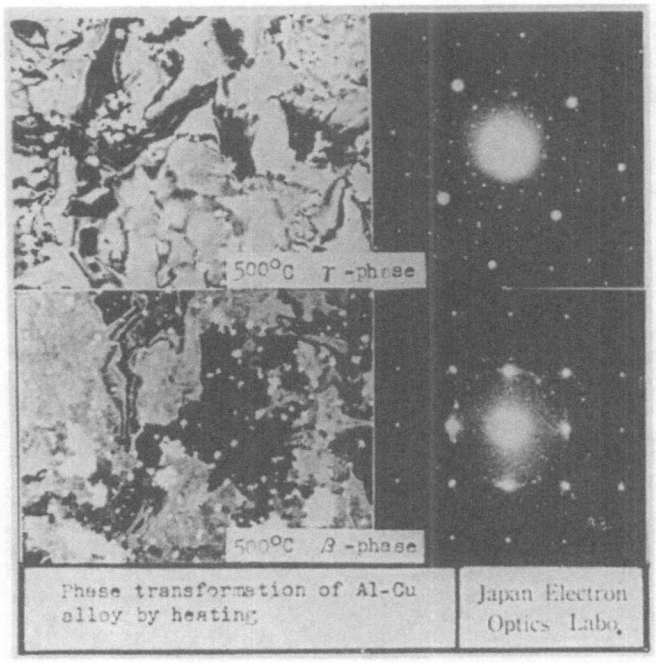

(c)

Fig. 15 (a) (b) (c)   50-50 Al-Cu alloy phase transformation
                      courtesy of N. Takahashi.

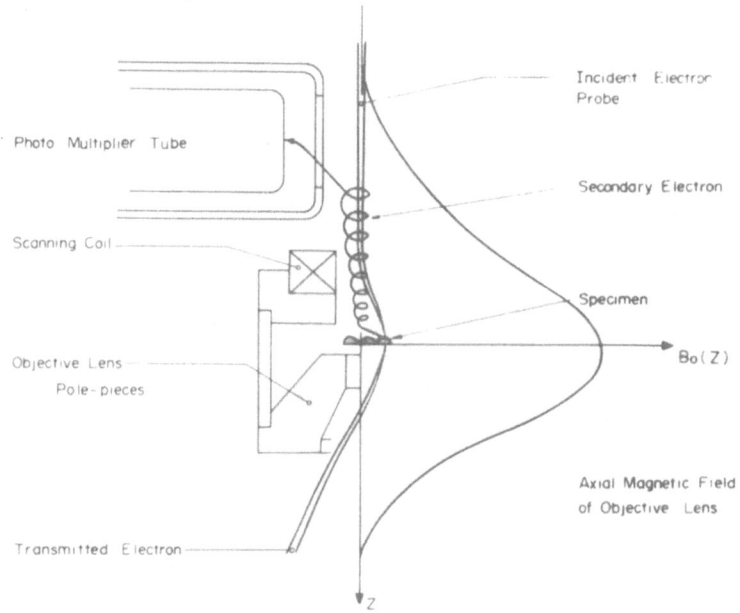

Schematic Illustration of Secondary Electron

Fig. 16 Schematic illustration of secondary electrons'
behavior in the objective lens.

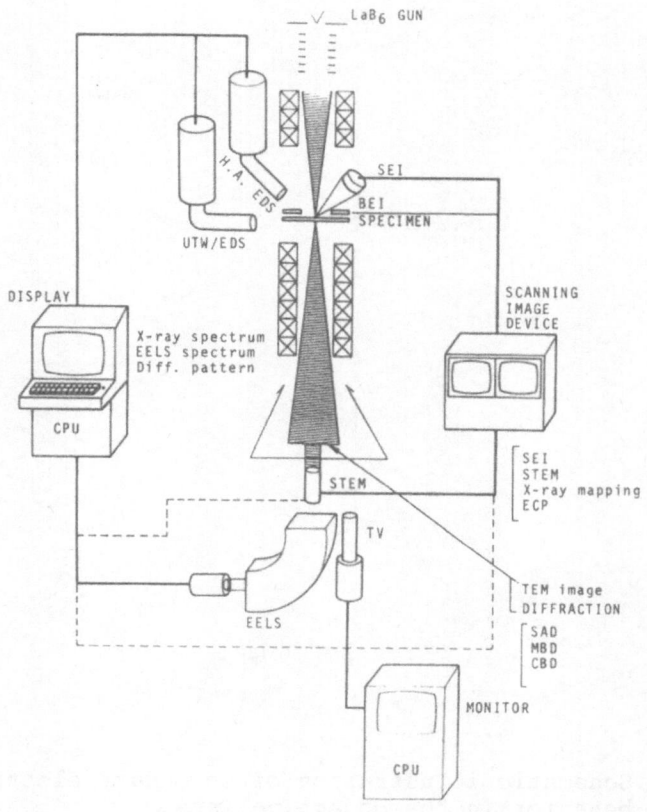

Fig. 17   Schematic illustration of analytical microscope.

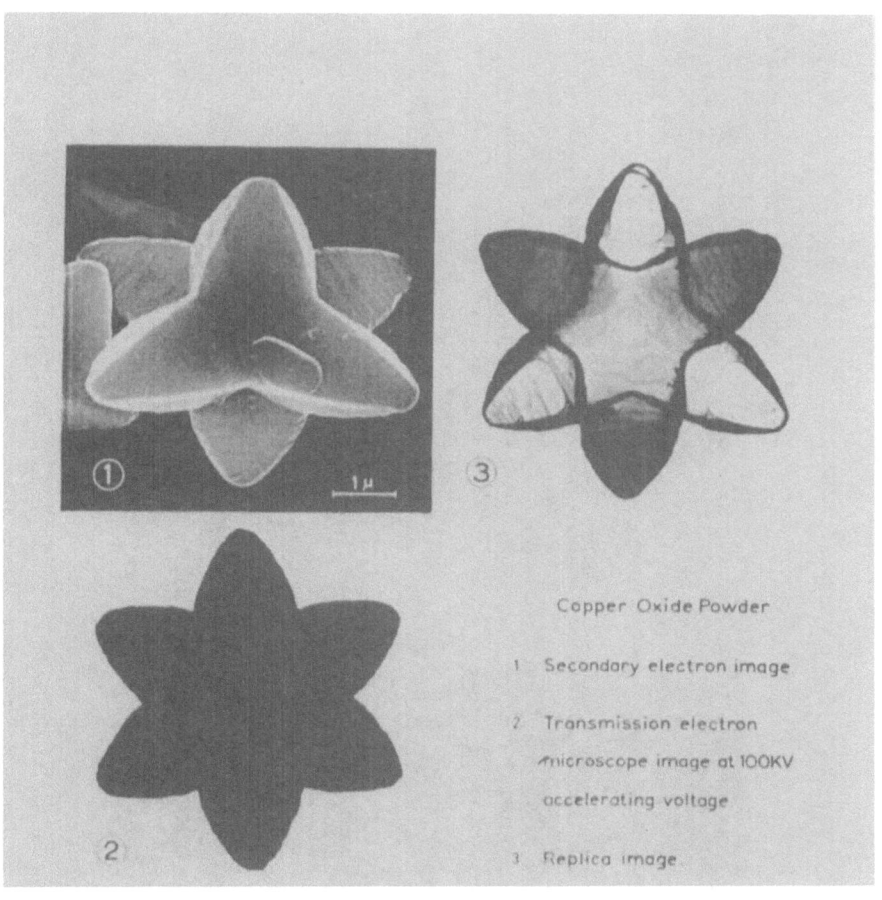

Copper Oxide Powder

1 Secondary electron image

2 Transmission electron
   microscope image at 100KV
   accelerating voltage.

3 Replica image.

Fig. 18 Comparison between TEM and SEM.

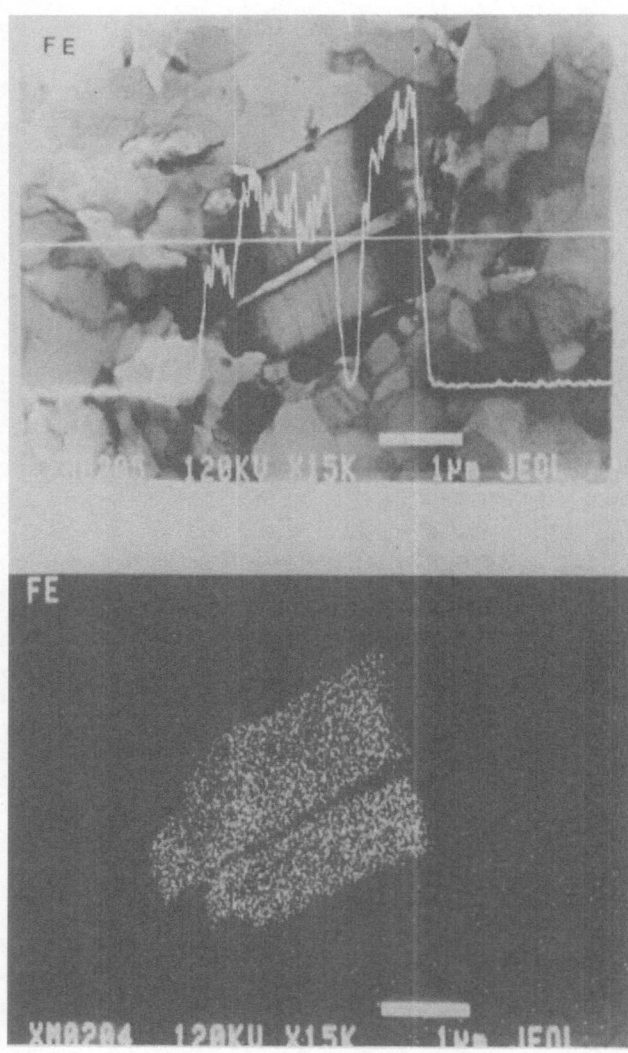

Fig. 19   Energy Dispersive X-ray spectrscopy together with electron microscope image.

41

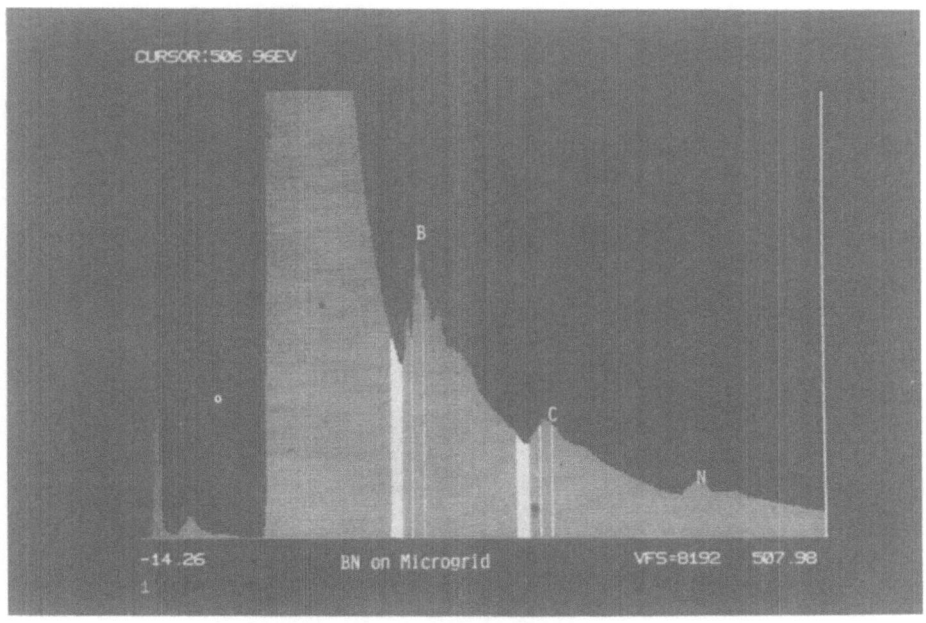

Fig. 20   One example of electron energy loss spectroscopy.

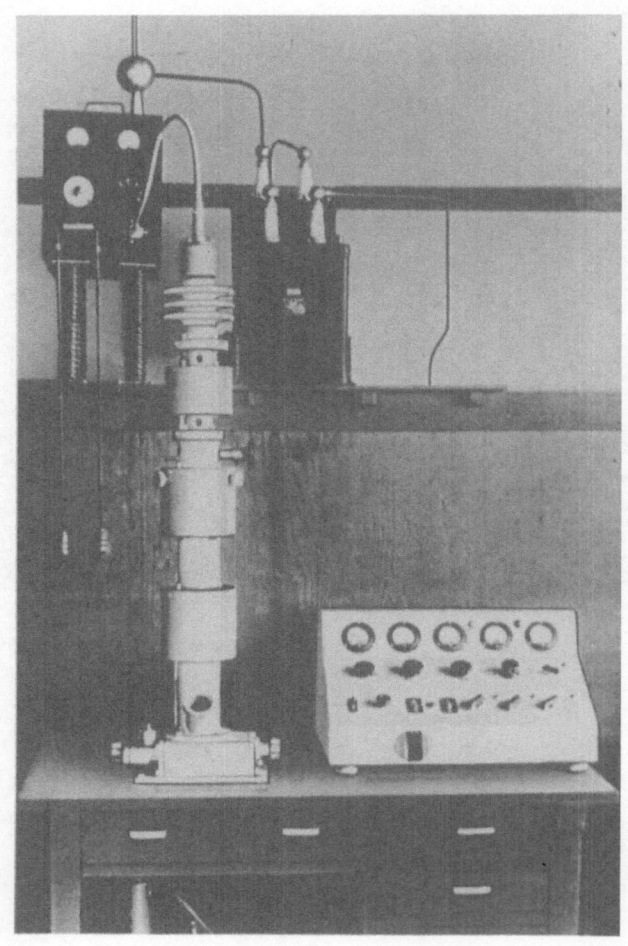

Fig. 21 The first electron micrsocope manufactured by us
in 1946.

Fig. 22   The latest microscope of 1,000 kV accelerating
          voltage.

Micro-biosensors for Clinical and Food Analyses

Eiichi Tamiya and Isao Karube

Research Center for Advanced Science and Technology,
University of Tokyo, 4-6-1 Komaba, Meguro-ku, Tokyo
227 (Japan)

ABSTRACT

Micro-biosensors for glucose and fish freshness were constructed by using silicon fabrication technology. A micro-H_2O_2 electrode was used to make a micro-glucose sensor. Glucose oxidase was immobilized on an organic membrane, prepared by vapor deposition of γ-aminopropyl-triethoxysilane (γ-APTES) and glutaraldehyde, and used for glucose determination in the range from 0.1 to 10 mg $\cdot$ dl^{-1}. The micro-oxygen electrode consists of two gold electrodes and agarose gel containing electrolyte covered with a gas permeable membrane. A negative photoresist was used as the gas permeable membrane which was directly cast onto the gel and submitted photochemical reaction. A 90% response time of the micro-oxygen electrode took approximately 3 minutes. A linear relationship was obtained between the response of the micro-oxygen electrode and that of a convinient Galvanic oxygen electrode. This sensor responsed almost linearly for glucose concentrations between 0.2 and 2 mM. The amorphous silicon ion sernsitive field effect transistor(a-ISFET) made by radio frequency plasma discharge was used as a pH-sensitive device. The size of the channel of the a-

ISFET is 10 μm long and 500 μm wide. The pH sensitive layer was silicon oxide evaporated over the amorphous silicon nitride layer. The pH sensitivity was about 46 mV/pH at 18°C in the range pH 5-10. The response times of this device to pH change are very rapid, being less than 30 sec to reach a steady state value. The fish freshness sensor was constructed by using an immobilized xanthine oxidase membrane and a-ISFET. Hypoxanthine was detected in the range 0.02-0.1 mM.

1. INTRODUCTION

Methods for the selective determination of organic compounds in biological fluids, such as blood, are very important in clinical and food analyses. Most analyses of organic compounds can be performed by spectrophotometric methods based on specific enzyme-catalysed reactions. These methods, however, involve complicated and delicate procedures and the assay times are rather long because of the several reactions involved. The spectrophotometric methods cannot be applied directly to colored samples or biological fluids.

Alternative systems based on electrochemical sensors have been developed. Electrochemical sensors employing immobilized biocatalysts have definite advantages. Namely, an enzyme sensor possesses excellent selectivity for biological substrates and can directly determine single compounds in a complicated mixture without need for a prior separation step[1, 2]. The development of a simple, inexpensive assay is therefore of interest and the miniaturization of enzyme sensors is especially important for clinical analysis. Semiconductor fabrication technology has permitted the development of ion selective field effect transistor and micro electrodes which have been utilized as pH and enzyme-based sensors [3,4,5,6].

In this paper, micro-electrodes prepared by silicon fabrication technology, are employed as micro-biosensor transducers. Micro-biosensors for glucose and fish freshness constructed from micro-transducers and immobilized enzyme thin membranes, are detailed and their characteristics are discussed.

2. MICRO-GLUCOSE SENSOR BASED ON H_2O_2 DETECTION

The determination of glucose in blood samples is important in the clinical field, and the development of bioelectrochemical devices considerably help routine laboratory work. Many enzyme sensors have been developed and some sensors are used for clinical analyses [2]. The development of miniaturized and implantable enzyme sensors employing micro-transducers are required in medical field. Recently, several amperometric micro-glucose sensors, which used wire electrodes [7], have been reported. Although a wire micro glucose sensor can be made small, its fabrication is not suitable for mass production. The reason is that each wire must be individually coated by a glucose oxidase immobilized membrane. On the other hand, a planar micro glucose sensor, which uses planar electrodes [8], will be suitable for mass production, because the glucose oxidase membrane can be deposited on all the individual electrodes simultaneously by using an IC process. Hydrogen peroxide sensor could be used for determination of glucose because glucose oxidase (GOD) catalyzes glucose oxidation according to the following reaction.

Glucose + Oxygen ⟶ Gluconolactone + Hydrogen peroxide

A micro-hydrogen peroxide (H_2O_2) sensor has been developed utilizing the currently available silicone fabrication technology [9]. The structure of micro-H_2O_2 sensor is shown in Fig.1. Micro-Au electrodes were created on the silicon nitride surface using the vapor deposition method and partially insulated by coating with Ta_2O_5. From the cyclic voltammogram shown in Fig.2, this unit works as a H_2O_2 electrode when the potential between both Au electrodes is set at 1.1 V. When the H_2O_2 electrode was placed in a sample solution containing

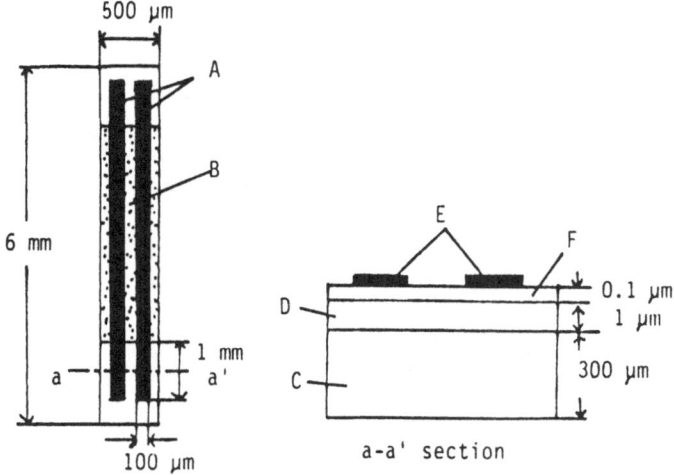

Figure 1. Schematic diagram of a microelectrode for H_2O_2.(A and E), Au;(B) Ta_2O_5;(C) Si;(D) SiO_2; (F) Si_3N_4.

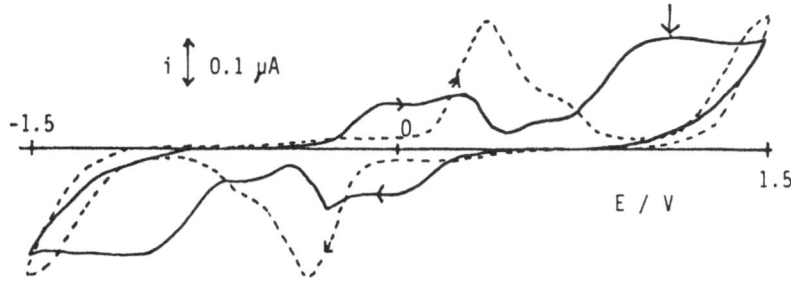

Figure 2. Cyclic voltammograms at the microelectrode. Scan rate is 100 mV sec^{-1} and the arrow indicates 1.1 V. (---)Phosphate buffer solution (——)Hydrogen peroxide solution (0.1 mM)

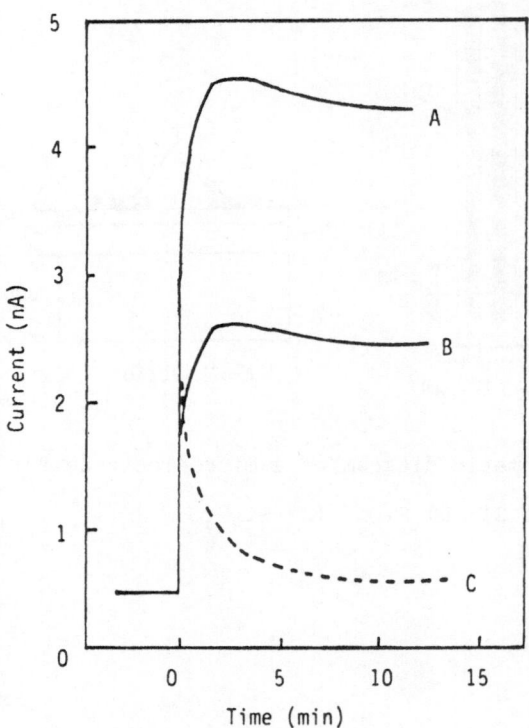

Figure 3. Response curves for glucose addition. Experiments were done at 37 C and pH 7.0. (A) 10 mg dl^{-1} glucose. (B) 5 mg dl^{-1} glucose, and (C) no glucose.

H_2O_2, the output current of the sensor immediately increased and reached a steady-state value within 1 min. A linear relationship was observed between the H_2O_2 concentration, over the range from 1 μM to 1 mM and the steady-state current. This electrode can be used as a transducer for a micro-glucose sensor.

The procedure for GOD immobilization onto the micro-electrode is as follows. Approximately 100 μl of γ-APTES was vaporized at 80°C, 0.5 torr for 30 min onto the electrode surface, followed by 100 μl of 50% glutaraldehyde vaporized under the same conditions. The modified micro-electrodes were then immersed in GOD solution containing bovine serum albumin and glutaraldehyde, the GOD becoming chemically bound to the surface of the micro-electrodes.

Figure 3 shows a typical response curve for the micro-glucose sensor. The output current increased after injection of a sample solution, steady state being reached within 5 min. Figure 4 shows a calibration curve for the micro-glucose sensor.

A linear relationship was obtained between the current increase (the current difference between the initial and steady state currents) of the sensor system and concentration of glucose ($0.1-10$ mg·dl^{-1}). Examination of the selectivity of the micro-glucose sensor indicated no response to other compounds such as galactose, mannose, fructose and maltose. Therefore, the selectivity of this sensor for glucose is highly satisfactory. The long-term stability of the sensor was examined with a sample solution containing 10 mg dl^{-1} glucose. The output current of the sensor system was almost constant for more 15 days and 150 assays (Fig. 5). Therefore, this micro-glucose sensor possesses both selectivity and good stability, its potential use as a micro-glucose sensor being very good.

2. FABRICATION OF A NOVEL GLUCOSE SENSOR BASED ON MICRO-OXYGEN
   ELECTRODE

Clark type oxygen sensing electrodes have been applied to various

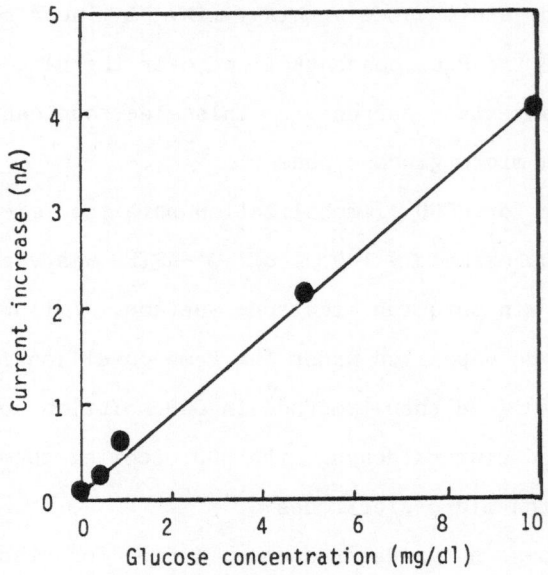

Figure 4. Calibration curve for glucose. Experiments were done at 37°C and pH 7.0.

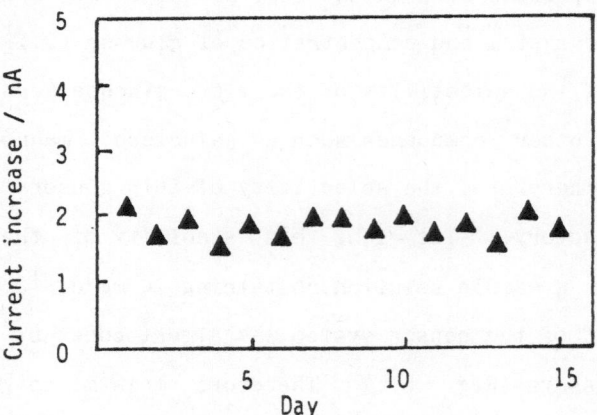

Figure 5. Stability of micro-glucose sensor.

biosensors by immobilizing either enzymes or microorganisms which catalyze the oxidation of biochemical organic compounds. But they have not yet reached the production line because they contain a liquid electrolyte solution, making adhesion of the gas permeable membrane to the substrate. Therefore, it is important to develop a disposable micro-oxygen electrode based on conventional semiconductor fabrication technology and use the micro-oxygen electrode in a biosensor. The key points of improvement were:(1) to use porous material (in this case agarose gel) to support the electrolyte solution, and (2) to use a hydrophobic polymer (in this case negative photoresist) as the gas permeable membrane, and submit it to direct casting over the porous material.

2.1 Construction of the micro-oxygen electrode

Construction of the micro-oxygen electrode is illustrated in Figure 6. The electrode has a U-shaped groove depth of 300 μm, and two gold electrodes over the SiO_2 layer that electrically insulates them. The agarose gel containing 0.1 M potassium chloride aqueous solution was filled in the groove followed by coverage of the gas permeable membrane. Only the pad areas of the two gold electrodes were exposed, while the other parts were covered with the same hydrophobic polymer used for the gas permeable membrane to insulate each electrode when used in a aqueous solution. The areas of the cathode and the anode were the same[10].

2.2 Characteristics of the micro-oxygen sensor

The response curve of the micro-oxygen electrode was examined after adding sodium sulfite to reduce the oxygen concentration. This response is the maximum response when oxygen concentration is reduced from the saturation point to the zero point. Although this curve is that of a micro-oxygen electrode width of 4 mm, the profile is similar regardless of its size. The ectrode responded as soon as sodium

52

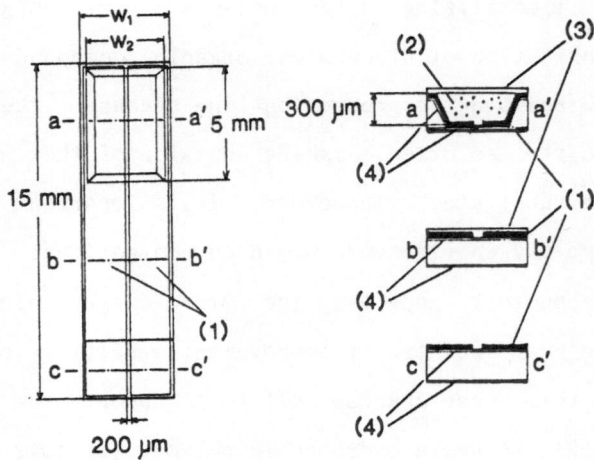

Figure 6. Schematic diagram of oxygen electrode. (1) Au, (2) agarose gel, (3) gas permeable membrane, (4) SiO_2 layer. W_1 is 2,3 or 4 mm. W_2 is 1.4,2.2 or 3 mm in each case. The cross sections on the right side corresponde to a-a', b-b' and c-c'.

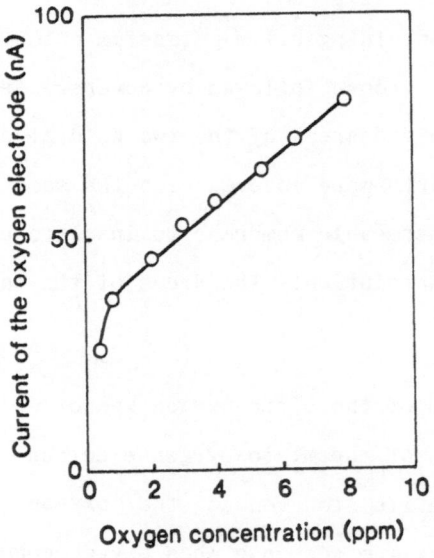

Figure 7. Caribration curve for the 2-mm-wide oxygen electrode. The terminal voltage was 0.8 V.

sulfite was added to the buffer solution, and stabilized 8 - 10 minutes after the addition. A 90% response time of the oxygen electrode took approximately 3 minutes, or three to four times longer than that of conventional oxygen electrodes. The response time was not affected by their cathode area but was considered to be dependent on the distance between the gas permeable membrane and the cathode, since the diffusion of oxygen through the gas permeable membrane and agarose gel seemed to be a decisive factor in the response time. The distance between the cathode and the gas permeable membrane can be shortened. It will be the next step of our improvement.

Output current obtained from a conventional Galvanic oxygen electrode was used as the references since this electrode uniquely responded to dissolved oxygen concentration controlled by adding sodium sulfite. As is shown in Figure 7, a linear relationship was obtained between the responses of the two oxygen electrodes when the terminal voltage between the two gold electrodes was 0.8 V. Therefore, the micro-oxygen electrode was found to work as an oxygen electrode.

Stability of the micro-oxygen electrode was tested using two micro-oxygen electrodes the width of which was 2 mm and 4 mm respectively. When a larger micro-oxygen electrode was used, its response decreased after a few times of successive use in experiments. But if it is stored in phosphate buffer solution or in distilled-deionized water for one day or two with no voltage applied between the two gold electrodes, its sensitivity recovers to the initial level. The smaller oxygen electrode could be stable used for more than 10 times. In this study, the smaller the oxygen electrode, the more stable it was. Bad stability of the larger oxygen electrode was thought to be due to accumulation of reaction products in the vicinity of each of the two gold electrodes.

2.3 Characteristics of the glucose sensor

The glucose sensor was fabricated by immobilizing glucose oxidase on a sensitive part of the oxygen electrode by cross-linking with bovine serum albumin(BSA) and glutaraldehyde(GA). The enzyme-immobilized membrane was formed by dipping the sensitive part into a mixture containing 2 mg of glucose oxidase, 20 µl of 10 % BSA solution, and 10 µl of 25 % GA solution.

Figure 8 shows a calibration curve for the glucose sensor at 30 °C and pH 7.0. As can be seen, the sensor responded almost linearly for glucose concentrations between 0.2 and 2 mM, which is comparable to conventional glucose sensors. The glucose sensor was a little sensitive to glucose at normal blood glucose concentrations(5 mM), but the sensitivity is easily shifted by adjusting the amount of immobilized enzyme. The stability of the glucose sensor was evaluated by performing the same experiments as were done to obtain the response curve. In subsequent experiments at 30 °C, its response gradually decreased, but it returned to the initial level when the sensor was stored with no voltage applied, as can be seen to be mainly dependent on the stability of the oxygen electrode used as the transducer.

3. FISH FRESHNESS SENSOR USING AMORPHOUS SILICON ISFET

3.1 Construction of amorphous silicon ISFET

The a-ISFET was mainly made by radio frequency (rf) plasma discharge. The type of grow discharge apparatus is the capasitatively coupled grow discharge deposition system. A 0.05 µm n^+ layer (3000ppm PH_3 in silane) was deposited over an evaporated aluminum layer on a glass substrate in order to ensure ohmic contact between the a-Si:H and aluminum. The evaporating apparatus (ULVAC Co.,TH-500A) was used to form the aluminum layer. The thickness of each layer was measured by tallysurf. After etching, the deposition of the amorphous silicon layer (0.3 µm) and the amorphous silicon nitride layer (0.3 µm) was performed successively in the same capacitatively coupled grow

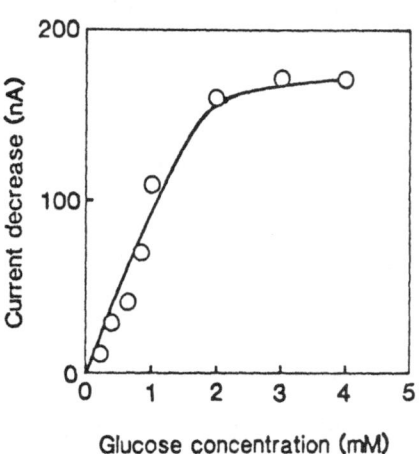

Figure 8. Caribration curve for glucose sensor

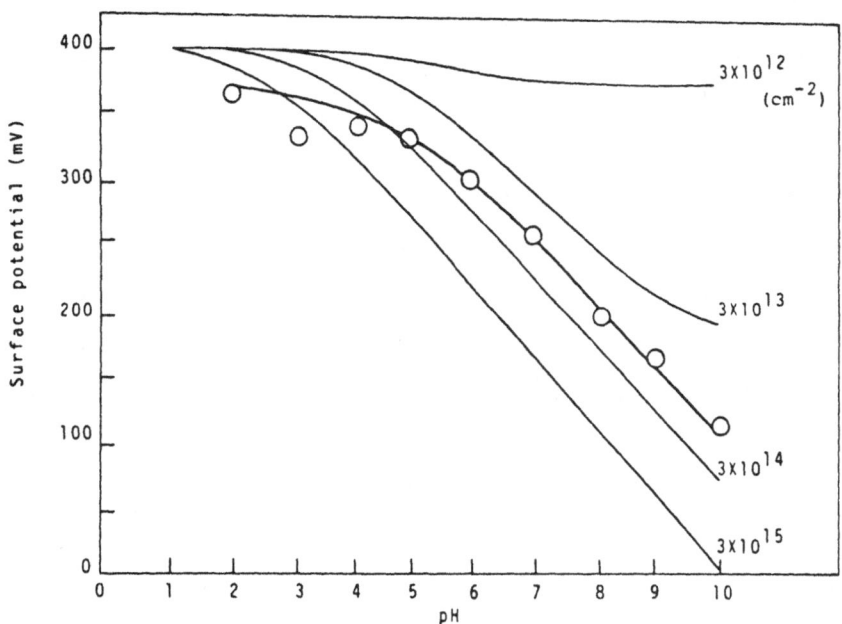

Figure 9.   pH characteristics of a-ISFET.   Circles revealed the data
points in 10 mM Tris-HCl buffer solution.

discharge deposition system operating at 13.56 MHz. The amorphous silicon layer was grown from a mixture of silane and hydrogen. The amorphous silicon nitride layer from silane and ammonia. All three layers are deposited at 300°C. The rf power level was 6 W net. Finally, a silicon oxide layer (0.2 µm) was evaporated over the amorphous silicon nitride layer. The size of the channel is 10 µm long and 500 µm wide[11].

3.2 Characteristic as field effect transistor

The characteristic as field effect transistor was measured by curve tracer. The source drain voltage (V_{SD}) / the source drain current(I_{SD}) characteristic of a-ISFET was investigated. The aluminum electrode was used as the gate. The gate voltage(V_{GS}) was applied to this device with two volts step. This device revealed characteristic as field effect transistor. When the gate voltage is zero, the source drain current is zero. The source drain current increase in accordance with the increase of the gate voltage. This device is the field effect transistor of the enhancement mode.

The Ag/AgCl reference electrode was placed in the same solution as the a-ISFET. The surface potential on the silicon oxide insulator of the a-ISFET is affected by the solution's pH, with a concominant change in the gate voltage proportional to the change in surface potential. Therefore, the surface potential change on the silicon oxide insulator of the a-ISFET, caused by a variation in pH, can be measured as the change in the gate voltage. In this case, the voltage between source and drain is held constant at 1.5 V, the current between source and drain also being constant at 0.1 µA.

Figure 9 shows the theoretical curve and the pH characteristics of the a-ISFET. Theoretical curves of the relation between the surface potential and pH was calculated with surface dissociated sites. In this figure, it is assumed that pKa is 5 and C_H is 20 µF/cm. The linear Vg/pH characteristic of a-ISFET was obtained over the pH range 5-10. The pH sensitivity was about 46 mV/pH at 18 ° C. The response times of a-ISFET to pH change are very rapid, being less than 30 sec to reach a steady state value.

3.4 Construction of fish freshness sensor

Hypoxanthine is considered to be a useful and reliable indicator of the fish meat freshness. The hypoxanthine sensor was constructed by using an immobilized xanthine oxidase polyvinylbutyral (PVB) membrane and a-ISFET. The PVB membrane was formed over the gate insulator of the a-ISFET as follows: 0.1 g of PVB and 1 mg of 1.8-diamino-4-aminomethyloctane were dissolved in 10 ml of dichloromethane. This polymer solution was spread over the gate insulator of a-ISFET. The a-ISFET was then immersed in 5 % glutaraldehyde solution at room temperature for 24 h to promote the cross-linking reaction of the amino group of 1,8-diamino-4-aminomethyloctane with glutaraldehyde. Xantine oxidase was covalently immobilized on the cross-linked membrane[12].

This sensor system was applied to the determination of hypoxanthine according to the following reactions. The uric acid formed was detected by this system.

$$\text{Hypoxanthine} \quad + \quad O_2 \xrightarrow{\text{xantine oxidase}} \text{xantine} \quad + \quad H_2O_2$$
$$\text{Xantine} \quad + \quad O_2 \xrightarrow{\hspace{3cm}} \text{uric acid} \quad + \quad H_2O_2$$

The characteristics of the sensor system were investigated. The optimum pH was approximately 7.0 at 37°C. The response reached a steady state for three minutes. A linear relationship was obtained in the range 0.02-0.1 mM between the initial rate of the gate output voltage change after hypoxanthine injection and the logarithmic value of hypoxanthine concentration. The sensor system revealed a response to 0.1 mM of hypoxanthine for more than 1 week. Over this period, the sensitivity decreased by about 30 %.

(Reference)

1)   I.Karube and S.Suzuki.   Amperometric   and   potentiometric
     determinations with immobilized enzymes and microorganisms.
     Ion-Selective Electrode   Review 6,(1984) 15-58

2)   I.Karube and S.Suzuki. Immobilized enzymes for clinical
     analysis, In Enzymes and immobilized cells in biotechnology
     (ed. A.I.Laskin)   (1985)   pp.209-226.   Benjamin/Cumming
     Publishing,London.

3)   S.Caras,   J.Janata.   Field effect transistor sensitive   to
     penicillin. Anal. Chem. 52,(1980) 1935-1937.

4)   I.Karube, E.Tamiya, J.M.Dicks and M.Gotoh. A microsensor for urea
     based on an ion-selective field effect transistor.
     Anal.Chim.Acta, 185,(1986) 195-200

5)   I.Karube, E.Tamiya, M.Gotoh and Y.Kagawa.   A microsensor for
     adenosine-5'-triphosphate pH-sensitive field effect transistor.
     Anal.Chim.Acta, 187, (1986) 287-291

6)   I.Karube, E.Tamiya, and M.Gotoh. Polyvinylbutyral resin membrane
     for enzyme immobilization to an ISFET microbiosensor.
     J. Mol. Catalyst, 37,(1986) 133-139

7)   S.J.Churchouse, C.M.Battersby, W.H.Mullen and P.M.Vadgama. Needle
     enzyme electrodes for biological studies.   Biosensors, 2, (1986)
     325-342

8)   T.Murakami, S.Nakamoto, J.Kimura, T.Kuriyama and I. Karube.
     A micro planar amperometric glucose sensor using an ISFET as a
     reference electrode. Anal.Lett, 19(1986) 1973-1986

9)   I. Karube and E. Tamiya.   Micro-biosensor for clinical   analyses.
     Sensors and Actuators, 15(1988) 199-207

10) H.Suzuki, E.Tamiya and I.Karube.   Fabrication of an oxygen

electrode using semiconductor technology. Anal.Chem., 60(1988)
1078-1080

11) M.Gotoh, S.Oda, I.Shimizu, A.Seki, E.Tamiya and I.Karube.
Construction of amorphous silicon ISFET. Sensors and Actuators,
16(1989) 55-65.

12) E.Tamiya, A.Seki, I.Karube, M.Gotoh and I.Shimizu. Hypoxanthine
sensor based on an amorphous silicon field-effect transistor.
Anal.Chim.Acta, 215(1988) 301-305

THE FUTURE OF ENGINEERING PLASTICS

Herbart G. Rammrath
President & Representative Director
GE Plastics Pacific Ltd.
Kowa 35 Bldg., 14-14, Akasaka 1-chome
Minato-ku, Tokyo 107, Japan

Plastics have evolved considerably since the first natural plastics, celluloid, was developed over a one hundred years ago in the United States as a substitute for ivory in billiard balls. It is safe to say that the consumer revolution of the past thirty years is basically a plastics' revolution. Without plastics, consumers could not afford or even get, many of the products that are on the market today.

Just think, last year in Japan almost the same volume of plastic resins were made and used than all of the steel, aluminum and copper combined. Total plastics produced in Japan in 1987 exceeded 10 million tons. Total production of steel, aluminum and copper was about 76 million tons. Using the different specific gravities of these materials, volume is almost the same. So I think you can see that plastics have become a very important material for use in Japan and throughout the world.

Unlike metal, plastics come in many different varieties and almost an infinite number of performance possibilities from rubber like materials to those having a stiffness to weight ratio approaching that of steel.

You already know that we divide plastics into two generic categories: Thermoset and Thermoplastic resins. Thermoset resins can be formed only once. While, Thermoplastics can be reheated and reshaped many times. Further classification of these two generic types would include commodities and engineering plastics. Commodity plastics serve non-structured applications, usually produced in high volumes such as shower curtains, trash containers, credit cars, ball point pens, cups and packages, etc. Among commodity plastics are polyethylene, polypropylene, polystyrene and PVC. In Japan commodity plastics comprise about 95% of the total volume of all plastics sold, but only about 80% of the total value, because of their generally lower prices.

In contrast to commodity plastics, engineering thermoplastics are high performance materials that offer prolonged use in more demanding and structured applications because they are outstanding in high stress chemical, electrical,thermal and mechanical environments.
They can also be produced to provide additional advantages in dimensional stability, high heat resistance, flame resistance and self-lubricating properties. Today, Engineering plastics represent less than 5% of all of the plastics being used in the world. But because these engineering plastics, more than any other materials, can help drive the major changes that are occurring in world industries, the use of these materials will grow enormously in the coming years. High performance engineering thermoplastics provide productivity design, productivity in manufacturing, and productivity in recycling. When properly formulated, engineering plastics can be fabricated into mechanically functional, precision parts with structures similar to metal --- but with other advantages that make them superior to metal in many important features. Compared to metals, engineering plastics do not corrode or rust. They are easier to fabricate and color. They can provide equal strength and durability and less energy is needed to produce a finished part versus a similar part made of metal. Throughout history metals have been extremely energy-intensive materials, requiring 00massive amounts of energy to mine, transport, refine and fabricate. Engineering plastics save energy.

Engineering plastics include a variety of high performance materials ranging from large volume plastics to relatively new , low volume, high cost specialities such as polyimide, liquid crystal polymers, polyetherketones and aramid fibers. These speciality materials will become more important in the future, because they are growing at more than 15% per year as end-user requirements increase and manufacturing techniques are altered or developed for their use. The needs of the aerospace and electronic industry are increasingly being met by these materials. At GE PLASTICS we are participating with ultem polyetherimide, supec polyphenylene sulfide, while LCPs and other new products are being developed in our laboratories.

Today, however, I will limit my discussion to higher volume engineering thermoplastics which will continue to grow at a rapid rate from a much larger volume base. Among these materials I include polyacetal, polyamide, polycarbonate, polyphenylene ether and molding grades of thermoplastics polyester, PBT and PET.

POLYACETAL

Acetal is a highly crystalline resin based on formaldehyde polymerization technology. The original polymer was introduced in the 1940's by Dupont.

Low friction and good wear resistance are major factors in selecting this product for dynamic applications such as gears and bearings. Acetal's toughness, abrasion resistance, and moisture, heat and solvent resistance have made it a preferred material in plumbing fittings.

Sales of Acetal resins are larger in Japan than in the U.S. In fact, Acetal is the highest volume engineering plastics in Japan with over 100,000 tons sold in 1987. This probably results from the fact that Acetal's low friction levels and lubricity enable it to be widely used for internal components in a wide range of consumer electronics products, a largely Japanese based industry. The leading producers of Acetals are Dupont, BASF, and Hoechst Celanese.

POLYAMIDE OR NYLON

Nylons are the oldest of the engineering thermoplastics, having been introduced in 1938 as the world's first synthetic fiber. It has the largest engineering plastic tonnage in Europe but in the U. S. It is second to polycarbonate. In Japan sales were about 80,000 tons in 1987. The most widely used Nylons are Nylon 66 and Nylon 6. Specialty grads, Nylon 11, Nylon 12, Amorphous Nylons, etc., have also been developed for specialty applications.

Nylons are mostly crystalline polymers which have excellent fatigue resistance, low coefficient of friction, good toughness and good chemical resistance. Nylons find their largest volume use in the automotive industry in mechanical and electrical hardware and under-the-hood applications. The introduction of impact modified (super though) Nylons in the 1970'S added to Nylon's growth and opened up new market opportunities. The major producers of Nylons are Dupont, Homechst Celanese, BASF, UBE and Asahi.

POLYCARBONATE

Polycarbonate has become the largest volume Amorphous Engineering Thermoplastic in the world. Its key properties are exceptional impact strength, toughness, transparency, high heat resistance, flame retardance, and dimensional stability. Applications range from Glazing of windows to compact discs to clear water bottles. The high degree of compatibility with other materials has resulted in many blends which have expanded market acceptance. Combinations with Polyesters have resulted in tough, chemical resistant products used in automobile bumpers. Ford Taurus and new Cadillac Models are using this material for bumpers in the U.S. In Korea, Hyundai has become a major user. Combinations with ABS are being used in business machines and automotive applications. About 80,000 tons Polycarbonate are General Electric, Teijin, Bayer and Mitsubishi Gas Chemical.

MODIFIED POLYHENYLENE ETHER

The combination of high impact polystyrene and polyphenylene ether produces an alloy with a broad property profile and good processability at reasonable cost. The properties vary with the quantities of each ingredient but generally, these products have excellent impact strength, toughness, dimensional stability and good hydrolytic stability, flame retardant grades find use in business machine applications, appliances, and electrical/electronic parts. Other grades are used in automotive applications due to excellent low temperature impact strength and retention of properties after repeated temperature variations.

Various fabricating processes, such as foam molding and blow molding, have expanded the use of these products, particularly in business machine applications. Over 70,000 tons per year are sold in Japan. Recent introduction of PPE/Nylon Blends, which have excellent dimensional stability, chemical resistance, and toughness, open new market opportunities such as automotive body panels. GE is the leading producer with over 80% share worldwide.

THERMOPLASTIC POLYESTERS

Thermoplastic polyester molding compounds are crystalline polymers with good chemical resistance (especially to organic solvents and oils), toughness, high heart resistance, and Dimensional stability. Polybutylene terephthalate (PBT) has received the most market acceptance compared to polyethylene terephthalate (PET) due to improved processing and electrical properties of the former material. However, the higher heat resistance of PET is desirable for some applications. Large markets include electrical/electronic application (such as intricate connectors) and automotive parts (such as distributor caps). About 70,000 tons per year are sold in Japan.

The title of my talk is "The Future of Engineering Plastics", so let me now describe what we at GE Plastics consider to be major trends for these materials in the future.

THE FIRST TREND IS THAT CONSUMERS AND MANUFACTURES ARE RECOGNIZING PLASTICS TO BE, IN FACT, HIGH QUALITY MATERIAL.

The old public image of plastics being cheap substitutes is giving way to a new image of plastics as high performance materials that enhance the design, function, manufacturability and general competitiveness of products. Because engineering plastics represent functional replacements of traditional materials, the polymers are providing high quality performance that equals and, in many cases, surpasses the materials being replaced.

For instance, with regard to applications most used by consumers, there are today in the marketplace products such as blenders and toasters made of plastics that are preferred over their metal counterparts.

Because consumers' perceptions of plastics are changing, their expectations of plastics are changing too. Being more comfortable with plastics, consumers are demanding, and will continue to demand, more sophisticated performance from plastics. In turn, the fulfillment of these demands by plastics suppliers will enhance the high quality image of plastics even more.

THE SECOND TREND IS THE INCREASING RECOGNITION BY MANUFACTURERS OF THE COMPETITIVE IMPORTANCE OF DESIGN.

A recent article in business week noted: "After relegating design to the backseat in the 1970s, U. S. manufacturers are once again discovering that it is key to industrial competitiveness. Design, they are learning, is more than skin deep. It's the very heart of a product. A good design appeals to the eye, but it also must be reliable easy, and economical to operate and service. It should also be simple to manufacture." In short, design not only enhances product appearance and function but it also drives manufacturing productivity.

Plastics, especially engineering thermoplastics, are especially suited to enhancing the power and scope of design. First, designing with engineering thermoplastics drives product quality advances. Quality is defined in one place only, the marketplace, and by one person only, the customer. Because the marketplace is constantly changing, because customers' needs are constantly changing, quality is constantly changing. So quality and change in the marketplace are not separate but interrelated. Using engineering thermoplastics offers the quickest and most efficient and effective means of adapting to and driving quality and change in the marketplace. Quality is enhanced through the use of engineering thermoplastics because designers have much greater freedom than when using metal, wood, glass or paper. Because suppliers can provide design and engineering expertise along with materials, new advances in materials performance and processing can be more rapidly incorporated into product development. New products, products directly linked to fulfilling consumer needs, can be more quickly cycled into the marketplace, usually at lower costs. Total cost incorporates not just materials' costs but manufacturing costs. Because the total manufacturing costs of a product usually breaks down to one third materials' cost and two thirds assembly costs, manufacturers must take a more comprehensive view of cost requirements. Engineering thermoplastics may be more costly on a per kilo basis, but their use can actually lower total manufacturing costs.

Designing with a broad range of performance offerings, designing for the combination of aesthetics and functionality, designing for serviceability, designing for manufacturability, designing with new materials and new processes, designing for quick cycling of products into the marketplace ----- all these design activities, that are enhanced by engineering thermoplastics ----- serve to make products less costly.

THE THIRD MARKETPLACE TREND THAT WILL ENHANCE THE GROWTH OF ENGINEERING THERMOPLASTICS IS THE MOVEMENT OF THE MARKET TOWARD LARGE PLASTIC PARTS.

For the first time since plastics were invented, the advent of new materials, new processing techniques and new conversion equipment is enabling engineering. Thermoplastics to be incorporated into large parts. In the early 1970s, the largest lexan polycarbonate injection molded part was a 10 kilogram snowmobile hood. Injection molding is an inherently slow process compared to metal stamping. Furthermore, the modulus of thermoplastics was not adequate for large part configuration. So large, infection molded parts, were impractical.

Before the development of high performance thermoplastic composites, like polyester/glass, manufacturers of large parts could choose either thermosetting composites, like polyester/glass, or stamped metal.
It was a limited choice. Both materials involve labor and time intensive processing. They require large amounts of floor space and very large machines. In addition, stamped metal requires expensive tooling and has limited functionality and design flexibility. But the development both of high modulus, engineering resin and glass mat composites and new conversion techniques, such as flow forming, is enabling engineering thermoplastics to be made into large parts. Composites combine a heterogeneous mixture of one or more polymers and reinforcements, providing a high modulus, though, lightweight material that is the functional equivalent of steel. They can be flow formed through a process that involves filling tools with preheated blanks of thermoplastic composites. Since the tools can incorporate functional components, large parts can be formed with such features as bosses, ribs, and inserts. Think of it: large, highly functional thermoplastic parts, flow formed in one operation.

These advances are leading to striking new innovations in the use of engineering thermoplastics in large parts, not just large parts for industrial applications, such as automobile seat shells, hoods, body and under body panels; large appliance fans, dryer tubs, bases and panels; construction modular flooring, to name just a few.

THE FOURTH TREND THAT WILL DRIVE ENGINEERING THERMOPLASTICS'
GROWTH IS MATERIALS' RECOVERY.

Recovery of materials, or recycling, is becoming a paramount
social issue, also of importance here in Japan. But equally
important are the business issues associated with materials'
recovery. Recycling can be a productive and profitable undertak-
ing. The metal industry has proven that point. Just as metals
can be re-used, so high performance engineering thermoplastics
can be profitably re-used as well. In a recent study, GE plastics found that engineering thermoplastics retain almost 100% of
their properties even after 10 years of weathering on first use
applications. Materials that ROT, RUST and CORRODE are simply
unproductive.

Indeed, Materials that cannot be used twice are also unproductive and are likely to become obsolete because they cannot
provide the economic advantages to be used once. For instance,
recycling ¥140 per kilo commodity plastics is not productive.
Cleaning them, reconverting them, repacking and reshipping them
is not worth the original cost. But recovering high value, high
performance engineering thermoplastics costing ¥500 or more per
kilo not only provides a significant value to recover, reconvert
and reuse them, it also provides materials stripped of original
manufacturing costs. In addition, using high performance engineering thermoplastics twice can drive out lower value materials
that are used only once. Just think of it, when plastics are
recycled, up to 60 percent of the energy required to make the
same product is saved.

THE FIFTH TREND IS THE MOVEMENT OF ENGINEERING THERMOPLASTICS
INTO THE BUILDING INDUSTRY.

As the cost of land, labor, materials, and capital continue
to rise around the world, engineering thermoplastics will provide
greater economies in manufacturing and building houses. Compared
to traditional materials of metal, wood and glass, engineering
thermoplastics can be more effectively designed for simplified
assembly procedures, they can be more effectively integrated into
automated and robotized production. And they can more effectively incorporate functional features. And houses built of engineering thermoplastics need not upset tradition. Such houses
will offer function, efficiency and beauty, with old and new
materials working side-by-side: the technology of the new combining with the wood, stone, and glass of tradition. GE plastics
is making a strong commitment to bring engineering thermoplastics
to the housing industry by building a polymer house, which will
open next September in Pittsfield, Massachusetts, U.S.A.

THE SIXTH TREND IN THE FUTURE OF ENGINEERING THERMOPLASTICS: OOBLENDS AND ALLOYS.

Polymer blends -- which are outgrowths of two other well-established composites, filled and reinforced plastics -- are achieving high growth for two reasons. One is that resin suppliers can longer afford to make the big investments needed in high research costs, long development times and marking uncertainties to develop and market new molecules. The only cost effective way to develop new materials is by blending.

The second reason is that suppliers, consumers, molders and extruders are increasingly developing new applications for engineering resins, requiring that materials be blended to fit specific market needs. Upgrading plastics performance is thus an optimization process that reinforces selected properties with due consideration to the established production and process specifications and to cost factors.

Blends are mixtures of two or more polymers that become a unique material. But blending is not simply mixing A & B to get C. Instead, it is a sophisticated effort that involves the utilization of a great deal of proprietary, specialized technology to create a chemical mix that will enhance the marriage of A & B. For example, Polymer "A" has excellent chemical resistant properties -- and Polymer "B" has high heat deflection temperature values -- but neither has both. By blending these resins, the polymer alloy will have both excellent chemical resistance and high heat deflection temperatures -- and consequently deliver the levels of performance the materials need to do the job effectively.

Today, there are numerous polymers alloy blends available in both commodity and engineering plastics. One of the original and most famous of the engineering blends is General Electric's MPPE or Noryl Resin. Polyphenylene ether itself had been of limited usefulness, until GE added polystyrene in the MID-1960s. The alloy was more processable than the basic resin. At the same time, the alloy retained the desirable physical properties.
With Noryl Resin, General Electric has served the low end of the engineering plastics market for years, filling the GAP between ABS and the higher-priced polycarbonate resins.

Examples of other blends produced by GE are: PC/ABS under the name cycolloy; PBT/ELASTOMER CALLED LOMOD. PC/PBT CALLED XENOY.

THE SEVENTH TREND IS IN PROCESSING TECHNOLOGY.

The principle method of forming thermoplastic materials is through injection molding. But recently some interesting new technologies and new ways of using old techniques have emerged.

For example, layering. In layering, different plastics each providing its own unique qualities, are combined during extrusion into a single structure capable of performing multiple tasks such as blocking the passage of light, flavor, moisture of gases such as oxygen or carbon dioxide. Engineering resins in general can be deficient in specific properties when used alone. For instance, Pet cannot be hot-filled; Polycarbonates have poor barrier and chemical resistance properties. Virtually all of these disadvantages of single polymer systems can be solved by layering technology. So a multi-layer film using PRT and Polycarbonate can have good barrier properties AND be hot-filled.

But advances in processing are not only keyed to new technologies but also to old technologies used in new ways. For example, Blow molding is a fifty year old method of forming plastics, mainly for bottles. Today, it has taken on a new aspect. That is compression blowmolding, a form of extrusion blowmoling accompanied by tacking off or squeezing together both walls of the parison, and producing parts with structurally enhanced characteristics. Structural blow molding produces double-walled configurations with extremely high stiffness-to-weight rations in large-sized parts. From an economy/performance standpoint, structural blowmolding fills a need for applications requiring fewer parts than would justify expensive tools for injection molding.

THE EIGHT TREND: THE INVENTION, DEVELOPMENT AND MANUFACTURING OF ENGINEERING RESINS REQUIRE SUBSTANTIAL INVESTMENT.

Despite the growth and profitability of engineering resins, it is quite difficult for new manufacturers to participate in this segment of the industry. First, the engineering resin producer must make strong commitments to a comprehensive marketing program and technology base from which product improvements will arise.

Second, the successful engineering resin producer must have the technological and financial strength to enter the market and the staying power to persist until financial results turn positive. These long payback periods result largely from the high cost of engineering resins capacity. The cost per unit capacity of an engineering resin plant is higher than that of the usual commodity facility for three reasons: First, engineering resin facilities are almost always smaller than commodity resin plants, and therefore, do not benefit from economies of scale to the same degree. Second, many engineering resins are produced VIA relatively complex syntheses which may require unusual or exotic and, therefore, expensive equipment. A third factor results from the typically rapid growth rates of engineering resins, which require that the manufacturing plants often be built to anticipate higher demand. This tends to cause initial low capacity utilization levels, which increase unit fixed costs.

So any company desiring to participate in the engineering plastics industry must be prepared to make very large investments, up front.

In conclusion, worldwide manufacturing is undergoing significant changes; changes in globalization, in technologies, and in materials. The most powerful change is that involving materials. And the materials that will be driving that change are high performance engineering thermoplastics. These materials provide productivity in design, manufacturing, cost reduction, lower energy requirements and materials' recovery. Engineering thermoplastics, thus far, have grown mostly through metal substitution. The fact that our industrial society is still predominately a metal-oriented society means that the growth of engineering thermoplastics has only just begun. Having only achieved a small fraction of its growth potential, the engineering thermoplastic industry is young, vibrant, and on the move.

Only strong companies can prevail in today's changing industrial environment. In addition, the successful producer of new materials must have a clear and comprehensive marketing vision and an abiding commitment to innovation and prudent risk that will turn that vision into profitable realities.

By the beginning of the next century, engineering thermoplastics will be so commonly used in so many different ways that their name will no longer be associated with exotic polymers, only playing a marginal role in manufacturing, but will be as common as the names: steel, glass and wood.

High—Temperature Oxide Superconductors :
Structural and Compositional Aspects
in the Manifestation of Superconductivity[*]

Kohji Kishio

Department of Industrial Chemistry, University of Tokyo,
Hongo, Bunkyo-ku, Tokyo 113, Japan

ABSTRACT

A series of recently discovered high-temperature cuprate superconductors are reviewed and compared mainly from the standpoint of structural and chemical aspects.

All the compounds are based on the perovskite structure and exhibit characteristic layered structures consisting of stackings of two-dimensionally connected copper-oxygen poly-hedra. The shapes and the combinations of the connected poly-hedra vary depending on the degree of stacking and on the nature of ordering of other constituent cations and seem to determine the approximate range of maximum T_c. However, appropriate concentrations of the charge carriers, either electrons or elec-tron holes, being accomplished by chemical doping, is quite important in the manifestation of superconductivity.

At this stage, the reason for the extraordinarily high-T_c observed in cuprate compounds has not been clarified, and it is still quite difficult to predict T_c values for new materials to be discovered in near future.

---

*) This paper is largely based on a review article "Crystal Struc-tures and Physical Properties of Ceramic Superconductors" that originally appeared in Japanese in "Ceramics Japan",24(8) 690-96 (1989).

INTRODUCTION

Almost three years have passed since high-T_c (critical temperature) superconductivity was discovered in a La-Ba-Cu-O system[1], and a considerable number of oxide superconductors with various chemical compositions as summarized in Table 1 have now been found. They are mostly based on rare earth or alkaline earth cuprates and possess crystal structures closely related to the perovskite-type. However, they commonly exhibit a distinct layered nature and show strongly two-dimensional characteristics of electronic structure as well as physical properties. In this respect, a single exception is the $(Ba,K)BiO_3$ system which has a three-dimensional ideal perovskite structure.

Quite interestingly, the first discovered high-T_c cuprate, $(La,Ba)_2CuO_4$ was initially found by Bednorz and Muller[1] by aiming at the perovskite composition, $(La,Ba)CuO_3$. This is because they selected the $Ba(Pb,Bi)O_3$ system[2] which had the highest T_c among the oxides at that time, as a prototype compound for the search for new materials. Whether this approach was the most effective way or not in search for new superconductors, all of the discovered cuprate superconductors do exhibit structures quite closely related to the perovskite. Moreover, a non-cuprate $(Ba,K)BiO_3$ with T_c=30 K[3], has been discovered through a close re-examination of the Ba-Bi-O system only after the birth of perovskite-related high T_c cuprates.

In this paper, I would like to review the crystal structures of various known cuprate superconductors and discuss correlations among composition, ionic nature of constituent elements and T_c. Although it would be desirable to extend the discussion also to non-cuprate superconductors, recent research efforts on those oxides based on other elements such as Ti, Nb, or W, have not been extensive enough to permit detailed comparison. Readers are referred to some review papers[4-6] in this direction.

Table 1. List of High-T_c Oxide Superconductors
(cationic and oxygen compositions are
ideal and vary in actual compounds.
T_c's are approximate maximum values)

"CuO$_x$ Single-Layered" Compounds

$(La_{1-x}M_x)_2CuO_4$ (x=0.08)
    M=Ba,Sr,Ca,Na,none   [20-40 K]

$(Nd_{0.8}Sr_{0.2}Ce_{0.2})_2CuO_4$ [30 K]

$(Ln_{1-X},M_X)_2CuO_4$ (x=0.08)
    Ln=Nd,Pr,Sm,Eu M=Ce,Th   [25 K]

"CuO$_x$ Multiple-Layered" Compounds

$Ba_2LnCu_3O_7$ [90 K]

$Ba_2LnCu_4O_8$ [80 K]

$Ba_4Ln_2Cu_7O_{15}$ [85 K]
    Ln= Y,La,Nd,Sm,Eu,Gd
    Dy,Ho,Er,Tm,Yb,Lu

$(Ba,Ln)_2(Ce,Ln)_2Cu_3O_8$ [43 K]
    Ln=Nd,Sm,Eu

"AO$_x$–CuO$_x$ Complex-Layered" Compounds

$(BiO)_2Sr_2Ca_{n-1}Cu_nO_{2n+2}$
    n=1,2,3   [7-110 K]

$(TlO)_1Sr_2Ca_{n-1}Cu_nO_{2n+2}$
    n=1,2,3   [75-100 K]

$(TlO)_2Ba_2Ca_{n-1}Cu_nO_{2n+2}$
    n=1,2,3   [80-125 K]

$(TlO)_1Ba_2Ca_{n-1}Cu_nO_{2n+2}$
    n=1,2,3,4,5   [20-120 K]

$Pb_2Sr_2(Y,Ca,Sr)Cu_3O_8$ [80 K]
    Ln= Y,La,Pr,Nd,Sm,Eu,
    Gd,Dy,Ho,Er,Tm

$Pb_2(Sr,Ln)_2Cu_2O_6$ [32 K]
    Ln=La

Non Cuprates

$Ba_{1-x}K_xBiO_3$ [30 K]

CuO$_x$ SINGLE-LAYERED COMPOUNDS

All of the cuprate superconductors exhibit characteristic layered structures. They consist of alternate stackings of two-dimensionally connected
Cu-O polyhedra (CuO$_x$ layers) and similarly connected AO$_x$ layers where A is
typically Bi,Tl, or Pb. Rare earth and/or alkaline earth ions are sandwiched
between these layers.

Among these materials, those having chemical compositions expressed as
(Ln,M)$_2$CuO$_4$ exhibit the simplest crystal structures; there are three such
structures. Each structure contains only a single type of connected CuO$_x$
layer without any AO$_x$ layers. Although their T$_c$'s are in the range of only 25
to 40 K and chemical compositions are quite similar, there are two classes of
compounds that are totally different in terms of their electronic transport
properties. Since this point seems to be the most essential and critical
issue in the current understanding of cuprate superconductors, I would like to
emphasize this group of materials in particular.

Structural Features

Fig. 1(a) shows the structure of (La,M)$_2$CuO$_4$, the first discovered high-
T$_c$ cuprate. The structure itself had been known as a K$_2$NiF$_4$-type (or the T-
type). The CuO$_x$ layers basically consist of interconnected corner-sharing
octahedra as in the cubic perovskite structure. However, the connections are
made only in two-dimensions, giving rise to a layered structure. Between the
two CuO$_x$ layers, La and M are situated with octahedral coordination of
oxygens, a situation quite similar to that found in the rock-salt structure.
When the total composition is designated as A$_2$BO$_4$, it may be interpreted as
A$_2$BO$_4$=ABO$_3$+AO. Because of this, the structure is sometimes described as

74

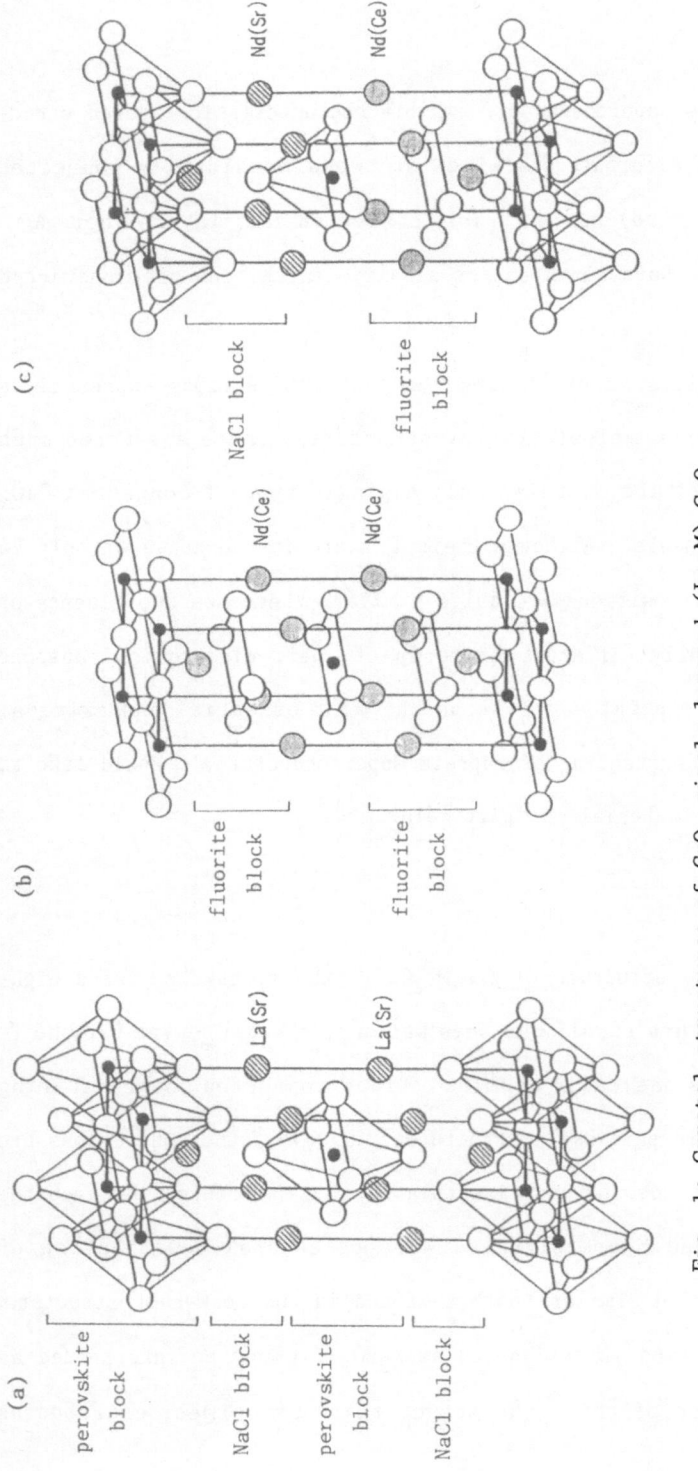

Fig. 1: Crystal structures of CuO$_x$ single-layered (Ln,M)$_2$CuO$_4$.

(a) K$_2$NiF$_4$-type (T-type); (La,Sr)$_2$CuO$_4$

(b) Nd$_2$CuO$_4$-type (T'-type); (Nd,Ce)$_2$CuO$_4$

(c) Intermediate structures of (a) and (b); T*-type; (Nd,Ce,Sr)$_2$CuO$_4$

In Figs. 1-5, black and white atoms reperesent copper and oxygen, respectively.

alternate stacking of a perovskite block, ABO_3, and a rock salt block, AO, although it is not possible to distinguish these A's since they are crystallo-graphically equivalent.

If the Cu-O bond lengths of the coordinated oxygen octahedra are careful-ly examined, it is found[7] that they are 1.89 Å within the horizontal plane and 2.43 Å for the vertical direction. Compared to 2.13 Å which is a simple sum of ionic radii of Cu^{2+} and O^{2-}, it is found that the former is shorter whereas the latter is appreciably longer. This suggests that the chemical bonding of vertical oxygens is much weaker than the bonding within the plane; the layered structure must be viewed as the strongly-interconnected square coordination of CuO_2.

When the Ln in $(Ln,M)_2CuO_4$ is replaced by the elements from Pr through Gd (the first half of the 4f elements), the structure transforms to the Nd_2CuO_4 (T'-type) shown in Fig. 1(b). The weakly coordinated apical oxygens in the vertical direction in the T-type structure are removed and the CuO_x layers become purely square-coordinated. The unbonded oxygens then move to the tetrahedral positions between two Nd layers, resulting in the local structure identical to that in the fluorite-type. If A_2BO_4 is interpreted as ABO_2+AO_2, then the present structure can be regarded as the alternate stacking of a defective perovskite block, ABO_2 and a fluorite block, AO_2.

Without the dopants M's, both of the above mentioned cuprates are merely semiconductors. They have clear energy gaps of basically a Mott-Hubbard type originating from strong correlations of electrons[8] in these systems. Super-conductivity appears only after partial substitution on the Ln^{3+} site by alio-valent dopant M's.

In this context it is interesting to see what kind of ions can be easily dissolved as dopants into the Ln sites and whether this is associated with the characteristics of each crystal structure. Experimental facts are that alkaline earth ions such as Ba, Sr or Ca are easily soluble in the T-type structure while tetra-valent Ce and Th are easily soluble in the T'-type. The reverse combinations of these are difficult. This may be understood as follows. The most stable form of pure Ce and Th oxides, as binary systems, is the fluorite-type structure; thus, these ions prefer the tetrahedral site in the oxygen sublattice of the T' structure. Alkaline earth ions, on the other hand, favor the octahedral sites of the rock salt-type structure as in their native binary oxides.

(Nd,Sr,Ce)$_2$CuO$_4$ shown in Fig. 1(c) is very interesting in that it exhibits a structure intermediate between 1(a) and (b). The unit cell consists of an upper half being the T-type while the lower half is identical to the T'-type. The resulting total structure is frequently referred to as the T*-type. The dopants, Sr and Ce, are confirmed by a neutron diffraction[9] to be partially substituting the Ln sites on the T- and T'-type sides, respectively. This distribution exactly follows the tendency of the dopant ions described above in Figs. 1(a) and (b). It should also be noted that, in this structure, the two-dimensionally connected copper-oxygen polyhedra are composed of pyramids because the apical oxygen on the T'-type side is missing. Quite recently, (La,Sr)LnCuO$_4$ where Ln=Sm, Eu and Gd, were confirmed[10] to exhibit the T* structure.

Manifestation of Superconductivity by Doping

Considering the Ln$_2$CuO$_4$ with Ln=La, the formal valence of Cu is increased by doping alkaline earth ions onto La sites, provided the oxygen composition is fixed at four. This corresponds to the doping of electron holes into the semiconductors in an electronic picture. However, these holes are not spatially distributed as in the free electron model, but enter a narrow energy band having a rather strong O-2p character[8]. In addition to this , the apical oxygens in the CuO_x polyhedra, octahedra or pyramids, were believed by many theorists to be particularly important in the manifestation of high-T_c superconductivity. However, these pictures were only valid until the end of 1988 when the T'-type superconductor was discovered.

Superconductivity in the T' structure was discovered by Tokura et al[11]. in $(Ln_{1-x}Ce_x)_2CuO_4$ with Ln=Nd, Pr, or Sm. The optimum Ce content, x, was found to be about 0.08, and T_c=25 K was obtained by quenching from high temperatures under reducing conditions. The valence of the Ce ion was confirmed to be +4 by XPS measurements. Doping of Th and F were also found[12-13] to be effective in making Nd_2CuO_4 superconductive. The Hall coefficient and thermoelectric power showed negative signs[14]. From all of these observations, it is clear that the doped carrier in this system is not electron holes, as has been observed in all other cuprate superconductors. The most significant aspects in the discovery of "electron-doped superconductors" are that neither holes in the O-2p band character nor the vertical apical oxygens around the Cu sites are indispensable to superconductivity.

While the T^*-type $(Nd,Ce,Sr)_2CuO_4$ $(T_c$=28 K)[15] had been a predecessor to T-type superconductors, it was a hole-doped conductor. It is now well recognized that the T-type $(La,Sr)_2CuO_4$ $(T_c$=40 K) is a typical hole-conductor.

It is interesting to see that all of these CuO_2 single-layered compounds, T, T^* and T' types, exhibit quite similar and rather low T_c compared to other multi-layered compounds, regardless of the local CuO_x structure. The most essential feature is that the two-dimensional planes consisting of interconnected layers of CuO_x are necessary in superconductivity, provided the doping of either electrons or holes is made in this plane. However, it should be noted that there is an optimum carrier concentration for maximum T_c as excessive doping makes the system more metallic but not superconductive[16].

In the previous section, the doping of ions in the three types of compounds was discussed from the viewpoint of local structures. The situation can also be considered from the perspective of carrier doping into the CuO_2 plane. Since negatively charged apical oxygen ions exist in the T-type structure, it is easier to dope, in the CuO_2 planes, electronically positive holes than electrons. Conversely, electrons are more easily doped than holes in the T'-type structure. More quantitative discussions along this line, using the calculation of Madelung energy on the various sites of these structures, can be found in the literature[17].

CuO_x MULTI-LAYERED COMPOUNDS

Structures with multi-layered CuO_x

A structure unknown before the birth of liquid nitrogen temperature superconductors but which immediately became one of the most famous ones in crystal chemistry, is $Ba_2YCu_3O_7$ (so-called 123 structure, shown in Fig. 2). The unit cell is based on triple unit cells of a perovskite, $(Ba,Y)_3Cu_3O_9$, from which oxygens are partially removed. Since two Ba ions and one Y ion line up in a regular manner, -Ba-Y-Ba-Ba-Y-Ba-, and oxygen becomes defective also in an ordered manner, two nonequivalent sites of Cu are formed.

Fig. 3 shows a closely related compound, $Ba_2YCu_4O_8$ (so-called 124 or 248). This 124 structure is quasi-quadruple in CuO_x layers, consisting of two interconnected pyramid CuO_x layers which face each other and two one-dimensional CuO chains. It should be noted that only a single chain was contained in the 123 structure. Another compound of this family is 247 (=123+124), $Ba_4Y_2Cu_7O_{15}$ which is made up by alternative stackings of the 123 and 124 structures. At first, these 123 related compounds were observed only in thin or thick film specimens, but recently they have been found to be a thermodynamically stable phase and can be synthesized[18] as bulk phases under high oxygen

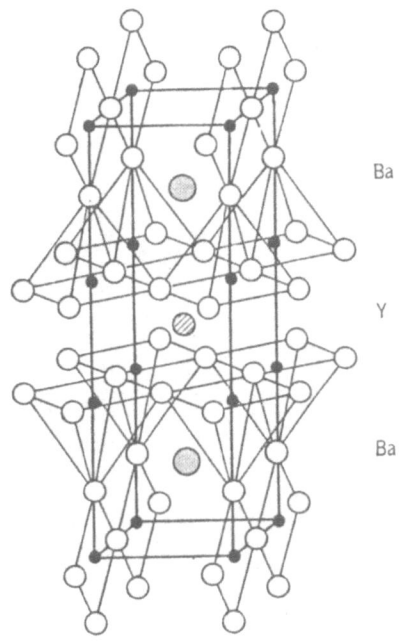

Ba

Y

Ba

Fig. 2: Crystal structure of "123-type" $Ba_2YCu_3O_7$.

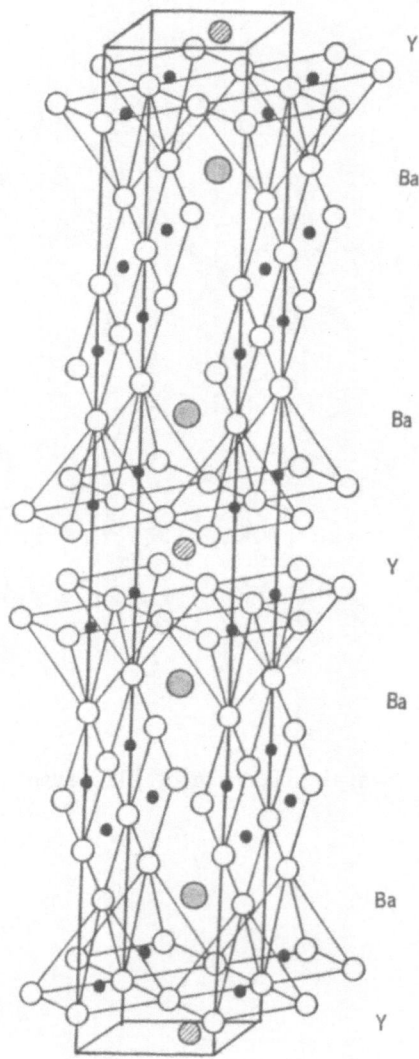

Fig. 3: Crystal structure of "124-type" $Ba_2YCu_4O_8$.

pressure treatments around 1000°C.

All of these three compounds exhibit T_c in the range of 80 to 90 K, irrespective of the number of Cu–O linear chains. This clearly suggests that the superconducting path is the interconnected CuO_2 plane of corner–sharing pyramids. The linear chains are thought to act as a "charge reservoir"[19] for electron holes. Because the oxygens in this chain are not strongly bonded, the oxygen nonstoichiometry of the 123 phase is appreciable and its T_c varies easily from 90 K, 60 K and to 0 K (non–superconductor) depending the oxygen composition. It is interesting from a materials processing view–point to note that the 124 compound is less susceptible to oxygen loss due to the stronger bonding of oxygen in the chain sites. Thus, the low temperature annealing of the specimens is not necessary as in the 123 phase.

Quite recently, Sawa et al.[20] discovered $(Ln,Ce)_2(Ba,Ln)_2Cu_3O_{10}$; (Ln=Nd, Sm or Eu) having T_c=43 K. The structure is shown in Fig. 4. Although the combination of the constituent elements in this system is analogous to T^*-type $(Nd,Sr)_2CuO_4$, the local structure resembles that of the 123 compound. The mutual arrangement of Cu ions is identical to that in the 123 compound, it has been reported that the oxygen sites around the middle Cu sites are about 50% defective similarly as in the 123 structure. However, the structure is tetragonal and no evidence of oxygen ordering has been observed by electron diffraction. The relatively low T_c compared to the 123 compound may perhaps be attributed to the fact that the local configuration around the pyramids is different from the 123 structure but equivalent to that in the T^* phase.

Compounds with complex multi-layered structures

The complex multi-layered family compounds[21] can be denoted as $(AO)_mM_2Ca_{n-1}Cu_nO_{2n+2}$ where A=Bi, Tl or Pb and M=Ba, or Sr. Each combination of elements contains a series of homologous phases with various m and n values

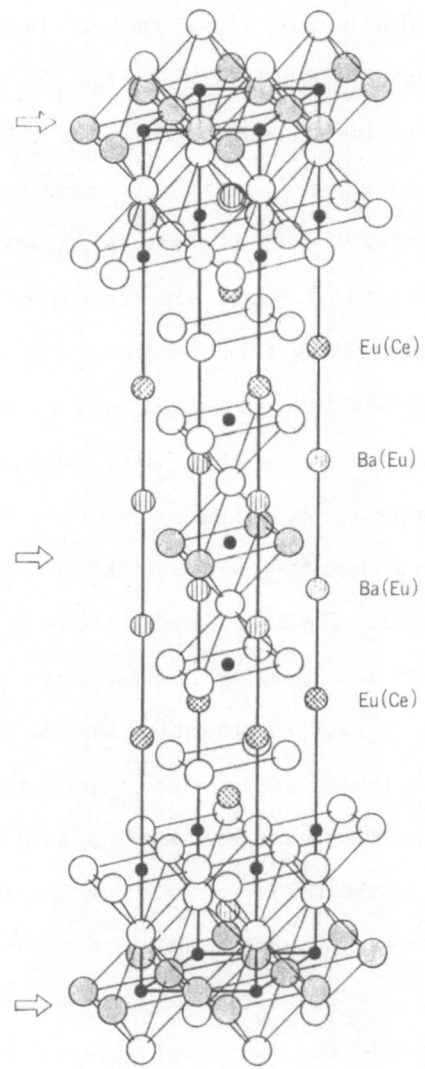

Fig. 4: Crystal structure[20] of $(Eu,Ce)_2(Ba,Eu)_2Cu_3O_z$.

It is reported that oxygens in the CuO_2 plane indicated by arrows

are 50% deficient.

and exhibits a wide range of T_c's. Among these, $(TlO)_2Ba_2Ca_2Cu_3O_8$ has been confirmed to have the highest T_c. Historically, the n=1 phases for A=Bi and Tl were discovered first and as the homologous phases having n=2 and 3 were found, the T_c gradually increased. Therefore, the synthesis of phases with n=4 and 5 were eagerly anticipated. However, these multiple layered phases, recently synthesized in Bi and Pb systems, showed decreasing T_c over n=3[22].

Since it is not possible to compare various compounds of this family within the space available here, only the structure of $Bi_2Sr_2CaCu_2O_8$ is shown in Fig. 5. The corresponding Tl-compounds exhibit essentially the same structure. The stacking sequence of the structure in Fig. 5 is expressed as -(BiO-BiO)-SrO-CuO$_2$-Ca-CuO$_2$-SrO-(BiO-BiO)-, where BiO double layers are arranged like in rock salt structures. The local structural configuration in which the bottom faces of pyramids face each other is similar to that in the 123 phase, and it is interesting to observe that the T_c (about 80 K) is in a similar range.

When compared to the corresponding Tl-compounds, the spacing within BiO double layers is much longer and Bi-compounds show characteristic cleavage. Another point to be noted is that the Bi-compounds exhibit a distinctive modulation structure which is not shown in Fig. 5 in which only an averaged structure is given. While this modulation has been interpreted as the combination of compositional and displacive modulations, it exhibits quite complex behavior[23] depending on the value of n and the preparation conditions of the specimens. The origin of the modulation could be related to the electronic structure of Bi^{3+} ions whose $6s^2$ electrons have a tendency to behave in a lone-pair-like manner[24].

There have been more than twenty compounds including non-superconductors discovered in this complex multi-layered compound family. Recently, Pb-systems such as $Pb_2Sr_2(Sr,Ca,Ln)Cu_3O_8$ (T_c=85K)[25] and $Pb_2(Sr,Ln)_2Cu_2O_6$

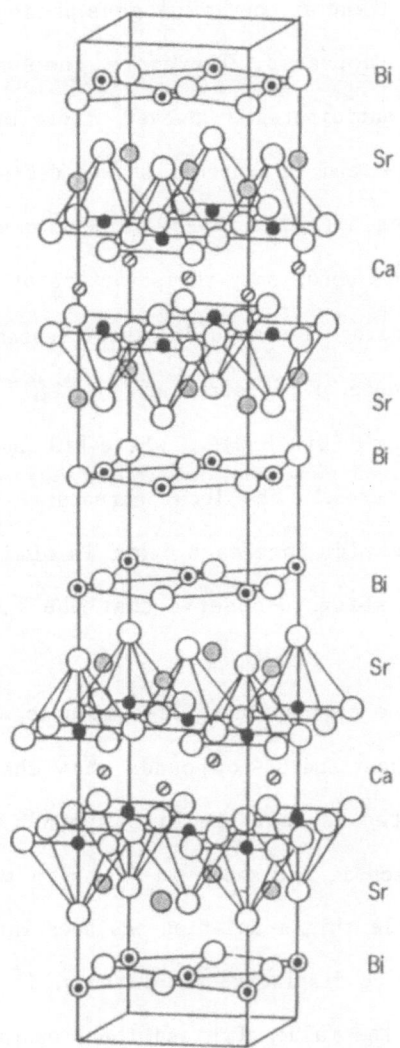

Bi

Sr

Ca

Sr

Bi

Bi

Sr

Ca

Sr

Bi

Fig. 5: Average crystal structure of $Bi_2Sr_2CaCu_2O_8$.

$(T_c=32$ K$)^{26}$ have been found, and the number of oxide superconductors in this family is expected to increase still very rapidly. However, as yet there is theoretically no rational basis for predicting the T_c of compounds exhibiting both new composition and structure. Unfortunately, we still have to take a relatively empirical path in our search for new high-T_c materials.

REFERENCES

1. J. G. Bednorz and K. A. Muller, "Possible High T_c Superconductivity in the Ba-La-Cu-O System", Z. Phys., B64 189-93 (1986).

2. A. W. Sleight. J. L. Gillson and P. E. Bierstedt, "High-Temperature Superconductivity in the BaPb$_{1-x}$Bi$_x$O$_3$ System", Solid State Commun., 17 [1] 27-28 (1975).

3. R. J. Cava, B. Batlogg, J. J. Krajewski, R. Farrow, L. W. Rupp Jr., A. E. White, K. Short, W. F. Peck and T. Kometani, "Superconductivity near 30 K without Copper: The Ba$_{0.6}$K$_{0.4}$BiO$_3$ Perovskite", Nature, 332 [6167] 814-16 (1988).

4. K. Kitazawa and S. Tanaka, "New Inorganic and Organic Superconductors", Hyoumen, 19 [7] 346-56 (1981), (in Japanese).

5. A. W. Sleight, "Oxide Superconductors: A Chemist's View", Mat. Res. Soc. Symp. Proc., 99 3-8 (1988).

6. K. Kitazawa and K. Kishio, "Structures and Properties of High-T_c Oxide Superconductors", Oyo-Butsuri, 57 [11] 1644-65 (1989), (in Japanese).

7. I. K. Schuller and J. D. Jorgensen, "Structure of High T_c Oxide Superconductors", MRS Bulletin, 14 [1] 27-30 (1989).

8. V. J. Emery, "Perspectives on the Theory of the New High T_c Superconducting Oxides", ibid, 67-71 (1989).

9. F. Izumi, E. Takayama-Muromachi, A. Fujimori, T. Kamiyama, H. Asano, J. Akimitsu and H. Sawa, "Metal Ordering and Oxygen Displacements in $(Nd,Sr,Ce)_2CuO_{4-y}$", Physica C158 [3] 440-48 (1989).

10. Y. Tokura, H. Takagi, H. Watabe, H. Matsubara, S. Uchida, K. Hiraga, T. Oku, T. Mochiku and H. Asano, "A New family of Layered Copper Oxide Compounds with Ordered Cations: Prospective High-Temperature Superconductors", Phys. Rev. B, in press.

11. Y. Tokura, H. Takagi and S. Uchida, "A Superconducting Copper Oxide Compound with Electrons as the Charge Carriers", Nature 337 [6205] 345-47 (1989).

12. J. T. Markert and M. B. Maple, "High Temperature Superconductivity in Th-Doped Nd_2CuO_{4-y}", Solid State Commun., 70 [2] 145-47 (1989).

13. A.C.W.P. James, S. M. Zahurak and D. W. Murphy, "Superconductivity at 27 K in $T'-Nd_2CuO_{4-x}F_x$", Nature, 338 [6212], 240-41 (1989).

14. H. Takagi, S. Uchida and Y. Tokura, "Superconductivity Produced by Electron Doping in CuO_2-Layered Compounds", Phys. Rev. Lett., 62 [10] 1197-1200 (1989).

15. J. Akimitsu, S. Suzuki, M. Watanabe and H. Sawa, "Superconductivity in the Nd-Sr-Ce-Cu-O System", Jpn. J. Appl. Phys., 27 [10] L1859-60 (1988).

16. J. B. Torrance, Y. Tokura, A. I. Nazzal, A. Bezinge, T. C. Huang and S. S. P. Parkin, "Anomalous Disappearance of High-T_c Superconductivity at High Hole Concentration in Metallic $La_{2-x}Sr_xCuO_4$", Phys. Rev. Lett. 61 [9] 1127-30 (1989).

17. J. B. Torrance and R. M. Metzger, Preprint.

18. J. Karpinski, E. Kaldis, E. Jilek, S. Rusiecki and B. Bucher, "Bulk Synthesis of the 81-K Superconductor $YBa_2Cu_4O_8$ at High Oxygen Pressure", Nature, 336 [6200] 660-62 (1988).

19. R. J. Cava, "Crystal Chemistry of the Copper Oxide Based High Temperature Superconductors", in "Advances in Superconductivity", Proc. 1st Int. Symp. on Superconductivity, Nagoya, 1988, pp.160-64, Springer-Verlag, Tokyo (1989).

20. H. Sawa, K. Obara, J. Akimitsu, Y. Matsui and S. Horiuchi, submitted to J. Phys. Soc. Jpn.

21. A. W. Sleight, M. A. Subramanian and C. C. Torardi, "High T_c Bismuth and Thallium Copper oxide Superconductors", MRS Bulletin, 14 [1] 45-48 (1989).

22. H. Ihara, R. Sugise, T. Shimomura, M. Hirabayashi, N. Terada, M. Jo, K. Hayashi, M. Tokumoto, K. Murata and S. Ohashi, "New Tl-Ba-Ca-Cu-O (1234, 1245 and 2234) Superconductors with Very High T_c", Ref. 19, 793-98.

23. Y. Matsui, H. Maeda, Y. Tanaka, S. Horiuchi, S. Takekawa, E. Takayama-Muromachi, A. Umezono and K. Ibe, "Applications of High Resolution Electron Microscopy to the Modulated Structures in Bismuth-Based Superconducting Oxides", JEOL News, 26E [2] 16-21 (1988).

24. B. Raveau, "The Superconductive Cuprates, a Large Structural Family, with Many Possibilities But Also with Many Defects", Proc. 1989 Spring Meeting of Materials Research Society, San Diego, 1989, to be published.

25. R. J. Cava, B. Batlogg, J. J. Krajewski, L. W. Rupp,Jr., L. F. Schneemeyer, T. Siegrist, R. B. van Dover, P. Marsh, W. F. Peck, Jr., P. K. Gallagher, S. H. Glarum, J. H. Marshall, R. C. Farrow, J. V. Waszczak, R. Hull and P. Trevor, "Superconductivity Near 70 K in a New Family of Layered Copper Oxides", Nature, 336 [6196] 211-14 (1988).

26. H. W. Zandbergen, W. T. Fu, J. M. van Ruitenbeek, L. J. de Jongh, G. van Tendeloo and S. Amelincks, Physica C159 [1] 81-85 (1989).

Advanced Ferrous Materials

Takuo Kohno

Deputy Director, Central R & D Bureau

Nippon Steel Corporation

1. Introduction

Man has been long associated with iron. Evidence shows that iron was used in Egypt and Mesopotamia in 3000 to 4000 B.C. The Old Testament, probably written in about 400 B.C., contains references to iron. Ferrous metals have been indispensable materials for the life of people and will assume ever increasing importance in the future.

I will start my presentation with the history of ferrous metals, touch on the uses of ferrous metals as conventional structural and functional materials, and introduce new ferrous metals as advanced materials.

2. History of Ferrous Metals

Traces are found of iron used by man in Egypt and Mesopotamia in about 3000 to 4000 B.C.

Japan entered the "Iron Age" with the introduction of ironware from mainland China into Kyushu in approximately 300 B.C. The Manyoshu, a collection of verse from the earliest times to the year 760, contains poems on iron and testifies to the antiquity of man's relation with iron.

In the seventeenth century, the "tatara (furnace)"

ironmaking process was developed, mainly in the Izumo region west of present Osaka, and reached its climax with the invention of "tenbin" and "fuigo (bellows)", as chronicled in Table 1.

In the middle of the nineteenth century, modern blast furnaces were built in Kamaishi, north of Tokyo, and marked the beginning of the steel industry in Japan. Since then, the Japanese steel industry has made phenomenal progress and now leads the world in ironmaking and steelmaking technology.

Table 1 History of ferrous metals

3000 to 4000 B.C.:
  Iron was used in Egypt and Mesopotamia.

400 B.C.:
  Use of iron was referred to in the Books of Isaiah and Micah in the Old Testament, as indicated by "they shall beat their swords into plowshares, and their spears into pruning hooks, ...".

300 B.C.:
  Ironware was introduced from the mainland of China into Kyushu, Japan.

100 B.C.:
  Ironmaking technology emerged in Japan. This can be inferred from a mythical story that the Ameno Murakumono Tsurugi (Rain-making Sword) was found in the dead body of the Yamata no Orochi (Eight-Headed and Eight-Tailed Dragon) slain by the Susanowo no Mikoto (Storm God).

607 A.D.:
  Iron nails were used to build the Horyuji, a Buddhist monastery.

759 A.D.:
  The Manyoshu was compiled. Of many poems mentioning iron, an example may be cited:
  "Muratamano Kuruni kugisashi Katametoshi Imoga kokorowa Ayokunamekamo"

> (Would my wife's lonely mind vacillate
> In my absence that I locked as securely
> As I locked the door hinges thrusting nails
> Into the ever freakish pivot sockets?)

14th century
Many charcoal-fueled blast furnaces were built and used in Europe.

1543:
Muskets were brought by the Portuguese to Tanegashima, a small island off the shore of southern Kyushu.

1580:
Oda Nobunaga built an ironclad ship.

1681:
With the invention of "tenbin" and "fuigo (bellows)," the "tatara" ironmaking process reached its climax in the Izumo region.

1735:
A. Darby, Jr. of England established the coke-fired blast furnace process.

1850:
Japan's first reverberatory furnaces were built in Saga, a province in Kyushu.

1853:
A reverberatory furnace was built in Nirayama, Izu, south of present Tokyo.

1855:
H. Bessemer of England invented the pneumatic steelmaking process.

1856:
Work was started on the construction of a blast furnace in Kamaishi, Japan.

1864:
P. Martin of England invented the open-hearth steelmaking process.

1878:
S.G. Thomas of England invented the Thomas converter.

3. Ferrous Metals as Structural Materials (for Towers and Buildings)

3.1 Uses of ferrous metals as structural materials are too numerous to mention. The Eiffel Tower that was built at the site of the World Exposition in Paris in 1889, just 100 years ago, is a famous steel structure.

The 300-meter tall tower was constructed with 7,300 tons of Bessemer steel produced in France. The Tokyo Tower, erected 70 years later, stands at 333 meters and used 4,000 tons of steel, graphically attesting to the improvements that had been made in the strength and toughness of steel in the intervening years.

The Hong Kong Head Office of the Bank of China, scheduled for completion in December of this year, will be 315 meters (70 stories) high and consume 15,700 tons of steel.

The data for the these structures is given in Table 2.

Table 2 Date for the two towers and one building

	Eiffel Tower Paris	Tokyo Tower Tokyo	Hong Kong Head Office Bank of China
Year constructed	1889 Built at site of International Exhibition in Paris (Centennial this year)	1958 (Completed in November 1958)	Scheduled for completion in December 1989
Height	Approx. 315 meters	333 meters	315 meters (70 stories)
Steel consumption	7,300 tons	4,000 tons	15,700 tons
Other	Designer: Alexandre Eiffel Type of steel: Bessemer steel	Designer: Kazunaka Naito	Engineering firm: Robertson, Fowler & Associates, USA

3.2 Bridges

The strength of galvanized steel wires for bridge cables has virtually remained at around 160 kgf/mm^2 for the past half century, as shown in Fig. 1. The Akashi Kaikyo Bridge now under construction is a long suspension bridge with a center span of 1,990 m, 1.4 times as long as that of the Humber Bridge, the world's longest suspension bridge to date. Cable wires of higher strength were essential for the structural stability and economy of the Akashi Kaikyo Bridge and the feasibility of the construction project depended on the development of such steel wires.

Wires with a strength of 180 kgf/mm^2, 20 kgf/mm^2 higher than that of conventional steel wires, were successfully produced from eutectoid steel containing 1% silicon, which is effective in strengthening the wire steel in the heat treatment stage before drawing and then softening the wire steel in the galvanizing stage after drawing. The high-strength galvanized steel wires have helped to allow the bridge roadway to be suspended with a single cable on either side, reduce the height of the towers, decrease the size of the substructure, and have contributed to the realization of the Akashi Kaikyo Bridge project (Fig. 2).

The Messina Bridge with a center span of as much as 3,300 m is being planned in Italy. The development of steel wires with a higher strength of 200 kgf/mm^2 is demanded for this bridge.

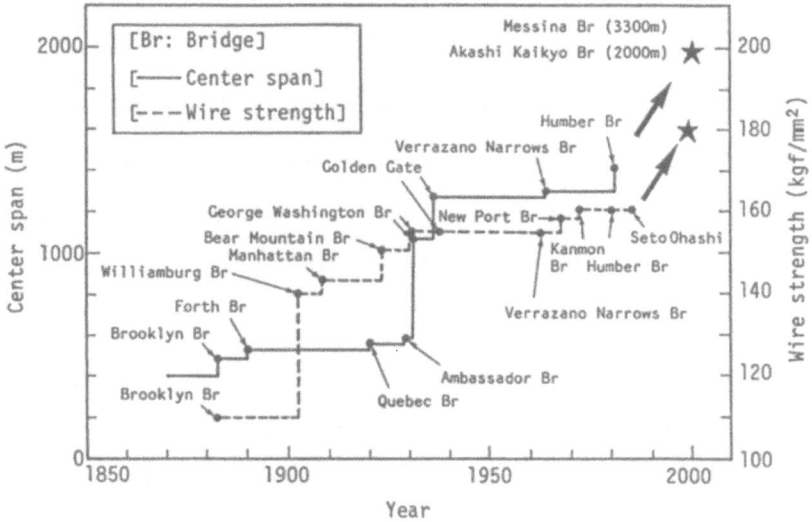

Fig. 1 Changes in center span and wire strength of Akashi
Kaikyo Bridge and other suspension bridges.

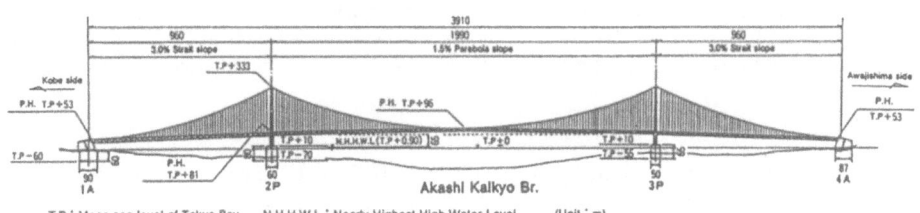

Fig. 2 Side elevation of Akashi Kaikyo Bridge
(Source: Honshu-Shikoku Bridge Authority).

3.3 Ultrafine Steel Wires

Steel wires for tire cord fabric have the highest strength of over 300 kgf/mm^2 among the ferrous metals in current use. They have a predominant position as tire reinforcement over competing fibers, such as aramid, glass and carbon fibers, in terms of tire performance, economy, and quality stability.

Research and development have made rapid progress in higher steel cord strength in recent years. Steel cord wires, measuring 150 to 300 μm in diameter and featuring a strength of 340 kgf/mm^2, are on the market. This level of strength is accomplished by the progress of the steelmaking technology that can eliminate inclusions and harmful elements that cause the breakage of ultrafine wires in the drawing and twisting stages. Much effort is also expended now in developing new steel grades and fabrication techniques to produce steel cord wires with a strength of 400 kgf/mm^2.

The construction of a steel-belted radial tire is illustrated in Fig. 3.

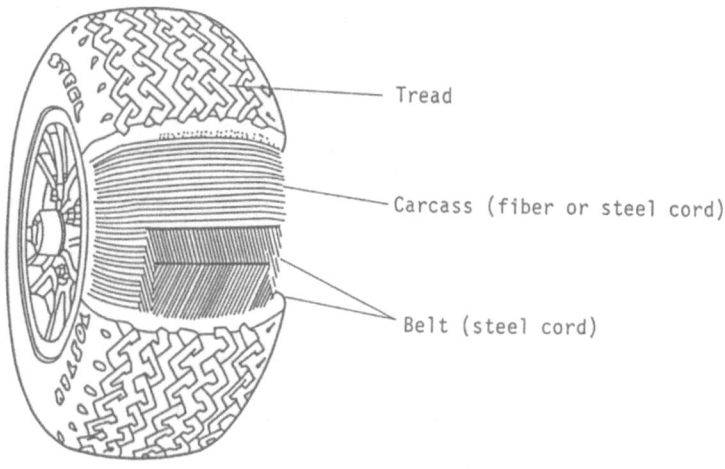

Fig. 3   Construction of a steel-belted radial tire
and typical use of steel cord.

Table 3   Properties of ultrafine fibers

Item of comparison	Steel fiber	Kevlar fiber	Glass fiber	Carbon fiber
Diameter (μm)	15–300	12	15	7
Tensile strength (kgf/mm^2)	340–450	300	220	360
Young's modulus (kgf/mm^2)	21,000	13,000	7,000	60,000
Density (g/cm^2)	7.8	1.5	2.5	1.8
Price (1) (¥/kg)	500–5,000	10,000–30,000	1,000–1,500	10,000–20,000
Price (2) (¥/kg)	19–186	112–336	36–55	30

96

4. Ferrous Metals as Functional Materials

4.1 Silicon Steel

Silicon steel is a representative example of steel used as a functional material.

Silicon-bearing steel has excellent magnetic properties and is employed as electrical steel in many electric appliances.

A new process was developed whereby the irradiation of the silicon steel surface with a laser beam refines the magnetic domains in the steel and sharply reduces the eddy-current loss that accounts for most of the core loss in silicon steel. The (C) process provides a core loss reduction of about 10% as compared with the best grade of grain-oriented silicon steel, as shown in Fig. 4.

Another domain-refining process that imparts heat resistance to silicon steel has also been developed (D).

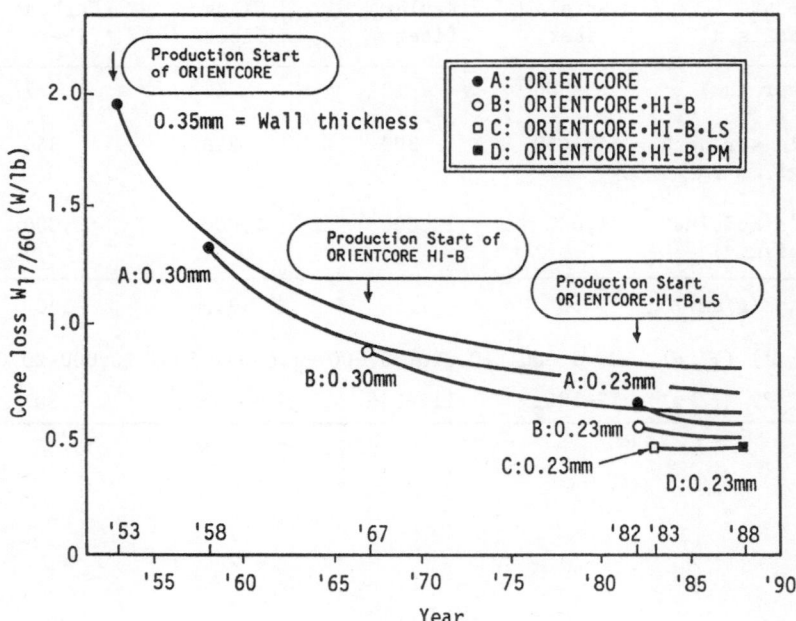

Fig. 4 Change in core loss of grain-oriented silicon steels in Japan.

5. Topics for Steel as an Advanced Material

5.1 Bake-Hardenable Steel Sheet

The first oil crisis of 1973 challenged the automobile industry to conserve materials and energy, and made it a most urgent issue to develop automobiles with greater fuel economy. Auto body weight reduction was devised as one measure for improving fuel economy and high-strength steel sheet was applied for this purpose.

One advantage of high-strength steel sheet is section thickness reduction (weight reduction). Production problems had to be solved to assure good press formability as well as high stretch stiffness, dent resistance, and impact strength.

After the solution of these problems, such bake-hardenable steel sheet was developed that has a tensile strength of approximately 40 kgf/mm^2 before press forming and provided a tensile strength of 60 kgf/mm^2 in the baking step after press forming. The basic concept of imparting bake hardenability is illustrated in Fig. 5.

The bake-hardenable steel sheet satisfied the demand of customers and has been extensively adopted since 1980.

High-strength steel sheet with a high retained austenite content and high-strength steel sheet of the copper solid solution type have been also developed.

a) After temper rolling   b) After press forming   c) After paint baking

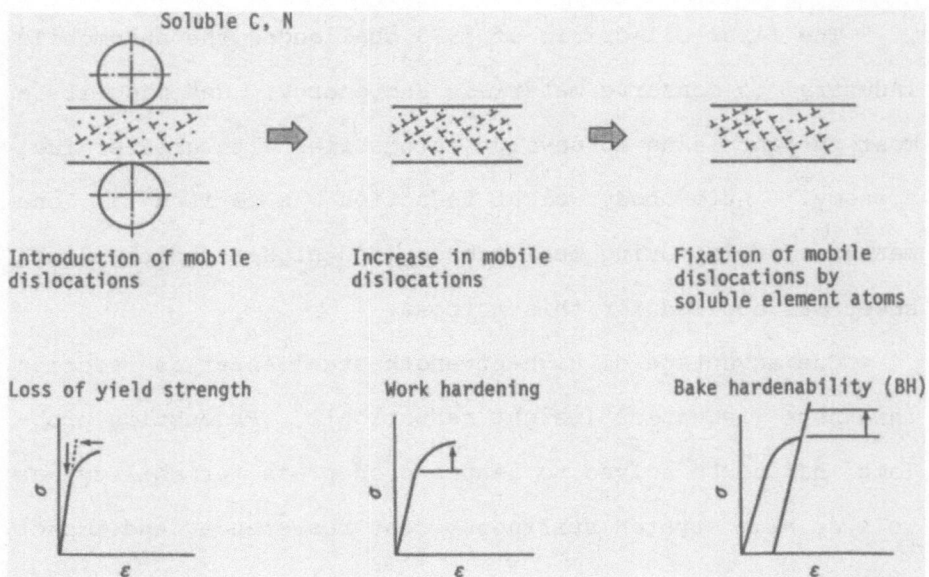

According to Fig. 5

a) Temper rolling introduces slight mobile dislocations and decreases yield strength.

b) Press forming introduces more mobile dislocations and causes the work hardening phenomenon to appear.

c) Paint baking helps soluble atoms to fix the mobile dislocations and provide a higher yield strength. This is called bake hardenability (BH).

Fig. 5 Basic concept of bake hardenability development.

5.2 Galvanized Steel Sheet for Automobiles

Highways are sprayed with large amounts of deicing rock salt in the cold regions of North America and North Europe. In this environment, automobiles are severely corroded and annually, increasing corrosion resistance has been required for coated steel sheet for automobiles.

Five and ten years of guarantee against surface corrosion and perforation, respectively, are the worldwide goals at present. These corrosion protection goals can be easily met by use of steel sheet coated with large amounts of zinc. Heavy zinc coatings, however, cause problems with press forming, welding, and painting.

Nippon Steel developed the new coated steel sheet "Excelite" jointly with an automaker. Excelite is a steel sheet electroplated with a zinc-iron alloy of higher corrosion resistance than pure zinc. It consists of an iron-rich upper layer and a zinc-rich lower layer as illustrated in Fig. 6. The two layers supplement each other in corrosion resistance and paintability as shown in Fig. 7. Thanks to its superior overall performance, Excelite is used in large quantities and is highly rated by customers.

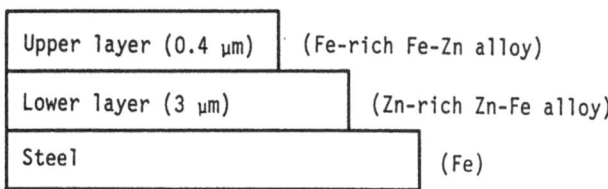

Fig. 6 Coating construction of Excelite.

Upper and lower layers of Excelite have different functions and display superior overall performance.

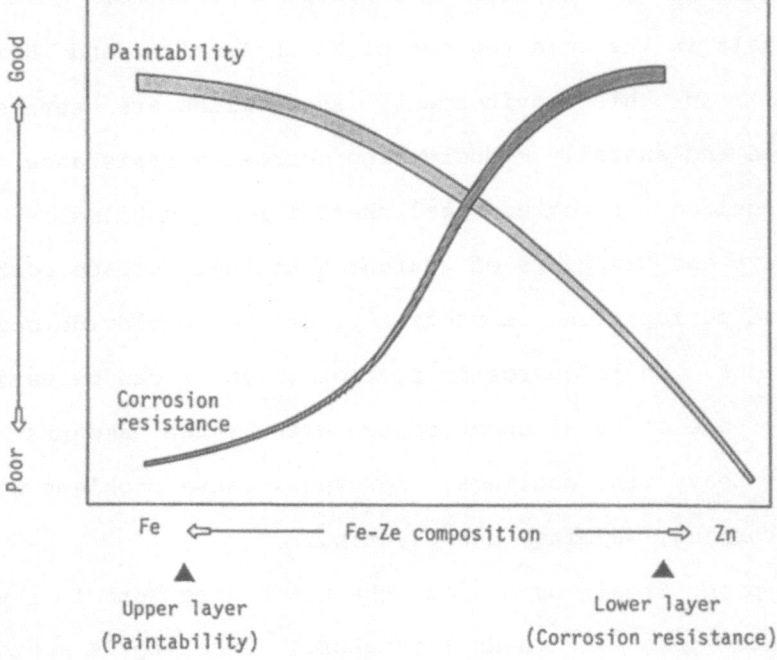

Fig. 7 Functions of Excelite surface layers.

5.3 Vibration-Damping Steel Sheet

Vibration-damping steel sheet is a plastic-steel composite of sandwish construction that has a viscoelastic resin layer of 40 to 80 m thickness enclosed between two steel sheets (Fig. 8). This material is characterized by a large vibration energy absorbing capability or an extremely large loss factor.

When the vibration-damping steel sheet is subjected to vibration, the resin layer undergoes shear deformation (slip deformation) and converts deformation energy into thermal energy. Vibration and noise are thereby abated.

The loss factors of different metals are shown in Fig. 9. Metallic materials generally have a low loss factor and are likely to vibrate and produce noise.

In recent years, noise control has come to be considered as an important social issue, as well as a critical
economical issue, because low noise enhances the commercial
value of machines and structures.

Vibration-damping steel sheet is used in various
industries, including automobiles, electric appliances,
building materials, and industrial machinery, and is
finding increasing use in many applications.

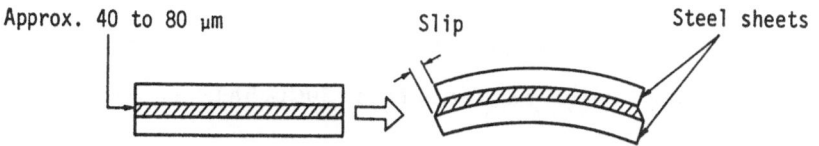

Bending vibration is attenuated by "slip deformation"
of viscoelastic resin.

Fig. 8   Construction of vibration-damping steel sheet.

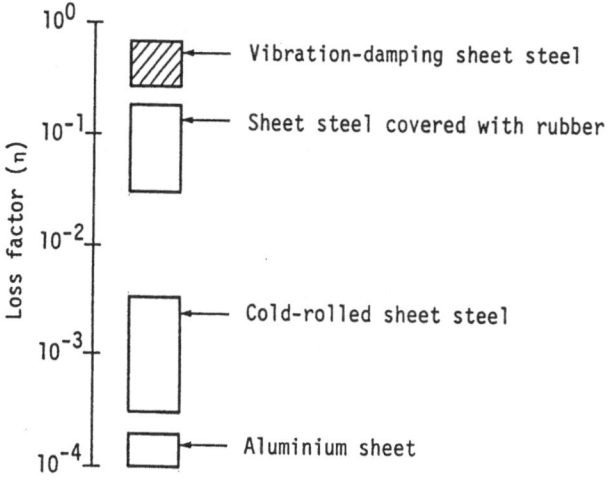

Fig. 9   Loss factor of various metallic materials
at room temperature.

5.4 Metallic Supports for Catalytic Converters

Ceramic honeycombs have been used in catalytic converters for automotive emission control.

Attempts had long been made to use metal support of heat resistant stainless steel, but they did not succeed mainly because of insufficient heat resistance of the steel foil.

Recent progress of material engineering has made it possible to stainless steel foils of higher heat resistance, and some European automobile models reportedly have actually started using this type of catalytic converters.

The steel foil is made from 20Cr-5Al heat-resistant stainless microalloyed with rare-earth elements.

In the hot exhaust gas of the automobile, the aluminum in the foil is selectively oxidized to create an aluminum oxide film which improves oxidation resistance. Furthermore this aluminum oxide film contributes to the better adhesion between the γ-alumina washcoat pregnated with platinum catalyst and the metal support.

The wall thickness of ceramic honeycomb is 120 - 200 μm, while it is only 50 μm for a metal support which offers a higher open cross section, resulting in lower exhaust gas back-pressure. The metal support does not need a knitted mesh cushion of expensive heat resistant alloys, as is necessary for the ceramic honeycomb, and it can also be used at higher temperatures, enabling the engine output increase.

Thus the metal support is anxiously awaited by automakers as a promising part that meets the recent trend towards higher engine output.

6. Conclusions

The history of ferrous metals has now been reviewed. It is known that man has used tools made of ferrous metals since ancient times.

The blast furnace ironmaking process was invented in the fourteenth century and the pneumatic steelmaking process was invented by Henry Bessemer in the middle of the nineteenth century. The two processes helped to establish the modern ironmaking and steelmaking processes and usher in the "Industrial Age."

History tells us that ferrous metals have made immeasurable contributions to the advancement of modern civilization.

To meet rapid changes in the conditions surrounding the more recent steel industry, iron and steel engineers have been opening up a "New Age of Steel" as an advanced functional material by making the most of refining, material engineering, and other related technologies.

Ferrous metals are expected to provide high economy with high functionality when combined with other materials and to play a more important role as advanced materials to support future technological innovations.

Materials other than ferrous metals are presently being developed as advanced materials. With the recognition, however, that ferrous metals belong to the family of advanced materials, we intend to continue our effort to develop better ferrous metals.

Design and Development of Superalloys in Japan

Michio Yamazaki

National Research Institute for Metals

(NRIM)

Abstract

We had two national projects in which superalloys were developed. The first one was "Advanced Gas Turbine"; in this project we treated conventionally cast nickel-base superalloys and directionally solidified nickel-base superalloys. Those alloy developments started in 1978 and ended in 1984. The second one, 1981 to 1988(fiscal), was "Advanced Alloys with Controlled Crystalline Structures"; single crystal nickel-base superalloys, superplastically workable nickel-base P/M alloys, and oxide dispersion strengthened nickel-base superalloys were treated. This paper briefly reviews the alloy developmental studies and process studies carried out in the two projects.

Introduction:

Agency of Industrial Science and Technology(AIST) of MITI planned and sponsored two national projects in which various types of nickel-base superalloys were developed. "Advanced Gas Turbine" was the first one and this project dealt with conventionally cast nickel-base superalloys (CC alloys) and directionally solidified columnar nickel-base superalloys (DS alloys). In this project the main theme was, of course, the development of an advanced gas turbine and the alloy developmental work started in 1978 and ended in fiscal 1984. "Advanced Alloys with Controlled Crystalline Structures" was the second project and three types of nickel-base superalloys were treated; they are single crystal alloys (SC alloys), superplastically forgeable P/M alloys, and oxide dispersion strengthened alloys(ODS alloys). This second project was carried out from 1981 to fiscal 1988.

National Research Institute for Metals designed alloys and proposed them as candidate alloys for process studies, which were chiefly carried out by companies participating in the two projects. More precisely, those companies performed their works as members of the especially established organizations, i.e. "Engineering Research Association for Advanced Gas Turbines" for the first project and "Research and Development Institute of Metals and Composites for Future Industries" for the second. Two other national research institutes took parts in the second project.

In the second project titanium alloys were also developed, in this paper, however, this subject will not be described.

Alloy Design:

The present author and his collaborators developed a computer-aided alloy design method for gamma/gamma-prime type nickel-base alloys(1) and later a revised version was reported(2). The essential part of these alloy design programs is made up of giving pairs of gamma and gamma-prime phase compositions in multi-component systems. Many pairs of analysed compositions of gamma and gamma-prime phase compositions were utilized to express phase relations. From the calculated phase compositions, by giving an arbitrary phase volume fraction, one can obtain an alloy composition together with some other factors for the alloy such as lattice parameters of the phases, lattice mismatch, density, creep rupture life, hot corrosion resistance, solution window, etc. The first version relied upon chemically analyzed phase compositions and the second utilized EPMA analysis values of the phases.

Conventionally Cast Alloys:

Many alloys with various gamma prime contents and various Cr concentrations were designed and examined. As a general trend, higher creep rupture strength alloys showed lower hot corrosion resistance. At a given hot corrosion resistance level, developed alloys gave creep rupture strengths higher than those of commercial alloys. Alloy TM-321 was proposed for the first stage blade of the Advanced Gas Turbine of the project. This alloy aimed a higher rupture strength at the sacrifice of hot corrosion resistance. The composition of TM-321 is as follows(in mass %).

TM-321

8.1Cr, 8.2Co, 12.6W, 5.0Al, 0.8Ti, 4.7Ta, 0.9Hf, 0.05Zr, 0.01B, 0.11C

For the second stage nozzle, Alloy TM-269 was proposed, which has a high melting temperature as well as a high strength. The composition of this alloy is as follows(in mass %).

TM-269

9.7Cr, 8.9Co, 13.2W, 4.3Al, 0.6Ti, 3.8Ta, 0.8Hf, 0.05Zr, 0.01B,0.11C

Melting stock manufacture was studied by Daido Steel Ltd. Investment casting of air-cooled blades was studied by Mitsubishi Metals Co. Ltd. and air-cooled second stage nozzles by Hitachi Metals Co. Ltd.

Aluminide coating and thermal barrier coating were studied by NRIM and Toshiba Co. Ltd; Y-doped (PVD) aluminide coating (NRIM) and automated plasma spray coating (Toshiba) were developed.

Directionally Solidified(DS) Alloy:

Grain boundary cracking is reduced in DS alloys due to columnar crystals. Cracking between columnar crystals in the solidification process is, however, a hazard of this type alloy. This is caused, just after the solidification, by expansion stress from a core to make an air-cooled hollow blade. To avoid this problem NRIM controlled gamma-prime contents in DS alloys. Alloy TMD-5 thus designed was proposed by NRIM to the project. The composition of the alloy is as follows(in mass %):

TMD-5

5.8Cr, 9.5Co, 13.7W, 4.6Al, 0.9Ti, 3.3Ta, 1.4Hf, 0.015Zr,0.015B,0.07C

Daido Steel Co. Ltd. again prepared melting stocks of this alloy. IHI studied manufacturing hollow DS blades and also took part in determining TMD-5 composition.

Single Crystal Alloy(SC):

The target for SC alloy was as follows:

Rupture life at 1040 C and 14 kgf/sq. mm; more than 1000 h.

Rupture elongation for the same condition as above; more than 10%.

NRIM applied their alloy design program modified to SC alloys. Various factors such as gamma-prime volume fraction, lattice mismatch of gamma and gamma-prime phases, W/Ta ratio in gamma-prime, solid solutioning degree of gamma-prime, and solution treatment temperature allowance(window). Some typical alloys developed are TMS-1, TMS-12, and TMS-26. The composition of TMS-26, the second proposal alloy for the project is as follows(in mass %):

TMS-26

5.6Cr, 8.2Co, 1.8Mo, 10.9W, 5.1Al, 7.7Ta

Near at the end of the project, the second version of alloy design program was developed, in which many pairs of gamma and gamma-prime compositions gained through intricate analyses by EPMA were utilized. In this program some sets of regression equations for alloy properties were also renewed. In this version the lattice mismatch played an important role and by running this program some high Mo alloys were indicated to show long creep rupture strengths. Experiments showed that this was true but unfortunately they gave small elongation values; some modifications to improve ductility will be done.

Melting stocks with low impurities and accurate concentrations of component elements were provided by Daido Steel Co. Ltd.

Making good cores is one of the important technologies for single crystal hollow blades. This was investigated by Government Industrial Research Institute, Nagoya. Cores must be held in molten alloy for about half an hour without damage and, after solidification, must be removed by leaching in an alkaline solution; this second condition requires that the base substance is silica which is not the highest heat resistant ceramics. The main remedies adopted were using fused silica of appropiate powder sizes, crystallization rate of fused silica being controlled less than 10% during sintering, dispersion of crystalline ceramics, and using injection molding. A good composition and process was proposed, and IHI made excellent cores for making experimental SC blades. The injection molding of ceramics parts is a new technology and is expected to be applied to other fields.

IHI, Hitachi, Ltd., and Hitachi Metals, Ltd. did experiments for producing SC blades and evaluating them. Plasma beam skull remelting was applied to an experimental SC making furnace to minimize the contamination during remelting of the melting stock. To another experimental furnace was applied a static magnetic field to reduce convection. This was expected to be effective for better SC qualities, but the result was not as expected.

Solidification simulation models were used to analyze temperature distribution during solidification. Flat solidification front is required to make an article made up of a single crystal; otherwise at the outer surface of the article other crystals would nucleate. The solidification models could show conditions to get a flatter solidification front.

Developed alloys and cores were tested to make hollow blades. Some improvements were further required in them; recrystallization at edges for the alloys and deformation for the cores were sometimes observed.

Superplastically Workable Ni-base Superalloys:

The target for this type alloy in the project was as follows. The UTS at 760 C is more than 160 kgf/sq.mm, tensile elongation at that temperature is more than 20 %, and the alloy must be superplastically forged at around 1050 C. After the project started it was found that this target, except superplasticity, was too high to be achieved.

The alloy is to be used for gas turbine disk materials. Preforms for superplastic forging were intended to be made without an extrusion process to avoid the usage of a big extrusion machine, which is not economical and practically can not be installed. Consequently, preforms were made through HIP-processing powders without extrusion.

For the alloy design, the above described alloy design program was applied to calculate gamma and gamma-prime compositions to be present in P/M Ni-base superalloy, RENE 95, and a series of alloys, including the original alloy, RENE 95, with various gamma-prime contents but with the calculated compositions of the two phases. An alloy, TMP-3 designed to have a gamma-prime content a little higher than the one in the original alloy showed better superplasticity than that of the original. NRIM proposed this as an official candidate alloy for the process research works, though the alloy does not satisfy the target.

NRIM continued research works to get alloys with higher strength and elongation values through composition modifications as well as heat treatments and some doping. Stronger alloys were developed. An improved version of alloy design program was then developed, and this showed that there would exist still higher strengths, which, however, could not reach the target values.

Melting stocks to be remelted for powder production were made by Daido Steel Co., Ltd. Two processes for powder making were employed. The first one was the argon gas atomization (by Kobe Steel, Ltd.) and the second the liquid helium cooling centrifugal atomization. The first one is rather conventional but the second proved to require much research work. Most isothermal superplastic forging experiments were done with powders made by the first method. For the second process a high speed rotating disc brought about difficulty.

Kobe Steel designed and constructed a superplastic forging equipment and made discs 400 mm in diameter from HIP preforms of gas atomized powders of alloy TMP-3. A computer calculation model utilizing the finite element method was applied to determine the shape of the preform to give uniform deformation. Kobe Steel also developed dual property discs made of two alloys.

Sumitomo Electric Industries, Ltd. treated powders in an attritor to give them strain. The strain induced in powders was expected to promote recrystallizaton of them and hence grain size reduction. The attritor treated powders, after HIP consolidation,showed improved superplasticity. Impurities introduced by this treatment sometimes reduced the mechanical properties after superplastic deformation but this can be avoided by careful attritor treatment. It was also shown that attritor-treated powders could be consolidated by CIP followed by sintering to make preforms for superplastic forging.

Hitachi, Ltd. made superalloy ribbons by melt spinning method. Those ribbons showed superplasticity and were used as inserts for diffusion bonding of cast superalloys.

Oxide Dispersion Strengthened(ODS) Ni-base Superalloys

ODS alloys, made up of gamma, gamma-prime and yittria particles, are stronger than SC alloys and expected to be used as materials for gas turbines. Mechanical alloying in an attritor, extrusion, forging, zone annealing, and bonding are usually applied to make ODS blades.

NRIM again proposed a candidate alloy; this was named TMO-2. A previously developed conventionally cast alloy, TM-220, which is one of the strongest alloys, was modified by the alloy design method to get TMO-2. The composition of TMO-2 is as follows(in mass %).

TMO-2

5.9Cr, 9.7Co, 12.4W, 4.2Al, 0.8Ti, 4.7Ta, 0.05Zr, 0.01B,

0.05C, 1.1yittria

This alloy, compared to MA 6000 is higher in W and gamma prime contents.

The target for ODS alloy was as follows:

Rupture life at 1100 C and 14kg/sq.mm; more than 1000h.
Rupture elongation at that condition; more than 5%.

Alloy TMO-2 gave a rupture life much longer than the target value and hence than that of MA 6000, but the elongation value was about 4% or less, being probably similar to that of MA 6000.

NRIM improved intermediate temperature strength of this type alloy by further increasing gamma-prime content of TMO-2, to get, for instance, TMO-20 which was designed to have a gamma-prime content of 75%.

Alloy TMO-2, and sometimes alloy MA 6000 as a reference material, were used for studying processes of ODS alloy.

Sumitomo Electric Industries, Ltd. took parts in mechanical alloying and extrusion. After trying to find appropriate conditions, they succeeded in making good extruded bars of TMO-2. The bars were 30-40 mm in diameter, high in hardness(as high as Hv 800), ready to be recrysallized, and isothermally forgeable.

Kobe Steel, Ltd. forged isothermally TMO-2 bar to give crude blade shapes. To make a hollow blade, the whole blade was divided into two parts or two sides. The two sides are to be bonded in later stage of the process; this type of blade was named a twin blade. It is known that high gamma-prime ODS alloy must be recrystallized for strengthening, and this forging process gives a bad effect to this recrystallization property of the alloy. Kobe Steel succeeded in forging blades, sound and recrystallizable, without platform portions, but some improvement was required to forge a blade with platform which can be recrystallized.

IHI was in charge of the final stage of the process. Zone annealing of forged bars is rather a difficult problem, because its shape is not uniform. One side of a twin blade before bonding was put into a divided mold to make the assembled article as if it were a solid round bar. This was zone annealed. This process was found effective if there was little clearance between the molds and the forged article inside. IHI did an experiment to give information for forging condition that is good for zone annealing recrystallization. Bonding of ODS alloys is known to be a difficult process. Solid state bonding without insert materials and TLP bonding were tried. The bonding strengths were not sufficient.

Concluding Remarks and Acknowledgements

Fig.1 shows temperature capabilities of developed alloys together with those of some reference alloys. Fig.2 shows CC, DS, and SC hollow blades cast by some of the participating companies using developed alloys.

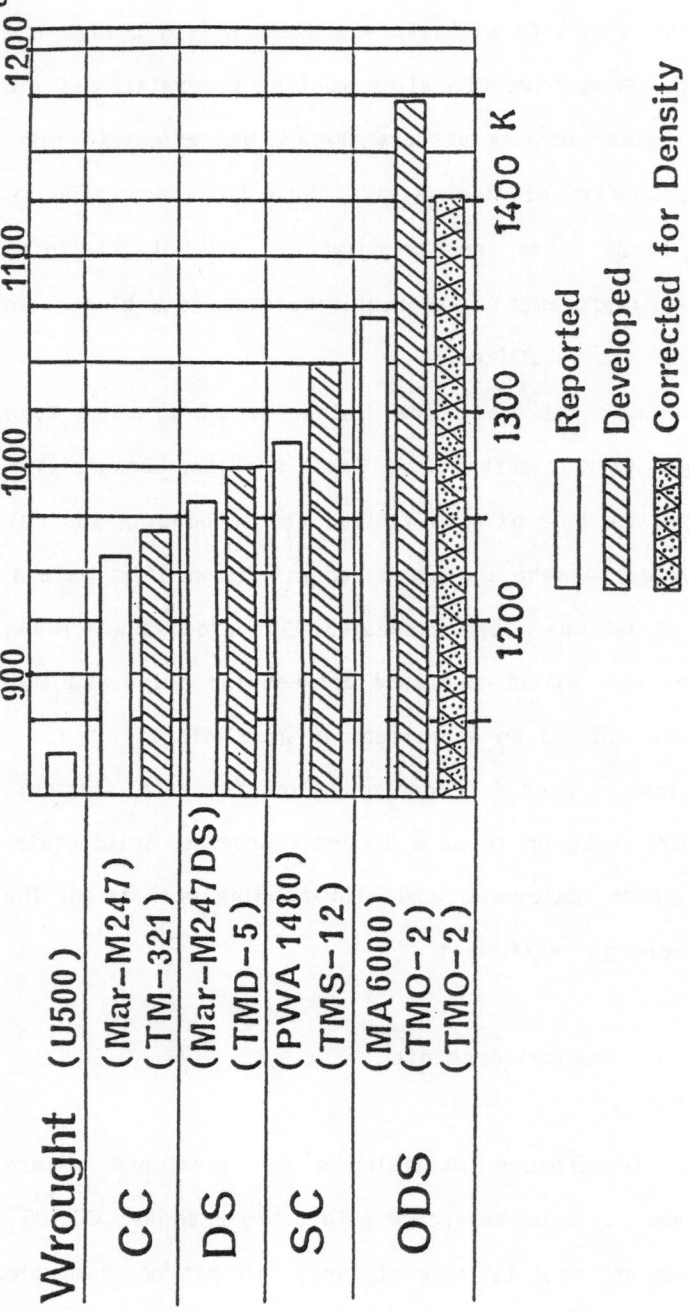

Fig.1   Temperature capability to give a 1000 h rupture life at 14

kg/sq.mm (137.3 Mpa) for Ni-base superalloys.

CC: Conventionally cast alloy, DS: Directionally solidified alloy,

SC: Single crystal alloy, and ODS: Oxide dispersion strengthened

alloy.

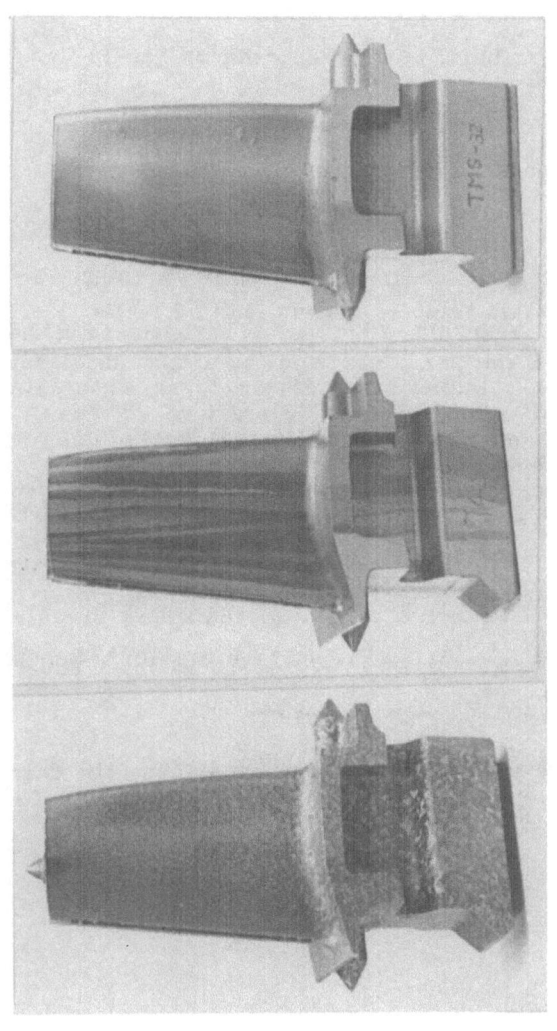

Fig.2 CC blade, DS blade, and SC blade (from left to right);all air-cooled hollow blades made using alloys developed by NRIM. Casting by Komatsu-Howmet, Ltd.(CC) and by IHI (DS and SC).

A symposium was held in March of 1989 for "Advanced Alloys with Controlled Crystalline Structures". All the participating organizations talked their results in the Symposium and the present paper, for the latter half, largely depends on the Proceedings of that Symposium(3).

References

(1) H.Harada and M.Yamazaki: Tetsu-to-Hagane, 65(1977), 1059, or(in English) H.Harada M.Yamazaki, Y.Koizumi, N.Furuya, and H.Kamiya : Alloy Design for Nickel-base Superalloys, Proceedings of International Conference on High Temperature Alloys for Gas Turbines, held in Liege, Belgium, Oct.,1981, D. Reidel Publishing Co., pp. 721-735

(2) H.Harada, K.Ohno, T.Yamagata, and M.Yamazaki:Phase Calculation and its Use in Alloy Design for Nickel-base Superalloys, Superalloys 1988, the Metallugical Society, 1988, pp.733-742.

(3) Proceedings of 6th Symposium for "Advanced Alloys with Controlled Crystalline Structures", held in Tokyo, 23-24 March, 1989.

Carbon Fibers and Their Composites

Kensuke OKUDA*

Abstract

A brief history of carbon fibers and their application is reviewed.

The existing situations of the types of carbon fibers on the market and the performance of typical products are summarized, and the technical issues on production introduced.

Major applications are outlined.

* Kureha Chemical Industry Co., Ltd., 1-9-11 Nihonbashi-Horidome-cho, Chuo-ku, Tokyo (103), Japan

1.   Introduction

Application fields of carbon fibers have expanded remarkably in the past fifteen years, through progress in the
production technology and composite materials.

There have been very significant improvements in the
performance of carbon fibers on the market, together with
the development of new types of carbon fibers. For example,
high strain-to-failure fibers with high strength and ultra-
high Young's modulus fibers have come into the market.
Moreover, activated carbon fibers, which are characterized
by excellent absorption power, have been manufactured on a
commercial scale.

Carbon fiber reinforced plastics (CFRP) have been widely
used for sporting goods, aircraft structures, machine parts
and so on. The construction industry has been applying
light-weight concrete fortified with chopped fibers to
large-sized buildings for several years.

In this paper, we report a brief history and the existing situation of carbon fibers and their composites laying
stress on recent advances in the performance of the fibers.

2.   A Brief History of Production and Application of Carbon
     Fibers

Carbon fibers which are used as reinforcements or fill-
ers are usually classified into two types. One is the

general-purpose type carbon fibers whose structure is isotropic. The other is the high performance type carbon
fibers which are characterized by a high modulus of elasticity. Besides these, a third type of carbon fibers, activated carbon fibers shall be added to the family. They are
distinctly different from the above ones in nature and application and were commercialized around ten years after high
performance fibers.

In 1959 Union Carbide Corp. (Amoco Performance Products,
Inc., presently) began commercial production of "graphite"
cloth as well as other fibrous forms by baking rayon cloth
and so on in an inert atmosphere to approximately 900°C,
followed by graphitizing to temperatures usually higher than
2,500°C [1] [2] (Fig. 1). These fibers belong to the general-
purpose type and were quickly applied to ablation materials
for space development.

Union Carbide also introduced the first high performance carbon fiber from rayon precursor to the market in
late 1965 [3]. These rayon-based high modulus "graphite"
fibers, however, disappeared a few years later because of
high processing cost in relation to their performance.

Just before this, Shindo developed a new process for
preparing a carbon fiber from polyacrylonitrile (PAN) [4],
and this invention led to the commercial production of PAN-
based general-purpose fibers by Nippon Carbon Co. in 1962.

After Shindo, Jonson et al. [5] succeeded in obtaining
high performance fibers from PAN by a method of improving

the superstructure. These fibers, which were immediately applied to the turbine blades of the Rolls-Royce RB.211 turbofan engines and now account for a large percentage of the present demand for carbon fibers, came into the market in 1967 in U.K. and in succession Toray Industries entered into production with advanced technology[6] in 1971.

Early in 1970, based on the pioneering studies by Otani[7], the first manufacturing plant for general-purpose carbon fibers derived form petroleum pitch sources began operation on a scale of 120 tpa at Kureha Chem. Ind.[8]. Meanwhile, Otani also published a paper on a high performance fiber derived from anisotropic pitch[9]. High performance fibers from pitch, however, were commercialized by U.C.C. in 1976[10].

Applications of CFRP to sporting goods such as golf clubs and secondary structures in military aircraft in the early 1970s caused PAN-based high performance fibers to take off as industrial products. Accordingly many enterprises have successively entered into the market.

Advanced composite materials (ACM), in particular carbon fiber based ACM, have also been used for secondary structures in large-sized passenger aircraft and even for primary structures in military aircraft and small-sized civil airplanes such as the Voyager in the 1980s, and moreover some of them have been qualified as materials for primary structures in next generation aircraft, which shows that the aerospace industry could be a huge market for high performance carbon fibers.

Just after the second oil crisis, many firms and institutes began to study pitch-based carbon fibers, in particular high performance fibers. Some of the firms have constructed commercial plants on a scale of the order of 100 tpa.

Applications of pitch-based carbon fibers are quite different from those of PAN-based high performance carbon fibers. Until quite recently, application of general-purpose types from pitch was limited to thermal insulating materials, sealing materials, reinforcements for engineering plastics and the like. In 1983, they were used for construction materials, which are expected to provide a growing market for them. Also, pitch-based high performance fibers whose major application is at present space equipment will be used in the same application fields as those of PAN-based high performance fibers.

In the mid-1970s, an activated carbon of fibrous form which was named activated carbon fiber was developed in Japan [11], in order to utilize more effectively the adsorption function of carbon. Activated carbon fibers have been pursued by several firms ever since and have been attracting public attention from the ecological point of view.

Over the last ten years, some enterprises have been developing new types of carbon fibers, which are prepared directly from low molecular weight hydrocarbon compounds in

the vapor phase and are named for vapor-phase-grown carbon
fiber.

3. Progress in Production Technology and Mechanical
   Property of Carbon Fibers

Most of the carbon fiber products on the market are
from PAN and pitch source, and rayon-based fibers are one
small portion. The performance of the products has been
improved significantly in response to materials requirements which advance with the expansion of application.

Before entering into this subject, I summarize the
basic process for manufacturing carbon fibers from PAN and
pitch below:

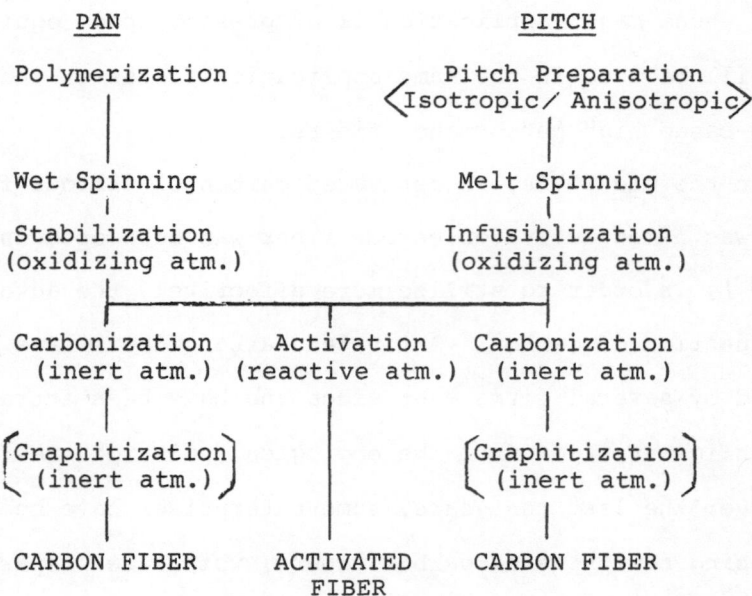

Prior to carbonization, precursor fibers, which are fabricated through conventional spinning technique, are stabilized (for PAN)/infusiblized (for pitch) in an oxidizing atmosphere. Stabilized PAN fibers/infusiblized pitch fibers are converted to carbon fibers by baking them in an inert atmosphere, and then graphitized if necessary.

The most important characteristic feature of both PAN and pitch is that the originally formed fiber structure of presursor fibers is kept through whole process, in other words, we can obtain the carbon fibers having preferred orientation which is one of the necessary conditions for giving excellent mechanical property.

In the case of pitch, isotropic pitch gives an isotropic fiber which belongs to the category of general-purpose carbon fibers, while if anisotropic pitch is used as precursor high performance fiber having the fiber structure is obtained.

Activated carbon fibers are manufactured by activating stabilized PAN fibers and infusiblized pitch fibers with for example steam at elevated temperatures, where activation means formation of a great number of open micropores which act as adsorption sites.

The tensile strength of the PAN-based high-strength type carbon fiber first marketed in 1967 was only 2,500 MPa with around 200 GPa of Young's modulus, as seen in Fig. 1, where the plotted value of Young's modulus of 400 GPa is for a high modulus type fiber. The mechanical pro-

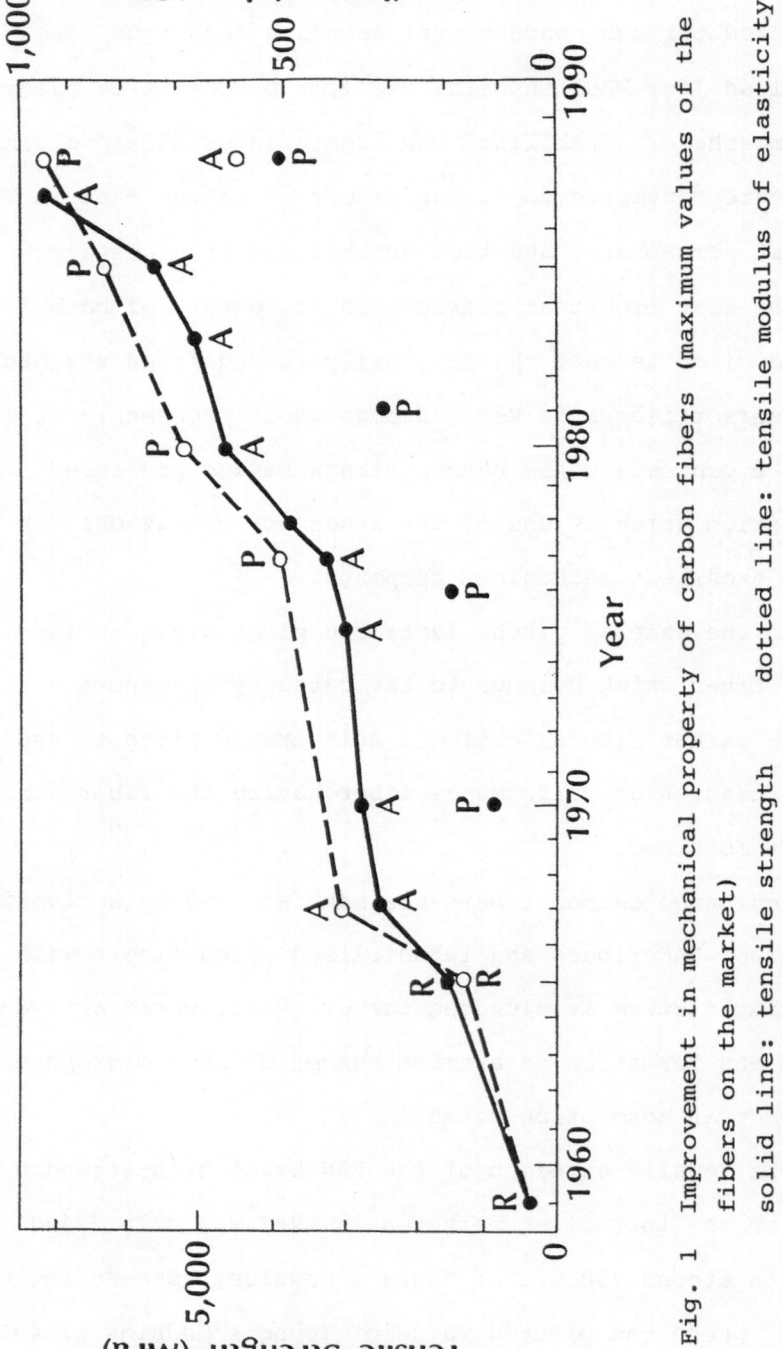

Fig. 1 Improvement in mechanical property of carbon fibers (maximum values of the
fibers on the market)
solid line: tensile strength   dotted line: tensile modulus of elasticity
A: PAN, P: pitch, R: rayon

perty of PAN-based fibers improved little till the middle
of 1970s, because their applications were limited to sporting goods, artificial satellites and so on. From that time
on, profound changes have steadily taken place in the production technology of carbon fibers, including pitch-based
carbon fibers (see Fig. 1).

Since the application of ACM to primary structures in
civil aircraft was projected at the beginning of 1980s,
related industries have been extensively developing advanced CFRP whose material requirements are, for example, design
strain equal to that of aluminum alloy for aircraft structures, high damage tolerance, higher specific modulus of
elasticity and so on. These material requirements of CFRP
can be, in terms of the reinforcing fibers, increase in
strain-to-failure with high modulus of elasticity, improvement in the interface property, increase in heat resistance
and so on, which improvement has been brought about through
the development efforts of the fiber manufacturers. For
instance, as illustrated in Fig. 1 and Table I, a fiber
T-1000 having strain-to-failure over 2% with modulus of
elasticity of 300 GPa came into the market in 1987. The
tensile strength of this fiber is twenty times higher than
that of the first fiber from rayon and twice as high as that
of the standard grade of PAN-based high strength type fibers,
e.g., T-300. Quite recently, new types of fibers have been
developed in the PAN-based family; an intermediate modulus
type fiber MRE 50 which is high in Young's modulus and also
thick in diameter compared with the same type fibers, and

128

Table I Physical Properties of Carbon Fibers on the Market

Type	Fiber Designation	Tensile Strength (MPa)	Tensile Modulus of Elasticity (GPa)	Elongation at Break (%)	Diameter (μm)	Manufacturer
GP	T-101S	720	32	2.2	14.5	Kureha Chem.
	T-201S	690	30	2.1	14.5	"
	S-210	784	39	2.0	13	Donac
		(686)	41	1.6		Ashaland
	GF-20	980	98	1.0	7 - 11	Nippon Carbon
HP (PAN)	T-300	3530	230	1.5	7.0	Toray
	T-400H	4410	250	1.8	7.0	"
	T-800H	5590	294	1.9	5.2	"
	T-1000	7060	294	2.4	5.3	"
	MR 50	5490	294	1.9	5	Mitsubishi Rayon
	MRE 50	5490	323	1.7	6	"
	HMS-40	3430	392	0.87	6.2	Toho Rayon
	HMS-40X	4700	392	1.20	4.7	"
	HMS-60X	3820	588	0.65	4.0	"
HP (pitch)	P-25	1400	160	0.9	11	Amoco
	P-75S	2100	520	0.4	10	"
	P-120S	2200	827	0.27	10	"
	E-35	2800	241	1.03	9.6	du Pont
	E-75	3100	516	0.56	9.4	"
	E-130	3900	894	0.55	9.2	"
	F-140	1800	140	1.3	10	Donac
	F-600	3000	600	0.52	9	"
ACF	FX-100	2 g/d	500 a)	18 b)	15	Toho Rayon
	FX-600	<1 g/d	1500 a)	50 b)	7	"
	A-10	245	1000 a)	20 c)	14	Donac
	A-20	98	2000 a)	45 c)	11	"

a) specific surface area (m²/g)
b) adsorption amount of benzene (%)
c) adsorption amount of acetone (%)

an ultra-high modulus type fiber HMS-60X with fairly high strength (see Table I). These results were achieved by improving the total process from precursor to finishing, such as increase in molecular weight and thorough purification, optimization of superstructure of PAN fibers, surface modification of carbon fibers, optimum process control and the like.

Besides reinforcements, high performance matrix resin which is tougher and more stable to temperature and humidity than conventional ones has been developed, because we cannot reach the material requirements mentioned above by means of improvement in the fiber performance only, as will be touched upon later.

Concerning pitch-based carbon fibers, significant advancement in the production technology has realized improvement in the mechanical property of high performance type fibers, as seen in Fig. 1 and Table I. For example, du Pont has introduced an ultra-high modulus type fiber E-130 whose Young's modulus is almost 900 GPa with 3,900 MPa of tensile strength. This Young's modulus value is over 80% of the theoretical value of graphite single crystal and twenty times higher than that of the rayon-based fibers commercialized first, and the tensile strength is improved two times or more higher than those of early pitch-based high performance type carbon fibers, for example, P-25, P-75S and others.

Four different types of basic superstructures or some

modifications of them exist in pitch-based carbon fibers because of anisotropic nature of the constituents of carbon. In principle, the mechanical property of pitch-based carbon fibers depends on the superstructure which is formed in the spinning process and maintained through the whole subsequent process, in other words, one of the necessary conditions for obtaining fibers that have higher tensile strength is to control the superstructure.

Since the superstructure can vary substantially according to the nature of precursor pitch and processing, in particular spinning conditions, the development work on pitch-based carbon fiber is concentrated on improvement or modification of precursor pitch, whose typical preparation methods are summarized in Fig. 2. The early products such as P-75S are from conventional mesophase pitch[10]. New types of pitches other than conventional mesophase pitch give higher tensile strength than the latter, though each of the former is different in nature depending on preparation method; neomesophase[12] belongs to the same category as conventional mesophase pitch, and domant anisotropic pitch[13] and premesophase pitch[14] have a rather naphthenic nature, which is caused by hydrogenation applied in any stage of the processes, and show good spinnability compared with mesophase pitch.

As understood from Table I, there is some correlation between the mechanical property of carbon fibers and the production process or nature of the precursor. Qualitatively summarizing, we can rather obtain carbon

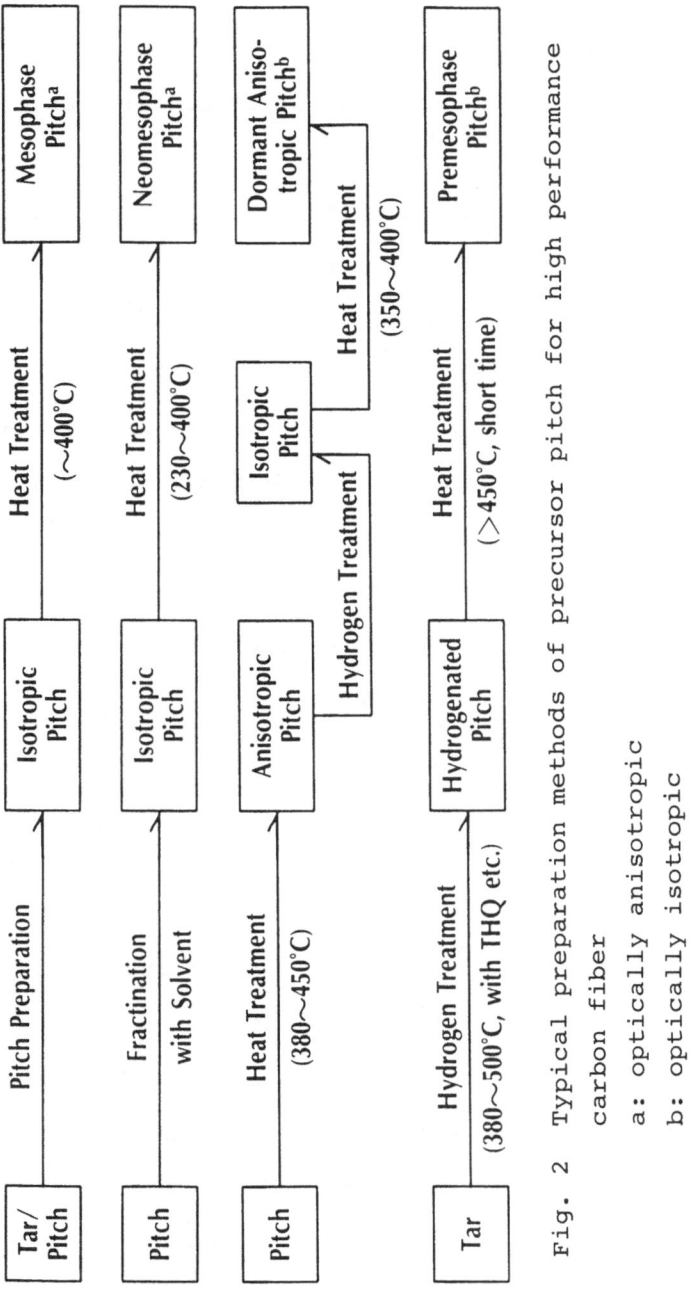

Fig. 2   Typical preparation methods of precursor pitch for high performance
carbon fiber

a: optically anisotropic

b: optically isotropic

fibers having higher strength from PAN precursor. This
tendency is due probably to non-graphitizable nature and
a suitable superstructure of the precursor fiber. While
pitch precursor tends to give higher modulus of elasticity,
which originates in that pitch is intrinsically converted
to soft-carbon.

After activated carbon fiber derived from rayon,
several products made from various kinds of precursors including PAN and pitch came into the market [15]~[17]. They
are different in surface property, in other words, the
adsorption performance as well as cost, as seen in Table I.

The invention of the seeding method with ultrafine
metallic catalyst particles [18] and the fluidizing catalyst
method [18] has remarkably advanced the production technology
of vapor-phase-grown carbon fibers, which are still under
development.

4.   Existing Application Situation

Carbon fibers are used in almost all industries and
most of the fibers are used as reinforcements for composite
materials, as seen in Table II, where the application fields
and related industries which use or will use products derived from carbon fibers are summarized.

Major application is, at present, resin matrix structural
materials with light weight and high rigidity for aircraft,
sporting goods etc, while application to construction mate-

Table II Application of Carbon Fibers

PRODUCT	APPLICATION	RELATED INDUSTRIES
FIBER	INSULATING MATERIALS (a)	ELETRONICS, CAR, AIRCRAFT, ATOMIC ENERGY
COMPOSITE — RESIN	SEALING MATERIALS (a, b)	CHEMICAL, PETROCHEMICAL
	FUNCTIONAL MATERIALS (a, b) (tribological, conductive, chemical-resistant materials and so on)	APPLIANCE, ELECTRONICS, COMMUNICATION, MACHINE, CAR, AIRCRAFT, CHEMICAL, MEDICAL
COMPOSITE — CARBON	STRUCTURAL MATERIALS (b) (light weight / high rigidity materials)	SPORT, MEDICAL, SPACE, AIRCRAFT, CAR, COMMUNICATION
	ABLATION MATERIALS (a, b)	SPACE, MILITARY
COMPOSITE — METAL	FRICTIONAL MATERIALS (a, b)	AIRCRAFT, CAR, RAILWAY, MACHINE
	CARBON / GRAPHITE (a)	STEEL, METAL
COMPOSITE — INORGANICS	CELL ELECTRODES (a, b)	ELECTRIC POWER, CAR
	CONSTRUCTION MATERIALS (a, b)	SHIP, BUILDING, HOUSING, PUBLIC WORKS

a : GP GRADE, b : HP GRADE, —— : ACTUAL USE, ... : DEVELOPMENT

rials has been growing in the past several years.

For example, the Voyager aircraft, which achieved a non-stop, unfueled, round-the-world flight in 1986, had a structure weight of 422 Kg, and its take-off weight was 5,153 Kg, some 4,065 of that being fuel, thanks to the use of CFRP for 90% of the aircraft's structure[20].

Although this is an extreme case and it cannot be concluded that CFRP will be immediately applied to primary structure in all kinds of passenger aircraft, we can confidently expect a bright future for the application of CFRP to aircraft, as illustrated in Fig. 3[21], which forecasts that 60% of the structure of passenger aircraft will consist of composite materials.

We can improve brittle hydraulic cement materials by reinforcing with carbon fibers, whose major reinforcing effects are: Remarkable improvement in tensile and flexural strength as seen in Fig. 4, where flexural strength to deflection curves of typical examples of concretes fortified with inorganic and organic fibers are summarized, Drastic increase in toughness and ductility (refer to Fig. 4), Significant improvement against impact force, High dimensional stability, Protection against crack formation due to drying shrinkage, Excellent durability, Good wear resistance, Anti-static charge, Light weight and so on.

Thanks to the above-mentioned advantages of carbon fiber reinforced concrete (CFRC) compared with conventiona precast concretes and other fiber reinforced concrete, CFRC has been

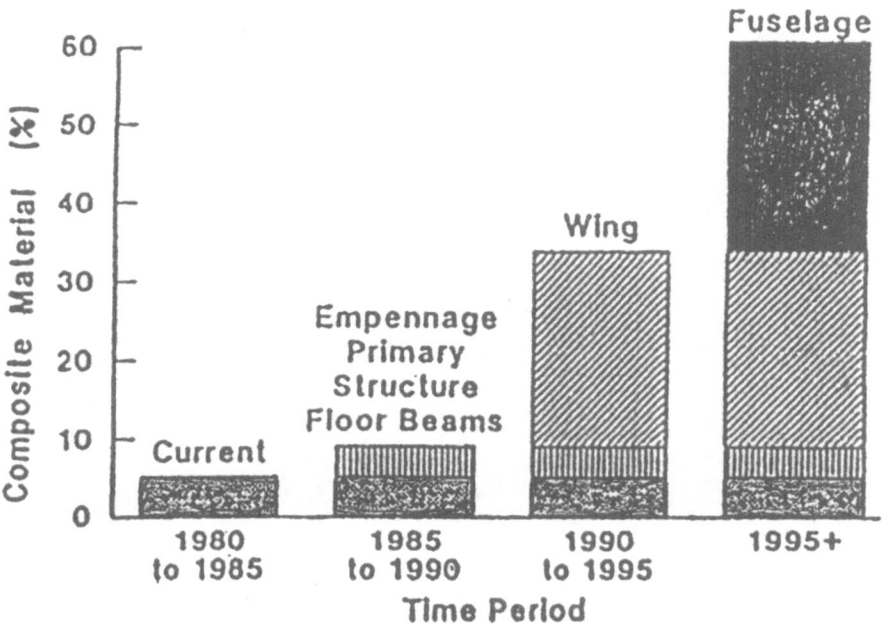

Fig. 3   Future application of composite materials
to passenger aircraft [21]

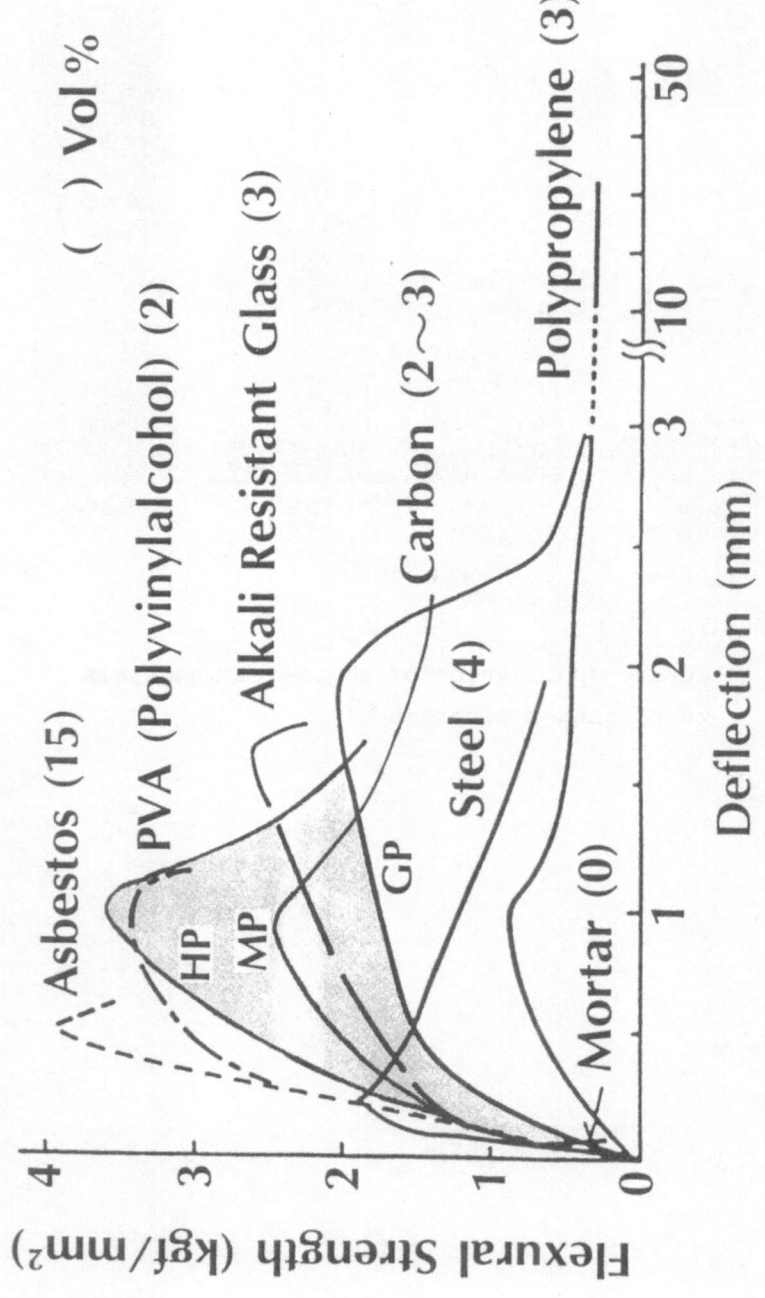

Fig. 4 Flexural properties of fiber-reinforced concrete

utilized for many constructions including the 37-story ARK Building completed in 1986 [22] and is expected to play an important role in the construction industry in the future.

5. Concluding Remarks

The production technology of carbon fibers has been significantly advancing for the past decade, though that of high performance fibers from pitch is still in its infancy. Accordingly, they have gained a position as the leading reinforcements and are used in almost all industries including the aerospace industry. Moreover, they are expected to grow at a rate of 10% or higher in the ACM market for some ten years and also a huge new demand such as for construction materials is confidently expected to be created, presupposing progress in the application technology of carbon fibers for manufacturing composite materials.

REFERENCES

1)   R. Bacon, "Carbon Fibers from Rayon Precursors", Chemistry and Physics of Carbon, Vol. 9, p. 1 ~ 102, Edited by P.L. Walker & P.A. Thrower, Marcel Dekker, Inc., New York, 1973.

2)   UCC, "Fibrous Graphite", USP 3107152 (1963) etc.

3)   UCC, "Process for Producing Fibrous Graphite", USP
     3454362 (1969).

4)   A. Shindo, "Method for Production of Carbon Products
     from Acrylonitrile-based Synthetic Polymers", JP S37-
     4405 (1962).

5)   W. Johnson, L.N. Phillips & W. Watt, "The Production
     of Carbon Fibres", BP 1110791 (1964).

6)   Toray, "Production Method for Heat Resistant Polymeric
     Materials", JP S46-35853 (1971) etc.

7)   S. Otani, "On the Carbon Fiber from Molten Pyrolysis
     Product", Carbon, 3, 31 (1965).

8)   Kureha, "Method for Production of Carbon Fibers from
     Molten Pyrolysis Material", JP S41-15728 (1966) etc.
     (GP), "Production of Carbon Shaped Articles Having
     High Anisotropy", JP S49-8634 (1974) (HP).

9)   S. Otani, Y. Kokubo & T. Koitabashi, "The Preparation
     of Highly-oriented Carbon Fiber from Pitch Material",
     Bull. Chem. Soc. Japan, 43, 3291 ~ 3292 (1970).

10) UCC, "High Modulus, High Strength Carbon Fibers
     Produced from Mesophase Pitch", USP 4005183 (1977).

11) Toyo Spinning, "Production Method for Fibers Having
     Adsorption Function", JP S49-133624 (OPI) (1974) etc.

12) du Pont, "Carbonaceous Pitch for Production of Carbon
     Fibers and Its Production Method", JP S63-5433 (1979
     OPI) etc.

13) S. Otani, A. Kikuchi & E. Ota, "A Study on Domant
     Mesophase Pitch", Meeting of the 117 Committee of

Japan Soc. for the Promotion of Science, 117-163-A-2,
Tokyo, 1981.

14)   Y. Yamada, "Mesophase Pitch-based Carbon Fiber",
      Seminar of the Carbon Soc. of Japan, P. 11 ~ 18,
      Tokyo, 1983.

15)   Toho Rayon, "Production Method for Activated Carbon",
      JP S51-132193 (OPI) (1976) etc.

16)   Nippon Kynol, "Production Method for Activated Carbon
      Fibers or Activated Carbon Fiber Structures", JP S55-
      7583 (OPI) (1980) etc.

17)   Osaka Gas, "Production Method for Carbon Fibers",
      JP S61-28020 (OPI) (1986) etc.

18)   M. Endo, T. Komaki & T. Koyama, "Vapor-grown Carbon
      Fibers by Seeding Method of Metal Ultra-fine Particles",
      Interntl. Symp. on Carbon, p. 515 ~ 518, Toyohashi, 1982

19)   M. Endo & M. Shikata, "Growth of Vapor-grown Carbon
      Fiber Using Fluid Ultra-fine particles of Metals",
      Oyobutsuri, 54, 507 ~ 510 (1985).

20)   D. Brewer, "Voyager's Composites 'Performed Superbly'",
      Adv. Composites, Jan./Feb., 1987, 58 ~ 60.

21)   M. Ohsumi, "Aircraft · Space", 87/3rd Joint Meeting of
      Research Groups, Soc. Polymer Sci., Japan, p. 1 ~ 4,
      Tokyo, Feb., 1988.

22)   S. Akihama, T. Suenage & N. Nakagawa, "Carbon Fiber
      Reinforced Concrete", Concrete Interntl., 10, No. 1,
      40 ~ 47 (1988).

Structural Change of Gels Prepared from Highly Acidic
Tetramethoxysilane Solutions in the Drying Process

Jun Yamaguchi, Hiromitsu Kozuka and Sumio Sakka
Institute for Chemical Research, Kyoto University
Uji, Kyoto-Fu 611

Introduction

Sol-gel reaction of tetramethoxysilane (TMOS) under highly acidic conditions with a limited amount of water provides highly porous monolithic silica gels composed of micrometer-sized particles[1-3]. Because of the continuous large pores and particles, these gels and heated derivatives may be suitable for porous materials which can be used as filters or supports for catalysts and enzymes. For these uses, surface characteristics of the gels should be controlled. Since the amount of water for hydrolyzing TMOS in the starting solutions is limited, some of alkoxy groups may be left to be hydrolyzed at the time of gelation and water in the atomosphere may affect the nature of surfaces of the large particles of the gels during drying. In the present study, the surface properties of the gels have been investigated by BET analysis and electeon microscopic observation as a function of drying time.

Experimental

Silicon alkoxide solutions having the mole ratio of TMOS : H_2O : CH_3OH : HCl = 1 : 1.53 : 2 : 0.40 were prepared by mixing the reagents at room temperature. 50 ml of the alkoxide solution was poured into a polypropylene container and gelled at

40 °C in the oven. Gelation took place in a very short time of about 1.3 h. The resultant wet gels were dried in the same container and at the same temparature of 40 °C for 2, 5, 7 or 13 days.

The specific surface area of the gels was measured by BET method using nitrogen gas and the pore size distribution of the gels was calculated on the basis of Cranston-Inkley model[4]. Gel powders of 75-106 μm in size were degassed at room temparature for 24 h and at 75 °C for 18 h before the BET measuremant.

Results and Discussin

Figure 1 shows the scanning electron micrographs of the gels dried for 2 days (Fig.1(a)) and dried for 13 days (Fig.1(b)). As seen in Fig.1(a), the surface of the large particles are very rough and composed of smaller particles, indicating that the micrometer-sized large particles are secondary particles. It is seen in Fig.1(b), however, that further drying reduces the roughness of the surface.

Figure 2 shows the pore size distribution curves of the gels. It can be seen from the figure that the micrometer-sized particles have small pores of the radii less than 20 A. These pores are regarded as voids among the primary particles. It is also seen that the number of detected pores decreaes with drying time.

Figure 3 shows the specific surface area of the gels. The data points for each drying time have been obtained by repeating the surface measurement with the same sample. For the drying

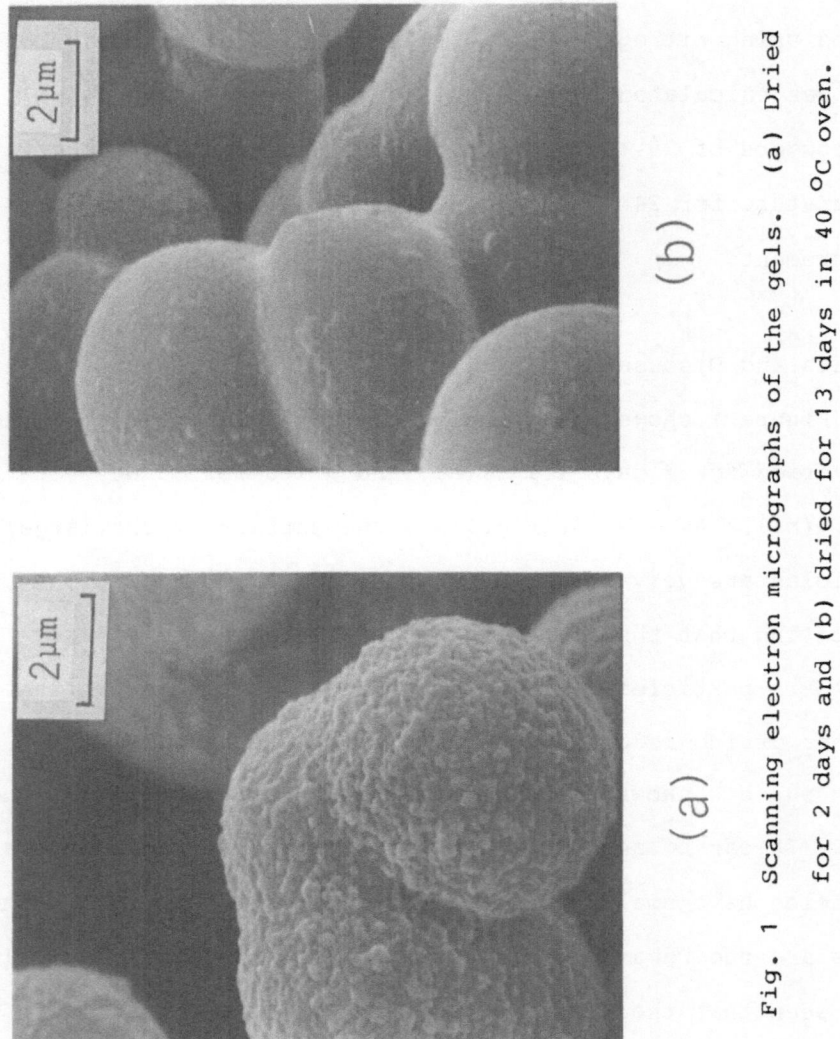

Fig. 1   Scanning electron micrographs of the gels.   (a) Dried
for 2 days and (b) dried for 13 days in 40 $^{\circ}$C oven.

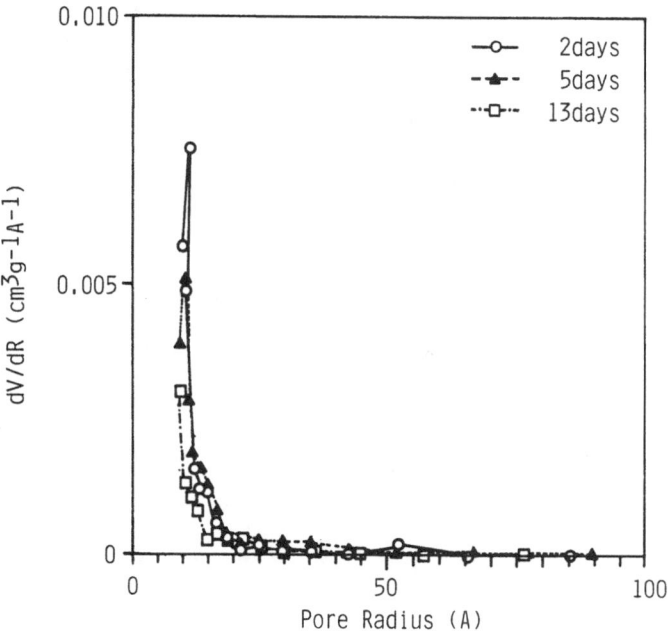

Fig. 2   Pore size distribution curves of the gels dried for 2

days (—o—), 5 days (— ▲ ··), 13 days (-·-□-·-).

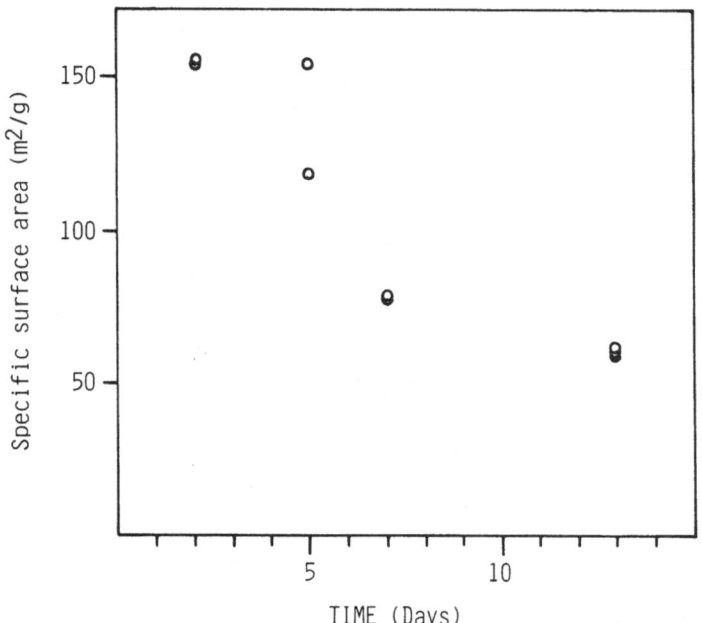

Fig. 3   Specific sruface area of the gels as a function of

drying time.

time of 2, 7 and 13 days, the repeated measurements give the
similar values, indicating that almost no change occurs in the
specific surface area during the measurement. A larger scatter
of the data for the drying time of 5 days indicates that in this
time periods, some change may occur during the treatment for the
measurement. The specific surface area decreases drastically
with drying time till 7 days, and then continues to decrease
more slowly. It is seen that the specific surface area
decreases down to about 1/3 (60 m^2/g) of the initial value after
13 days. It is assumed that this decrease in surface area
reflects the collapse of a portion of voids among the primary
particles whitin the secondary particles. Assuming taht the gel
consists of 5 μm dense silica particles with smooth surface, the
specific surface area of the gel is estimated to be 0.5 m^2/g.
This value is much lower than the value of 60 m^2/g obtained by
the BET method, indicating that there remain still some voids to
be collapsed whitin the secondary particles even after drying in
the air atomsphere.

The theoretical amount of water required for the complete
hydrolysis and polycondensation is 2 moles per mole of TMOS.
Taking into account that the amount of the water added to the
starting solution is 1.53 mole per mole of TMOS, it is presumed
that not all of the alkoxy groups may be hydolyzed and some of
the alkoxy groups may remain non-hydrolyzed when gelation takes
place. These non-hydolyzed alkoxy groups would be attacked by
water vapor in the atomosphere during drying. Thus, the
decrease in the specific surface area and the number of pores
can be attributed to further hydrolysis and subsequent

polycondensation reaction leading to bonding of primary
particles whitin a secondary large particles. The mechanism of
formation of large micrometer-sized secondary particles, has
been discussed eleswhere[5].

Conclusion

The change of the surface characteristics of the gels with
micrometer-sized large particles prepared under highly acidic
conditions with a limited amount of water during drying have
been investigated. It has been found that the specific surface
area and the number of pores decrease with drying time. It has
been discussed that the micrometer-sized particles of the gels
are densified by the collapse of voids among the primary
particles due to water vapor in the drying atomosphere causing
hydrolysis and polycondensation reactions.

References

1. H. Kozuka and S. Sakka, "Formation of Highly Porous Opaque
   Gel from Alkoxysilane Solution," Chem. Lett., 1987, 1791-
   1794 (1987).
2. S. Sakka and H. Kozuka, "Properties and Sintering of Gels
   Derived from Metal Alkoxide," Sintering '87, ed. by
   S. Somiya, M. Shimada, M. Yoshimura and R. Watanabe,
   Elsevier Applied Science, New York, 145-50 (1988).
3. H. Kozuka, J. Yamaguchi and S. Sakka, "Effect of Solution
   Composition, Aging and Exposure to Water Vapor on the
   Structure and Properties of Highly Porous Silica Gels,"
   Bull. Inst. Chem. Res., Kyoto Univ., 66, 68-79 (1988).

146

4. R. W. Cranston and F. A. Inkley, "Determination of Pore Structures from Nitrogen Adsorption Isotherms," Adv. in Catalysis, 9, 143-54 (1957).

5. H. Kozuka and S. Sakka, " Formation of Particulate Opaque Silica Gels from Highly Acidic Solutions of Tetramethoxysilane," submitted to Chemistry of Materials.

SYNTHESIS OF SiC-AlN CERAMIC ALLOY BY THE COMBINATION OF REACTION SINTERING AND HIP

JING-FENG LI and RYUZO WATANABE
DEPARTMENT OF MATERIALS PROCESSING
FACULTY OF ENGINEERING
TOHOKU UNIVERSITY,SENDAI,980,JAPAN

ABSTRACT

Ceramic alloys were synthesized by the combination of reaction sintering and hot isostatic pressing (HIP) in this work. The starting powders were commercial silicon nitride and aluminium nitride and graphite powders. A small amount of calcia was added in the form of $Ca(NO_3)_2 \cdot 4H_2O$, and the reaction sintering was performed in a nitrogen atmosphere in a temperature range of 1670 to 2170K. The reaction process was well investigated and found to be as following: at first, α-sialon forms from 1770K to 1870K, then at above 1973K, graphite reduce α-sialon to produce SiC-AlN solid solution. When AlN content was less than 35mol%, a fine-grained uniform 2HSiC-AlN solid solution was synthesized, and its grain size was three times finer than that of the starting powders. On the other hand, when an excess amount of AlN was added, two-phase composite was formed. After HIP-sintered at 2070 to 2120K under a pressure of 200MPa, porous SiC-AlN ceramic alloy specimens were fully densified.

INTRODUCTION

It is well known that silicon carbide is a strong candidate for high temperature structure ceramics due to its excellent high temperature strength, but its fracture toughness is too low. Alloying has been proved to be an effective method to improve the mechanical property of metal materials, so it is also necessary to attempt alloying SiC ceramics.

A solid solution forms in 2H-SiC and AlN system in a wide range of composition because both have the same structure and their lattice parameters are very close.[1] So SiC can be alloyed with AlN, and a little of work about the synthesis of SiC-AlN solid solutions has been done, for example, Ruh et al. hot pressed SiC and AlN powder mixtures at 2173 to 2473K and obtained solid solutions of the hexagonal 2H structure, but their bending strength values were much low, due to inhomogeneous mixing.[2] SiC-AlN alloys were also prepared by the carbothermal reduction of silica and alumina or their mixture.[3,4]

Recently, it was reported that SiC-AlN solid solutions can be formed by the reaction of Si_3N_4, AlN, CaO and Carbon, but the reaction process is not clear and only porous specimen was obtained.[5] In the present work, it is the objective to synthesize uniform SiC-AlN alloys by this method and clarify the reaction process. And in order to obtain dense specimens, densification condition of HIP-sintering was also investigated.

EXPERIMENTAL

Commercial Si_3N_4[#], AlN[$], graphite powders[*] were mixed in ethylalcohol according to the composition ratio given in Table 1, 4mol% CaO was added in the form of $Ca(NO_3)_2 \cdot 4H_2O$[+] which is dissolvable in ethylalcohol. After ball milled and dried, the mixtures were heated at 1073K for 15 minutes in the flow of high-purity H_2 gas to decompose $Ca(NO_3)_2 \cdot 4H_2O$ to CaO, then compacted to a columnar green body of 10mm in diameter, using a metal die (100MPa) and a cold isostatic pressing machine (200MPa). The sintering was performed in a nitrogen atmosphere at 1673K to 2173K.

The synthesized specimens with high porosity were coated with BN powders to prevent the reaction between specimen and glass capsule, and encapsulated in a vacuum ($<10^{-4}$ Torr.) into a vycor glass tube, then HIP-sintered at 2070 to 2120K under a pressure of 200MPa, using the equipment made by Nippon Steel Corporation (Model HIP2000S).

In order to investigate the reaction process, X-ray diffraction analysis was done on the reaction products sintered at various

#α-Type Si_3N_4,99.0% pure, major impurity: free C(1.0w/o),by Toshiba Co.
$TOYALNITE, 99.0% pure, major impurity: oxygen (1.0w/o),by Toyo Aluminium Co.,Ltd.
*Graphite, flake shape, special grade, by Junsei Chemical Co.,Ltd.,
+Calcium Nitrate Tetrahydrate, 99.9%, by Wako Pure Chemical Industries Co.,Ltd.

temperature using the commercial X-ray diffractometer made by Rigaku Denki Co.,Ltd., which is equipped with a personal computer for data processing. In all case, Ni-filtered CuKα radiation was used. The microstructure was observed using the scanning electronic microscope (SEM) made by Hitachi Corporation (Model S-530).

RESULTS AND DISCUSSION

The X-ray diffraction patterns of the specimens sintered at different temperature are shown in Fig.1. After the sample was sintered at 1873K for 90 minutes, a new compound which can be indexed on the basis of α-sialon (according to ASTM card No.33-260) was formed, and added graphite existed as unreacted. But after it was sintered at 1973K, almost only the peaks of 2H-SiC were detected out, and a small quantity of β-SiC coexisted. Furthermore, the sample was sintered for various times at the same temperature of 1973K: after sintered for 5 minutes the major phase was α-sialon, but for 15 minutes almost only 2H-SiC solid solution were obtained,

From the X-ray diffraction results, the reaction process is thought to be as shown in the model of Fig.2, at first, the liquid phase appeared on the powders surface due to the existence of oxides, then α-sialon formed with increasing temperature or sintering time, but graphite powders remained in an unreacted state, when the sintering temperature was higher than 1873K, graphite reduced sialon to produce SiC-AlN solid solution. So α-sialon whose molecular formula can be written as $CaSi_{12-x}Al_xO_yN_{16-y}$ was a transient phase. The reaction formulae can be written as following:

$$Si_3N_4 + xAlN + CaO \rightarrow CaSi_{12-x}Al_xO_yN_{16-y}$$
$$CaSi_{12-x}Al_xO_yN_{16-y} + C \rightarrow 3SiC:xAlN + Ca\uparrow + CO\uparrow + N_2\uparrow$$

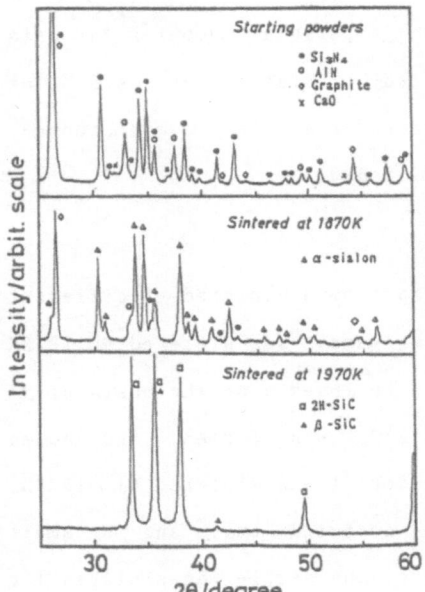

Fig.1 The X-ray diffraction profiles of starting powders and reaction products

Table 1 The composition of starting powders and designed reaction products.

| sample | composition/mol% | | | | | |
| | starting powders | | | | reaction products | |
	Si₃N₄	AlN	C	CaO	SiC	AlN
SNAN01	18	6	72	4	90	10
SNAN02	16	16	64	4	75	25
SNAN03	14.5	23.4	58.1	4	65	35
SNAN04	12	36	48	4	50	50
SNAN05	8	56	32	4	30	70

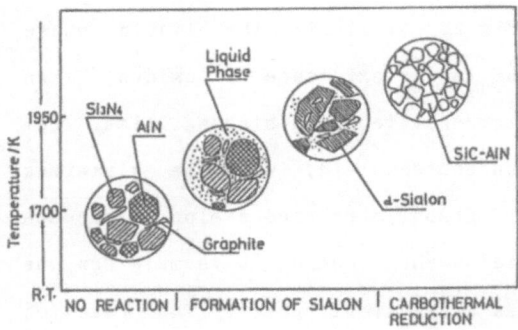

Fig.2 The schematic drawing of reaction process.

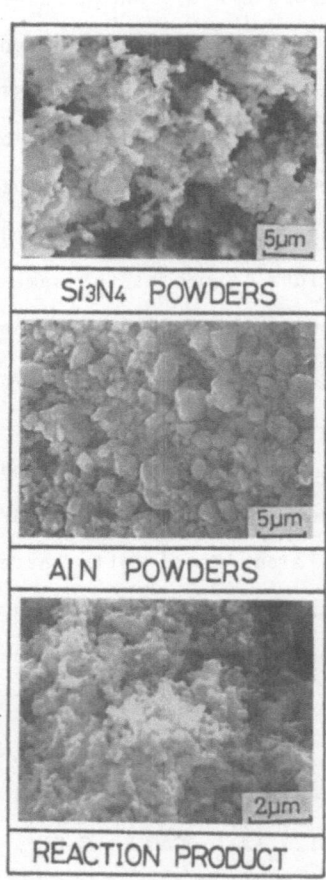

Fig.3 The SEM photographs of the starting powders and the reaction products of sample SNAN02.

But at a few place where no liquid phase existed, Si_3N_4 reacted directly with graphite powders, so that a small quantity of β-SiC was formed, as shown in Fig.1.

The SEM photographs of the starting powders and one sintered specimen are shown in Fig.3. It should be noticed that the grain size of SiC-AlN solid solution is three times smaller than that of the starting powders. It is very meaningful that ultrafine SiC-AlN alloy can be synthesized by reaction sintering.

Because 2H-SiC and AlN have the same crystalline structure and their lattice parameters are close, their X-ray diffraction peaks are too near to decide whether SiC-AlN solid solution was exactly formed or not by the ordinary operation of X-ray diffraction, so more careful diffraction was done on the (002) plane. As shown in Fig.4, only one peak was detected out when less than 35mol% AlN was added, this indicates that the single phase was formed below 35mol% AlN content, but the two phase of SiC-AlN solid solution and AlN (or AlN-SiC solid solution) coexisted when 50mol% or 70mol% AlN was added. This result coincides with the phase diagram presented in reference(5).

As described above, SiC-AlN solid solution or composite was able to be synthesized by the reaction sintering method, but because a large amount of gas formed due to the reaction mentioned above, no densification occurred during the reaction sintering. To measure the mechanical property of this new material, dense specimens are required, so the specimens were HIP-sintered.

Fig.5 shows the change of bulk density, because weight loss occurred during the reaction, the density of reaction product was small than green density, but after HIP-sintered for two hours at 2070K, the specimens containing more than 25mol% AlN were fully densified, on the

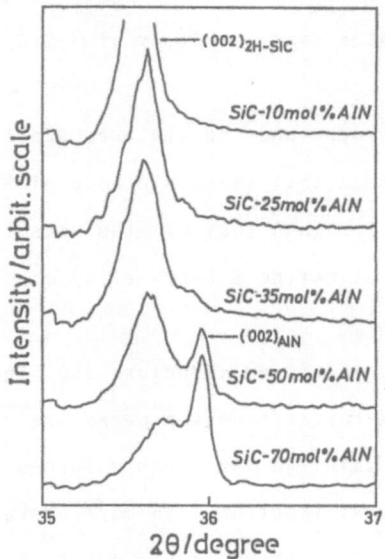

Fig.4 The X-ray diffraction
results on (002) plane.

Fig.5 The change of bulk density.

Fig.6 The fractography of SiC-AlN alloys.

contrary, relative density of pour SiC powders was about 60%. It can be concluded that addition of AlN effectively accelerates the densification of SiC.

The SEM photographs of the fractured surfaces were shown in Fig.6. the ultrafine grains are SiC-AlN solid solution, and the larger grains are AlN (or AlN-SiC solid solution). there is only one single phase in the specimens containing 10mol% and 25mol% AlN,but two phases exist in those containing 50mol% and 70mol% AlN.

CONCLUSION

SiC-AlN ceramic alloys were synthesized by the reaction sintering at rather low temperature. When AlN content was less than 35mol%, a fine-grained uniform solid solution was formed. On the other hand, when an excess amount of AlN was added, two-phase composite was obtained. The reaction process includes two steps: the formation of α-sialon and the carbothermal reduction. It can be described by the following reaction formulae.

$$Si_3N_4+AlN+CaO \longrightarrow CaSi_{12-x}Al_xO_yN_{16-y}$$
$$CaSi_{12-x}Al_xO_yN_{16-y}+C \longrightarrow 3SiC-xAlN+Ca\uparrow+CO\uparrow+N_2\uparrow$$

The synthesized SiC-AlN alloy specimen with high porosity was fully densified after HIP-sintered at 2070 to 2120K under a pressure of 200MPa.

REFERENCE
(1). I.B.Culter, P.D.Willer, W.Rafaniello, H.K.Park, D.P.Thompson and K.H.Jark, "New Materials in the Si-C-Al-O-N and Related System", Nature, 275[5679]434-435(1978)
(2). R.Ruh, A.Zangvil, "Composition and Properties of Hot-Pressed SiC-AlN Solid Solutions", J. Am. Ceram. Soc., 65[5]260-265(1982)
(3).W.Rafaniello, K.Cho, A.V.VirKar, "Fabrication and Characterization of SiC-AlN alloys", J. Mater. Sci., 16[12]3479-3488(1981)
(4). Y.Sugahara, K.Sugimoto, K.Kuroda and C.Kato, "The Formation of SiC-AlN Solid Solution By The Carbothermal Reduction Process Of Montmorillonite", J. Mater. Sci. Lett., 7[7]795-797(1988)
(5). M.M.Dobson, "Silicon Carbide Alloy", Research Reports in Materials Science, Series One, (1986)

ENERGY EQUILIBRIA AND RATE PROCESS

IN BRITTLE FRACTURE

Yoshizo Inomata

National Institute for Research in Inorganic Materials

1-1, Namiki, Tsukuba-shi, Ibaraki, 305

So as to treat brittle fracture by an atomic model, problems in the energy equilibria on the crack stability and those in the rate process are discussed, and new expressions are proposed. Newly derived critical condition of crack propagation for Griffith type through micocrack in brittle material is given as follows,

$$\sigma_0 = 2(E\gamma_s/1)^{1/2}$$

where, σ_0; critical tensile stress (threshold stress of subcritical crack growth) operated normal to crack surface and parallel to specimen length, E; Young's modulus, γ_s; surface free energy of the material, 1; specimen length.

And it is shown that the following formulation of rate equation is useful to explain the velocity of crack propagation.

$$v = f\{\exp(-\Delta G_a/RT)\} \cdot \psi \cdot \{1 - \exp(-\Delta G_0/RT)\}$$

Where, v; velocity of the reaction like crack propagation, f; frequency factor, ΔG_a; activation energy necessary for the reaction remarked, ψ; measure unit of the reaction rate (in case of crack propagation, i.e. a lattice parameter), ΔG_0; excess free energy stored in the system which is expected to be finally dissipated through the reaction.

The term expressed in the parenthesis, { } corresponds to an effective

probability of the transformation and it becomes zero under the critical

(i.e. equilibrium) condition. A pair of these equations makes it possible

to treat the equilibrium and the rate process continuously.

PREFACE

Stability of a crack in brittle materials under load is a problem of

thermodynamic equilibrium and the velocity of crack propagation is a

problem of reaction rate. In the present report, both problems are

discussed basically.

STABILITY OF CRACK

According to thermodynamics, an equilibrium of two different states of

same material is expressed by the equality of their free energy level.

Consider the two different states as shown in Fig. 1. If the crack size

2C satisfies a condition $2C \ll t < w \sim 1$, and the product, $(1 \cdot t \cdot w)$ is

equal to molar volume V, choosing stress free and crack free state of same

size of material as the reference state, the free energy level of the

state I can be well approximated by following equation,

$$\Delta G_I = (\sigma^2 V / 2E) \tag{1}$$

where the σ is uniform tensile stress operated at the boundary normal to

the crack surface and along the direction of l.

Since the state F is free from stress and free from kinetic energy, the

excess free energy stored in the system is only the surface free energy

necessary to separate the specimen in two pieces. That is,

156

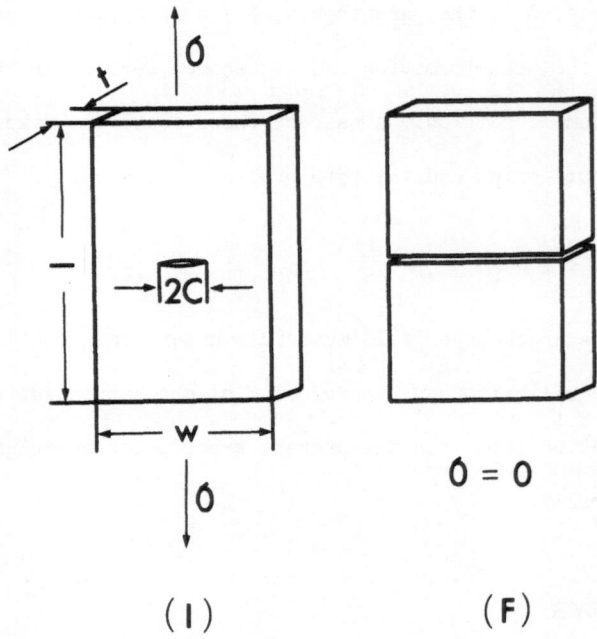

Fig. 1

Defined initial state (I) and final state free from kinetic energy (F).

$$\Delta G_F = 2\gamma_s tw = (2V\gamma_s/l) \tag{2}$$

The thermal equilibrium between state I and F is given by a condition, $\Delta G_I = \Delta G_F$, and following relation is obtained [1].

$$\sigma_0 = 2(E\gamma_s/l)^{1/2} \tag{3}$$

Where, σ_0 is the critical stress of crack extension, i.e. the threshold stress of subcritical crack growth. If additional works are concerned with γ_s, one can add those works to the γ_s.

Though, similar condition to Eq. (3) is already discussed by Griffith [2,3], in case of his system, the same treatment with Eq. (1) - (3) gives a result, $\sigma_0 = 0$. Since his system is characterized by infinite value of l and w.

RATE EQUATION TO EXPRESS REACTION OCCURRED IN HETEROGENEOUS SYSTEMS

There have been proposed many experimental and theoretical formulae
[4-7], but all of these proposals have not succeeded to take into account
the threshold condition of the crack propagation. This problems can be
solved by application of a new proposal for the generalized expression of
the rate process [8]. That is,

$$v = f\{\exp(-\Delta G_a/RT)\} \cdot \psi \cdot \{1 - \exp(-\Delta G_0/RT)\} \tag{4}$$

Where, v; velocity of the reaction, f; frequency factor, ΔG_a; activation
energy of a reaction remarked, ψ; measure unit of the reaction rate (in
case of crack propagation, i.e. lattice parameter), ΔG_0; excess energy
stored in the system which is finally dissipated through the reaction (in
case of crack propagation, i.e. the energy difference, $\Delta G_I - \Delta G_F$ in Fig.
1), R; gas constant, and T; absolute temperature.

The term expressed in the parenthesis, { } corresponds to an effective
probability of transformation of the reaction and it becoms zero under the
equilibrium state expressed in the former section. The expression of Eq.
(4) is useful to treat a rate process occurred in heterogeneous system
such as crack propagation, and it makes possible to treat the equilibrium
and the rate process continuously. Similar expressions to Eq. (4) have
successfully applied for the explanation of diffusion controlled mass
transport in heterogeneous systems [9].

RATE OF CRACK PROPAGATION

Consider a system illustrated in Fig. (1)-I, and assume 2C and the
stress satisfying 2C << t < w ~1 and $\sigma > \sigma_0$. Under the condition, the
detailes of Eq. (4) can be shown as follows,

$$v = (n'/n)(RT/Nh)\left[\exp\{(\Delta G_{\sigma m} - \Delta G_{th})/RT\}\right]\cdot\lambda\cdot\{1 - \exp(-\Delta G_0/RT)\} \quad (5)$$

where, n; number of atomic pairs which must be broken to advance crack an

atomic spacing, n'; number of atomic pairs bearing maximum stress at the

crack tip, N; Avogadoro's number, h; Planck's constant, $\Delta G_{\sigma m}$; molar strain

energy stored in the atomic pair bearing the maximum stress at the crack

tip [10], ΔG_{th}; molar strain energy supposed to be stored in the atomic

pair under maximum theoretical stress [11], λ; lattice spacing along the

direction of crack propagation, and ΔG_0; excess free energy stored in the

system (in case of the example shown in Fig. 1, i.e. $\Delta G_I - \Delta G_F$). In

general, n = n' may be preserved.

The ΔG_{th} increases by the introduction of additional works such as

dislocation formation and by crack branching, but $\Delta G_{\sigma m}$ decreases by these

processes and by crack blunting. Chemical reactions at the crack tip such

as moisture effect, generally decrease ΔG_{th}.

LIFE PREDICTION

The life prediction of precracked specimen by K_I mode becomes possible

by the following formula, using Eq. (5),

$$\Delta t = \int_{2C}^{W} dC/v(C,\sigma) \quad (6)$$

and if $\Delta G_{\sigma m}$ is linearly propotional to the crack length [10], under a

condition w > 20C, the result of integration of Eq. (6) is given as

follows, assuming n = n',

$$\Delta t = \{\exp(-\Delta G_C/RT)\}/(\Delta G_{\sigma m}/RTC)(RT/Nh)\lambda(\Delta G_0/RT)\exp(-\Delta G_{th}/RT) \quad (7)$$

where, Δt; time to failure from crack length C to a catastrophe, ΔG_c; molar strain energy stored in the atomic pair bearing maximum stress at a crack tip of a crack in length 2C.

REFERENCES

1) Y. Inomata, J. Surface Science Society Japan, in Japanese, $\underline{9}$, 591 -594 (1988); Critical Stress in Brittle Fracture.

2) A.A. Griffith, Phil. Trans. Roy. Soc., $\underline{221}$, 163-198 (1920); The Phenomena of Rupture and Flow in Solids.

3) ibid., Proc. 1st Int'l Cong. of Appl. Mech., (Delft, 1924) pp.55-63; The Theory of Rupture.

4) J.E. Sinclair, Phil. Mag., $\underline{31}$, 647-671 (1975); The influence of the interatomic force law and kinks on the propagation of brittle cracks.

5) A.G. Evans and S.M. Wiederhorn, Int. J. Frac., $\underline{10}$, 379-292 (1974); Proof testing of ceramic materials - an analytical basis for failure prediction.

6) S.M. Wiederhorn, H. Johnson, A.M. Diness and A.H. Heuer, J. Am. Ceram. Soc., $\underline{57}$, 336-341 (1974); Fracture of Glass in Vacuum.

7) A.G. Evans and T.G. Langdon, Progress in Material Science, Vol. 21, "Structural Ceramics" Edited by B. Chalmers, J.W. Christian and T.B. Massalski, (Pergamon Press, 1976) pp. 300-319

8) Y. Inomata, J. Surface Science Society Japan, being in print.; Rate Process in Brittle Fracture.

9) ibid., J. Surface Science Society Japan, in Japanese, $\underline{5}$, 308-312 (1984); Free Energy Theory of Material Transport Controlled by Diffusion.

10) C.E. Inglis, Trans. Inst. Nav. Archit., 55, 219-241 (1913); Stress in
    a plate due to the presence of cracks and sharp corners.

11) E. Orowan, Rep. Prog. Phys., 12, 185-232 (1949); Fracture and
    Strength.

ATOMIC CRACK TIPS IN COVALENT CRYSTAL

Hidehiko Tanaka, Yoshio Bando,

Mamoru Mitomo and Yoshizo Inomata

National Institute for Research in Inorganic Materials

1-1 Namiki Tsukubashi Ibarakiken 305 Japan

ABSTRACT

Crack tip geometries in covalent crystals were studied by high resolution (HR) TEM. The materials used in the study were SiC, Si crystals and 15R-sialon grain. Cracks were introduced by Vickers indentation in SiC and Si. In 15R-sialon, crack was naturally introduced. It was found that the crack tips were atomically sharp and there was no macroscopic yielding zone. The results of TEM observation are summarized here.

INTRODUCTION

According to continuum fracture mechanics, crack tip is imaged as an elliptical shape and stress becomes infinite at the tip. This large stress accumulation at the tip is modified by mechanism of small scale plastic zone or healing mechanism by cohesive force between two crack walls. But actual materials have discrete and non-elastic natures in an atomic scale, and the crack tip does not seem to be elliptical.

S.M.Wiederhorn et al.[1], B.J.Hockey et al.[2] first observed crack tips which propagated in ceramic materials such as SiC, Al_2O_3 by TEM. They found that there was no macroscopic deformation around the tips and discussed that spontaneous closure and healing behind the tip were an essential mechanism of crack propagation in highly brittle materials.

Another approach to the investigation of crack tip geometry is computer simulations using simplified lattice model. The calculations revealed various aspects of crack propagation. J.E.Sinclair et al.[3] obtained the results showing complete cleavage crack propagation in a diamond structural model. J.H.Weiner et al.[4] calculated dislocation generations at the tip in a two-dimensional crystal model. The features of the calculated crack tips depend largely on bonding potential and lattice structure used in the models. In order to understand the fracture behavior of brittle materials, it is important to observe directly crack tips in the atomic level.

The authors have studied crack tip geometry which propagated in ceramic materials by HR-TEM. This paper summarizes the results of the TEM observations on the cracks in SiC, Si crystals and 15R-sialon grains[5,6].

EXPERIMENTAL PROCEDURE

6H-SiC was obtained from ingot which was industrially produced by Acheson furnace (Taiheiyourundum Co.Ltd.,3N pure). Si was semiconductor grade crystal which was commercially obtained (Nihonsilicon Co.Ltd.,11N pure). The crystal structure of SiC and 15R-sialon is hexagonal, and that of Si is cubic.

The SiC and Si crystals were cut into small disks, cracks were introduced by Vickers indentation. After thinning by ion bombardment, the samples were observed by TEM (JEM-4000FX,JEOL). 15R-Sialon($SiAl_4O_2N_4$) was prepared by hot-pressing of Si_3N_4, Al_2O_3 and AlN powders. The sample was crashed. It was placed on a holey carbon grid and observed by the TEM. The cracks were found at the edge of the crashed grains.

RESULTS OF CRACK OBSERVATION

1. SiC crystal

Cleavage cracks propagated preferentially in the (0001) and {1120} planes in SiC. HR-TEM photographs of the cracks are shown in Fig.1. It is revealed that the crack is atomically sharp. The crack opens between two adjacent layers. At the tip, lattices of the crystal do not show any disorder. Crack tip blunting does not occur. Another important feature which Fig.1 pointed out is that there were no dislocation generations from the tip. Macroscopic plastic yielding zone does not exist. The fracture of SiC exhibit completely brittle manner.

By close examination of the tip, the lattices are found to be bent or distorted. This distortion extends for a few atomic spacings (about 2-3 nm) in front of the tip. The lattice appears to be partially debonded and the two adjacent debonded lattice planes close here.

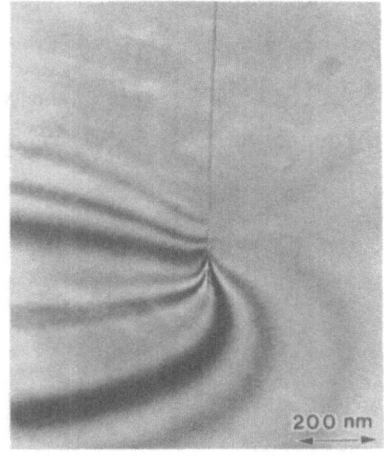

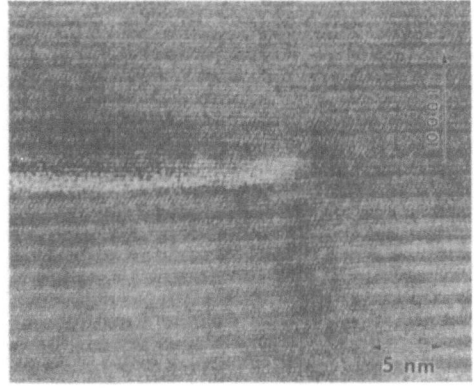

Fig.1 Atomically sharp crack tips in α(6H)-SiC[6].

2. 15R-sialon grain

Crack tips in 15R-sialon were shown in Fig.2. Completely brittle crack tip feature, that is, no evidence of plasticity, was observed again in 15R-sialon grain. At the tip, lattice spacing of 15R-sialon expands slightly. Fig.2 shows more clearly the atomically sharp cracks in 15R-sialon than in SiC.

3. Si crystal

Cleavage cracks in Si always propagate in <111> plane (Fig.3). Although the crack tip in Si crystal is not so clear as those of SiC and 15R-sialon, it is also atomically sharp. In Si, however, small dislocation segments were found around the periphery of the crack and the tip. It is considered that the dislocations were formed owing to the stress around the tip concentrated when the crack propagated and the successive ion-bombardment of Si crystal.

DISCUSSION

Atomic fracture process has been studied by many computer

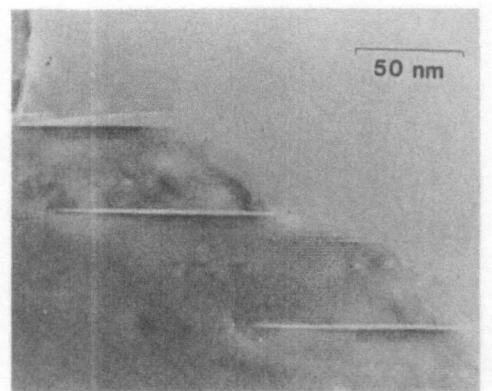

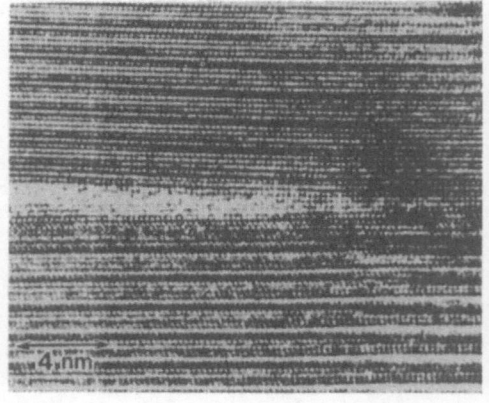

Fig.2 Crack tip in 15R-sialons[5]. The crack propagates in C plane.

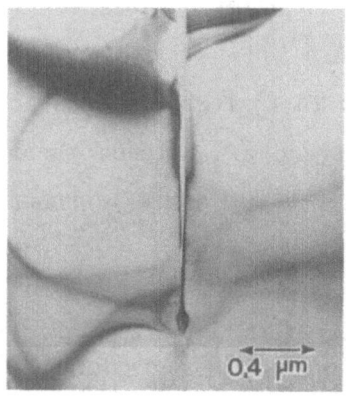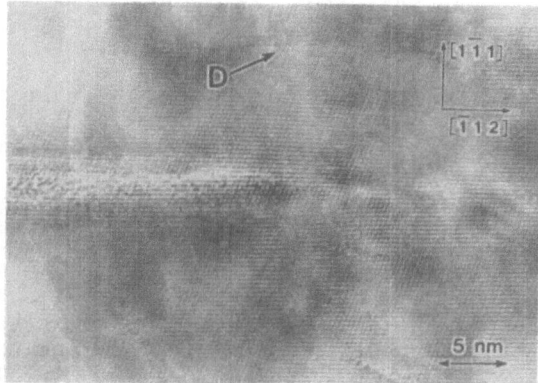

Fig.3 Crack tips in Si crystal[6]. Crack propagates in (111) plane. D:dislocation segment around the tip.

calculations[3,4]. J.E. Sinclair et al.[3] simulated a completely brittle fracture. They used a model which had diamond lattice structure and covalent bonding potential. The calculation showed that the crack propagated in a cleavage manner. Non-linear lattice deformation was limited in a very small region at the tip, dislocation generation was less stable than cleaving of the lattice, and there was no cohesive interaction between two crack planes behind the tip.

The results of TEM observation (Figs.1,2 and 3) suggest that the crack tip features in the covalent crystals correspond basically to the result of Sinclair et al.'s calculation[3]. This means that the crack tips in the covalent crystals are atomically sharp and the crack blunting or the macroscopic yielding around the crack tip do not occur. In Si crystals, however, dislocation segments, a wake of fracture, were found around the tip. It should be noticed that Si crystal is not completely brittle. The atomically sharp crack tip feature observed in this work leads to conclusion that fracture behavior must be considered not only in terms of energy balance theory but of kinetics of debonding of lattice, as S.M.Weiderhorn[1] pointed out.

Acknowledgment: The authors thank the American ceramic society for the permission of reproducing the figures 1,2 and 3 from refs.5 and 6, and thank also Mr. H.Kawabata, Taiheiyourundum Toyama-shi Japan, for preparing SiC single crystals.

REFERENCES

[1] S.M.Wiederhorn, B.J.Hockey and D.E.Roberts, "Effect of Temperature on the Fracture of Sapphire," Phil. Mag., 29, 783-96(1974).

[2] B.J.Hockey and B.R.Lawn,"Electron Microscopy of Microcracking about Indentations in Aluminium Oxide and Silicon Carbide," J. Mater. Sci. 10, 1275-84(1975).

[3] J.E.Sinclair and B.R.Lawn, "An Atomistic Study of Cracks in Diamond-Structure Crystals," Proc. R. Soc. Lond. A, 329, 83-103(1972).

[4] J.H.Weiner and M.Pear, "Crack and Dislocation Propagation in an Idealized Crystal Model," J. Appl. Phys., 46[6], 2398-405(1975).

[5] H.Tanaka, Y.Bando, Y.Inomata and M.Mitomo, "Atomically Sharp Crack in 15R-Sialon," J. Amer. Ceram. Soc., 71[1], C32-C33(1988).

[6] H.Tanaka and Y.Bando submitted to J.Amer.Ceram.Soc.1989.

THEORETICAL CALCULATION OF THE RATE OF
CRACK PROPAGATION AND TIME TO FAILURE IN SiC SINGLE CRYSTAL

Yoshizo Inomata

National Institute for Research in Inorganic Materials

1-1, Namiki, Tsukuba-shi, Ibaraki, 305

Using new proposals regarding the stability condition of cracks in a brittle material and the rate equations for heterogeneous system, numerical calculations are performed on the rate of crack propagation and the time to failure in silicon carbide single crystal, assuming Griffith type atomically sharp crack and linear elasticity in the material.

A technological expression of K_I - v relation is proposed as follows,

$$v = A \left[\exp\{(B\ K_I^2 - C)/RT\} \right] \cdot (K_I^2/C)$$

where, v; verocity of crack propagation, C; half length of Griffith type crack, K_I; stress intensity factor near by the crack tip, R; gas constant, T; absolute temperature, and A, B and C are material constants.

PREFACE

In the present report, the rate of crack propagation is calculated on silicon carbide single crystal, using a new proposal to treat the rate process in heterogeneous system [1]. The basic equation is,

$$v = (n'/n)(RT/Nh) \left[\exp\{(\Delta G_{\sigma m} - \Delta G_{th})/RT\} \right] \cdot \lambda \cdot \{1 - \exp(-\Delta G_0/RT)\} \quad (1)$$

where, v; velocity of crack propagation, n; number of atomic pairs which must be separated to advance crack an atomic spacing, n'; number of atomic pairs bearing maximum stress at the crack tip, N; Avogadoro's number, h;

Planck's constant, $\Delta G_{\sigma m}$; molar strain energy stored in the atomic pair

bearing the maximum stress at the crack tip [2], ΔG_{th}; molar strain energy

supposed to be stored in the atomic pair under maximum theoretical stress

[3], λ; lattice spacing along the direction of the crack propagation, and

ΔG_0; excess free energy stored in the system which is expected to be

finally relaxed through the separation of material in two pieces (in case

of Griffith type microcrack, that is the strain energy finally turning

into kinetic energy by the separation).

ASSUMPTIONS AND NUMERICAL DATA USED IN THE CALCULATION

Linear elasticity is assumed, and the atomic force between adjacent

atoms is supposed to be expressed as follows,

$$\text{if } a_0 \leq a \leq a_{th}; \qquad \sigma = E(a - a_0)/a_0$$
$$\text{and if } a > a_{th}; \qquad \sigma = 0 \tag{2}$$

where, a_0; atomic distance between atoms under stress free state, a; the

distance under stress σ, a_{th}; the distance under maximum theoretical

stress.   The maximum theoretical stress, σ_{th} which corresponds to a_{th},

is supposed to follow Orowan's expression [3],

$$\sigma_{th} = (E\gamma_s/a_0)^{1/2} \tag{3}$$

where, E; Young's modulus.

The maximum stress realized in the specimen which has a Griffith type

atomically sharp crack, is assumed to be expressed by the result of Inglis

[2]. Assuming the radius of curvature at the crack tip being $a_0/2$, and

using V for molar volume, the stress σ_m is,

$$\sigma_m = (8C\sigma^2/a_0)^{1/2} \tag{4}$$

From Eq. (2)-(4), assuming a Griffith type microcrack in the specimen, ΔG_{th}, $\Delta G_{\sigma m}$ and ΔG_0 in Eq. (1) are given as follows,

$$\Delta G_{th} = (E\gamma_s/a_0)(V/2E) \tag{5}$$

$$\Delta G_{\sigma m} = (8C\sigma^2/a_0)(V/2E) \tag{6}$$

$$\Delta G_0 = (\sigma^2 V/2E) - (2V\gamma_s/l) \tag{7}$$

Physical data used in the next section for an actual calculation of the rate of crack propagation in (111) plane of β-SiC are, V; 1.261×10^{-5} m^3/mol, γ_s; 3.17 J/m^2, a_0; 2.518×10^{-10} m, and E; 4.6×10^{11} Pa.

RATE EQUATION AND EQUATION FOR THE TIME TO FAILURE

Since n = n' may be realized for the Griffith type through crack, Eq. (1) can be rewritten, using Eq. (5)-(7),

$$v = (dC/dt) = (RT/Nh)\left[\exp\{(V/2ERTa_0)(8C\sigma^2 - E\gamma_s)\}\right]\cdot\lambda\cdot$$
$$\cdot\left[1 - \exp(1/RT)\{(2V\gamma_s/l) - (\sigma^2 V/2E)\}\right] \tag{8}$$

and under the condition, $\Delta G_0 \ll RT$ and $(\sigma^2 V/2E) \gg (2V\gamma_s/l)$, following equation gives an apploximation to Eq. (8).

$$v = (RT/Nh)\left[\exp\{(V/2ERTa_0)(8C\sigma^2 - E\gamma_s)\}\right]\cdot\lambda\cdot(\sigma^2 V/2ERT) \tag{9}$$

The result of calculation of Eq. (9) at 293 °K is shown in Fig. 1. If one takes into account a nonlinear elastic model for the estimation of ΔG_{th} and $\Delta G_{\sigma m}$, v may become smaller than that given by Eq. (9).

Since the time necessary for the crack propagation from crack length C_1 to C_2 is given by the following expression,

$$\Delta t = \int_{C_1}^{C_2} dC/v(C,\sigma) \tag{10}$$

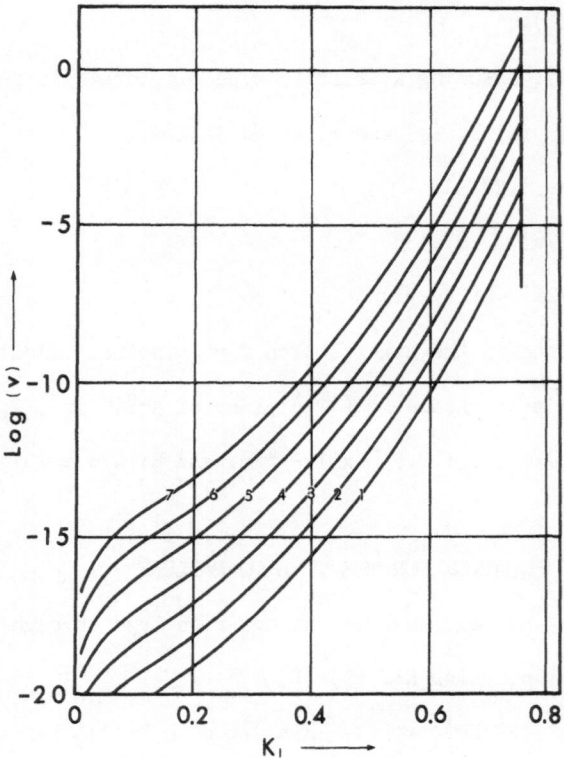

Fig. 1

Calculated $C(m)-K_I(MN/m^{3/2})-v(m/sec)$ relation at 293 °K in (111) plane of
β-SiC single crystal by Eq. (9). The number attached to each curve
corresponds to the power n in the relation $C = 10^{-n}$ (m).

where, C_1; half length of the initial crack, C_2; half length of the crack
after the propagation by a stress σ for a period, Δt.

Integration of Eq. (10) using Eq. (9) for $v(C,\sigma)$ leads to the folowing
results.

$$\Delta t = \frac{\left[\exp(-8C_1\sigma^2 V/2ERTa_0)\right] - \left[\exp(-8C_2\sigma^2 V/2ERTa_0)\right]}{(8\sigma^2 V/2ERTa_0)(RT/Nh)\cdot\lambda\cdot(\sigma^2 V/2ERT)\exp(-VE\gamma_s/2ERTa_0)} \tag{11}$$

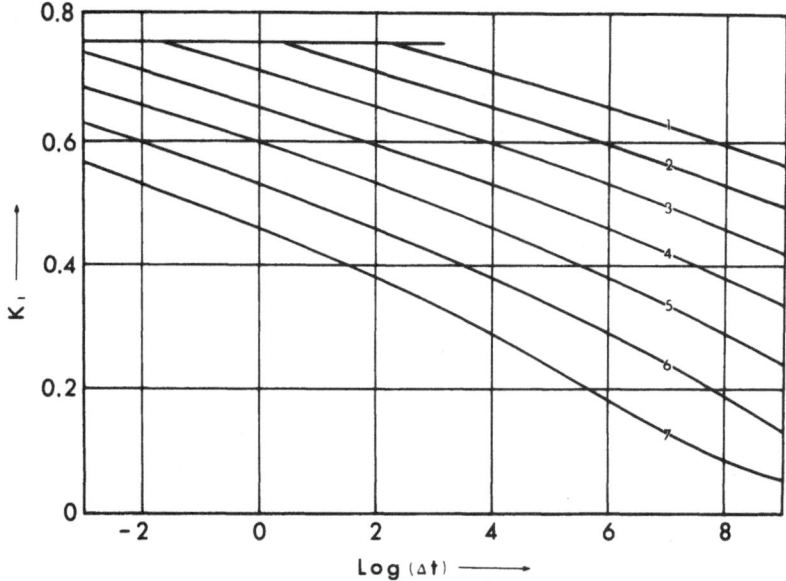

Fig. 2

Calculated $C(m)-K_I(MN/m^{3/2})-\Delta t(sec)$ relation at 293 °K in (111) plane of β-SiC single crystal by Eq. (11), assuming $C_2 > 10\ C_1$. The number attached to each curve has the same meaning with that in Fig. 1.

The second term of the numerator in the Eq. (11) can be neglected, if $C_2 > 10C_1$ is preserved. Figure 2 is a result of actual calculation performed under the same condition with Fig. 1. In the calculation, T is choosen at 293 °K and $C_2 > 10\ C_1$ is assumed.

TECHNOLOGICAL EXPRESSION OF K_I - v RELATIONS

The Eq. (8) and (9) leads to a technological expression of K_I - v relations as follows, using a definition, $K_I = \sigma(\pi C)^{1/2}$,

$$v = A\left[\exp\{(B\ K_I^2 - C)/RT\}\right] \cdot (K_I^2/C) \tag{12}$$

where, K_I; stress intensity factor near by the crack tip, **A**, **B** and **C** are material constants, and **B** and **C** are affected by the atomosphere especially in the region of subcritical crack growth [4]. If K_{IC} is concerned with certain velocity of crack propagation, Eq. (12) predicts that the K_{IC} may be affected by crack length.

REFERENCES

1. Y. Inomata, a paper submitted in this Symposium, entitled "Energy Equilibria and Rate Process in Brittle Fracture".
2. C.E. Inglis, Trans. Inst. Nav. Archit. 55, 219-241 (1913); Stress in a plate due to the presence of cracks and sharp corners.
3. E. Orowan, Rep. Prog. Phys. 12, 185-232 (1949); Fracture and Strength of Solids.
4. S.M. Wiederhorn, "Fracture Mechanics of Ceramics, Vol. 2", Ed. by R.C. Bradt, D.P.H. Hasselman and F.F. Lange, Plenum Press, (1974) p. 613-646

FORMATION OF INTERMETALLIC COMPOUND LAYER FORMED BY DIFFUSION TREATMENT OF THERMAL SPRAY COATINGS, AND ITS APPLICATION.

SACHIO OKI, KAZUYOSHI KAMACHI and SUSUMU GOHDA

Faculty of Science and Technology,

Kinki University,

3-4-1 Kowakae, Higashi-osaka, 577, Osaka, Japan

ABSTRACT

In order to develop high temperature oxidation resistive coatings, multi-layer thermal spray coatings containing Ni-Cr-Al and Al layers were applied to austenitic stainless steel substrate. The Ni-Cr-Al layers were coated with Ni-20%Cr-5%Al pre-alloyed powder or mixtures of Ni-20%Cr powder with 5% of pure Al powder. Then, pure Al was sprayed on the Ni-Cr-Al coated surface. Prior to high temperature oxidation test, conducted at 1273 K in air, sprayed specimens were heat treated at 1123 K for 3.6 ks and 1423 K for 3.6 ks in air. The oxidation resistivity of coating sprayed with pre-mixed powder is superior to that made with pre-alloyed powder. This result is attributed to the difference in the distributions of the Al rich alloy phases formed by the diffusion between each layer.

INTRODUCTION

Thermal spraying is one of the practical method of surface coating to improve wear resistance, heat resistance and corrosion resistance of metals and alloys. Despite of the widespread use of the thermal spraying, the cohesive properties have not been researched from a point of view of the basic cohesion mechanism. The cohesivity expected to be enhanced through the mutual diffusion between the sprayed coating and the substrate. The authors have been aiming to improve the cohesive strength by appropriate diffusion treatment after spraying, and to develop the novel method of the surface modification by the use of the resulting intermetallic compound layers[1, 2, 3].

In this paper, the attempt to develop the high temperature anti-oxidation coating formed by the diffusion treatment of spray coating is described. This method was applied to the Ni-Cr-Al and Al multi-layer coatings on the austenitic stainless steel substrates, and the formation of the intermetallic compounds phases by diffusion treatment and the high temperature oxidation properties of these coatings were investigated.

It is well known that the Al-rich alloy layers formed on the surface of base-metals improve their oxidation properties by the resulting Al_2O_3 protective films[4]. However, Al coatings sprayed on stainless steel substrates did not form stable compound layer at high temperature[3].

The Ni-Cr-Al alloys were developed for high temperature protective coatings, and also have applied to the under-coat for ceramics coatings[5]. But the Ni-Cr-Al alloy spray coatings are available only by the Low Pressure Plasma Spraying method[6]. Hence, they have a limited field of use.

In view of these facts, we developed anti-oxidation coating available under oxidizing environment at about 1273 K. The prerequisites of the development are to form Al-rich intermetallic compound layers which have high cohesive strength with substrate and is stable at high temperature. In addition, it is an important requisite that the thermal spraying and the heat treatment can be conducted in air.

EXPERIMENTAL PROCEDURE

The Ni-Cr-Al alloys and pure Al multi-layers, as shown in Table 1, were coated on the commercial austenitic stainless steel substrates containing 18%Cr and 8%Ni (JIS SUS 304) by plasma spraying method in air. Plasma spraying was carried out using Bay State PG-100 torch. The stainless steel substrates were glit blasted prior to thermal spraying. The Ni-Cr-Al alloy layers were coated with pre-alloyed powders or powder mixtures of Ni-Cr pre-alloyed powders and pure Al powder. Then pure Al was sprayed on the Ni-Cr-Al of Ni-Cr sprayed surfaces. In table and figures, - and + show the pre-alloyed and the pre-mixed powders, respectively. The properties of

material powders are listed in Table 2. The thickness of the Ni-Cr-Al layers
and Al layers were 200 ~ 250 μ m and 50 ~ 100 μ m, respectively.

Table 1. Constitution of multi-layer coatings

Designation	1st layer	2nd layer
NC10A/A	Ni - 20%Cr + 10%Al	Al
NC5A/A	Ni - 20%Cr + 5%Al	Al
NCA/A	Ni - 20%Cr - 5%Al	Al
NCA	Ni - 20%Cr - 5%Al	-
NC	Ni - 20%Cr	-

Table 2. Properties of material powders

Powder	Chemical composition / mass%						Particle size / μ m
	Al	Ni	Cr	C	Si	Fe	
Ni-Cr-Al	4.6	Bal.	18.12	1.1	-	-	53 - 10
Ni-Cr	-	Bal.	19.16	0.014	1.26	-	45 - 10
Al	Bal.	-	-	-	0.06	0.11	106 - 32

The sprayed specimens were heat treated at 1123 K for 3.6 ks and 1423 K
for 3.6 ks in air.

High temperature oxidation properties were estimated by their mass
gains during heat cycle tests conducted at 1273 K in air. The oxidation
products were also determined by X-ray diffractometry (SHIMADZU XD-5A).
Monochromated Cu K α radiation was used as the X-ray source.

In order to determine the intermetallic compound phases formed by the
heat treatment and during the heat cycle test, some metallographical
examinations, i.e. metallography, electron probe microanalysis (JEOL
JXA-8600), and X-ray diffractometry were conducted.

RESULTS and DISCUSSION

The result of high temperature heat cycle test was shown in Fig. 1. As shown in this figure, the multi-layer coating made from Ni-Cr-Al pre-alloyed powder (NCA/A) was peeled in a few number of heat cycles. The multi-layer coating which did not contain Al in the first layer (NC/A) shows the same tendency. On the other hand, no peeling can be detected in the coatings made from Ni-Cr+Al pre-mixed powders (NC5A/A and NC10A/A). Mass gain of Ni-Cr single layer coating (NC) during the heat cycle test were less than that of Ni-Cr+Al pre-mixed multi-layer coatings. This fact is attributed to the high vapor pressure of Cr_2O_3 formed on the Ni-Cr coated specimen during high temperature heat cycle test. The same tendency was observed in the case of the Ni-Cr-Al single layer coating (NCA).

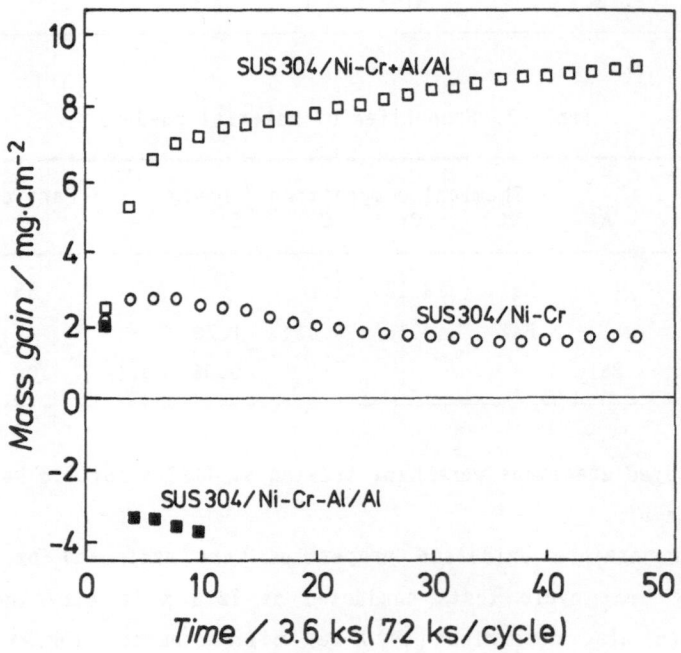

Figure 1. Effect of constitution of multi-layer coatings on high temperature oxidation properties at 1273 K.

The results mentioned above shows that the oxidation resistivity of the coating sprayed with pre-mixed powder is superior to that made with pre-alloyed powder, although their compositions are the same. Then, the microstructures of the specimens coated with pre-alloyed powder and pre-mixed powder were examined for revealing to the effect of the method of Al addition on the oxidation properties.

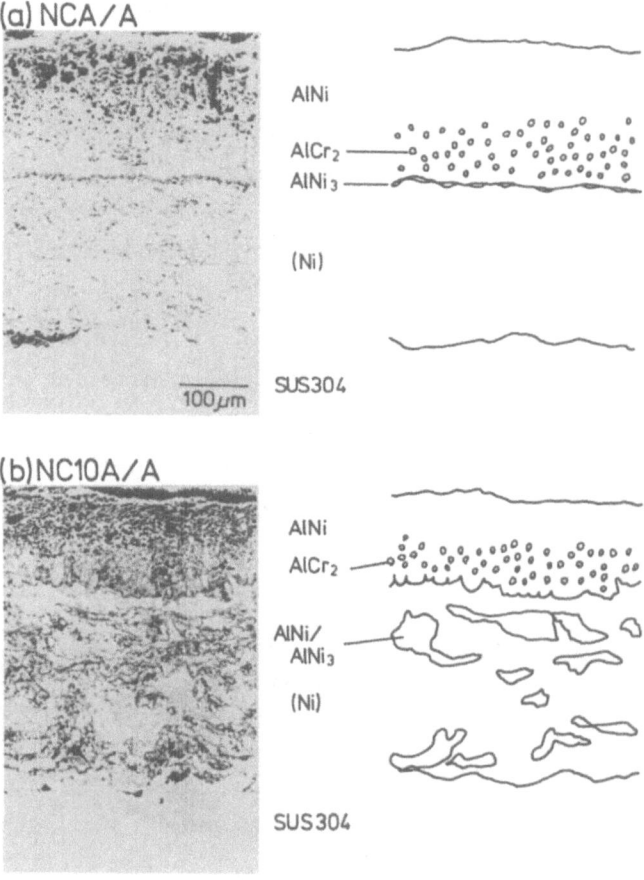

Figure 2. Microstructures and their schematic illustration of multi-layer coatings made from (a) pre-alloyed powder and (b) pre-mixed powder, after heat treatment.

The microstructures of the multi-layer coatings made from pre-alloyed powder (NCA/A) and pre-mixed powder (NC10A/A) are shown in Fig. 2. They consisted of AlNi phase and Ni solid solution, and islands of $AlCr_2$ phases were dispersed in the AlNi layers. $AlNi_3$ phases were formed in the interface between AlNi and Ni layers of the coating made from pre-alloyed powder (Fig.

2 (a) NCA/A). On the other hand, the Ni solid solution layer which made from pre-mixed powder (Fig. 2 (b) NC10A/A) had dispersion of Al rich phases consisted of AlNi and AlNi₃.

Furthermore, the secondary images and characteristic X-ray images of Ni, Cr, Al and O were shown in Fig. 3. There were some differences in distributions of alloying elements between pre-alloyed and pre-mixed coatings. In the case of the coatings made from pre-alloyed powder (Fig. 3 (a) NCA/A), distributions of Al and O formed a continuous net-work. This fact could be attributed to the oxidation of the surface of the molten particles and the interface of the deposited particles during spraying, and/or heat treatment. On the other hand, in the case of the pre-mixed coating (Fig. 3 (b) NC10A/A), the pre-mixed Al particles were oxidized preferentially, and the cohesive strength between deposited particles were consequently higher. The same tendency has been observed in the coating sprayed from Ni-Cr+5%Al pre-mixed powder (NC5A/A), which has the same composition as the Ni-Cr-Al pre-alloyed powder.

(a) NCA/A

(b) NC10A/A

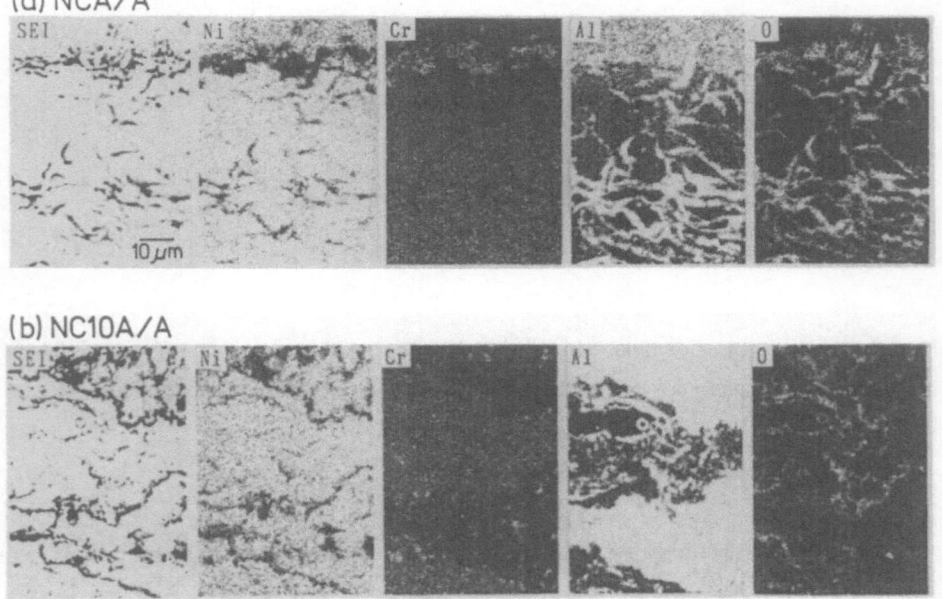

Figure 3. Secondary electron images and characteristic X-ray images of sprayed and heat treated multi-layer coatings made from (a) pre-alloyed powder and (b) pre-mixed powder.

Therefore, the difference of the oxidation properties between pre-alloyed and pre-mixed coatings could be ascribed to their distinct morphology of oxides in the spray coated layers. We can conclude that the pre-mixed Al addition is available in an attempt to develop the high temperature oxidation resistive coatings by means of thermal spraying method.

CONCLUSIONS

In order to develop high temperature oxidation resistive coating, Ni-Cr-Al and Al multi-layer thermal spray coatings were applied to austenitic stainless steel substrate. It has been proved that the oxidation resistivity of the coating sprayed with pre-mixed powder was superior to that made from pre-alloyed powder. This fact is attributed to their distinct morphology of oxides in the sprayed coatings.

REFERENCES

1. K. Kamachi, S. Oki, S. Gohda and G. Ueno, Reaction Diffusion between Thermal Sprayed Aluminium and Copper Substrate, J. Japan Inst. Metals, 52, 995 - 998 (1988).

2. S. Oki, K. Kamachi, S. Gohda and Y. Hirato, Formation Process of Alloy Phases by Reaction Diffusion between Thermal Sprayed Aluminium and Substrate of Armco Iron and Carbon Steels, J. Japan Inst. Metals, 52, 999 - 1005 (1988).

3. S. Oki, K. Kamachi and S. Gohda, J. Japan Inst. Metals, to be published.

4. N. Birks and G. H. Meier, *Introduction to High Temperature Oxidation of Metals*, Edward Arnold Ltd., translated by K. Nishida and T. Narita, Maruzen, Tokyo, 1988, p. 133.

5. K. Shimotori and T. Aisaka, The Trend of MCrAlX Alloys for High-temperature-protective Coatings - On the Effects of Alloy Compositions - , ISIJ, 69, 1229 - 1241 (1983).

6. K. Shimotori, Thermal Spraying Materials, in *Thermal Spraying Handbook*, ed Japan Thermal Spraying Society, Tokyo, 1986, pp. 216 - 225.

EFFECTS OF Al$_2$O$_3$: SiO$_2$ RATIO OF THE STARTING MIXTURE ON THE COMPOSITION ,MICROSTRUCTURE AND MECHANICAL PROPERTIES OF MULLITE PRODUCED BY REACTION SINTERING.

P.D.D. Rodrigo, Department of Materials Engineering, University of Moratuwa, Sri Lanka.

P. Boch, E.N.S.C.I., UA C.N.R.S. 320, F-87065 Limoges, France.

ABSTRACT

Composition of mullite formed by reaction sintering mixtures of α-alumina and amorphous silica varies from 70.4 to 74.8 wt% Al$_2$O$_3$ depending on the bulk composition of the starting mixture.   3 Al$_2$O$_3$: 2 SiO$_2$ stoichiometric mullite results in the best densification and finest grain size. Mullite rich in alumina (ie. 74.8 wt% Al$_2$O$_3$) exhibits rather poor densification and excessive grain growth. A small amount of silica in excess of the solubility limit, which is present in the form of a vitreous phase, has no significant effect on either the densification or the grain growth whereas a small amount of excess alumina resulting in α-alumina precipitates reduces the excessive grain growth and improves the densification of mullite rich in alumina. Mullite begins to form with a composition close to 70.4 wt% Al$_2$O$_3$ irrespective of the Al$_2$O$_3$: SiO$_2$ ratio of the starting mixture. Alumina dissolves into mullite progressively after all free silica is consumed.

Mechanical strength of mullite is sensitive to its composition. Presence of a small amount of vitreous silica leads to a higher flexural strength at elevated temperatures (up to 1200 $^{\circ}$C) and α-alumina precipitates slightly improves fracture toughness.

INTRODUCTION

Mullite is a solid solution of alumina and silica[1]. Good chemical and thermal stability, high creep resistance and refractoriness, low thermal expansion and thermal conductivity and medium strength at high temperatures are some of the attractive properties of this material. It can be synthesised at temperatures above 1300 $^{\circ}$C using mixtures of fine alumina and silica of different crystalline forms [2,3,4,5]. Sintering of mullite is rather difficult. Pressureless sintering of fine   (1 μm) mullite powders at about 1650 $^{\circ}$C for several hours results in densifications of the order of 90% [6,7] . Therefore it is difficult to produce dense mullite by sintering of synthetic mullite powders. This shows the necessity of studies on the possibilities of producing dense mullite by reaction sintering.

The aim of this work was to study the effects of the Al$_2$O$_3$: SiO$_2$ ratio of the starting mixture on the composition, microstructure and mechanical properties of mullite produced by reaction sintering mixtures of α-alumina and amorphous silica.

EXPERIMENTAL

The characteristics of amorphous silica and α-alumina which were used as starting materials are given in Table 1. The powder mixtures were prepared by attrition milling in ethanol using zirconia balls. After drying, crushing and mixing with the organic binder, powder mixtures were sieved through a 200 um sieve. Disc shape specimens (diameter : 30mm, thickness : 4mm) were prepared by uniaxial pressing under 150MPa in a floating die which is subjected to an ultrasonic vibration to improve compaction homogeneity[8].

After burning off the binder by firing up to 800 $^{\circ}$C, the powder compacts were subjected to different predetermined heat treatments. Fired specimens were characterized by measurement of density, observation of microstructure and both qualitative and quantitative analyses of crystalline phases using X-ray diffraction techniques. Mechanical strength was measured by biaxial flexure of discs[9]. Fracture toughness was measured according to SENB method using rectangular bars (25mm x 4mm x 4mm).

RESULTS AND DISCUSSION

The effects of the Al_2O_3 : SiO_2 ratio of starting mixture on densification and microstructure of the final product were studied using mixtures of α-alumina and amorphous silica of different compositions, namely 68.0, 71.8, 75.0, 77.3 and 80.0 wt% Al_2O_3. According to X-ray diffraction patterns given in Figure 1 corresponding to specimens fired at 1600 $^{\circ}$C for 10h, mullite and α-alumina are present in all samples which initially contained 75.0 wt% or more Al_2O_3. This observation is confirmed by the corresponding micrographs (Figure 2). Mullite is the only crystalline phase present in fired samples of other two compositions.

Samples of all starting compositions, fired at 1600 $^{\circ}$C for 10h were finely ground and leached with an aqueous solution of 10% HF for 3h in order to reveal the presence of any glassy phase. A loss in weight was recorded only in the case of samples of 68.0 wt% Al_2O_3 starting composition, which indicates that there was no significant amount of glassy phase in other sintered samples. The glassy phase present in sintered samples containing 68.0 wt% Al_2O_3 transformed into cristobalite when it was crushed, ground and fired again at 1350 $^{\circ}$C for 2h (Figure 3). This indicates that a certain amount of free silica is present in samples containing 68.0 wt% Al_2O_3 even after sintering at 1600 $^{\circ}$C for 10h. Quantities of different phases present in sintered samples of all compositions were determined using X-ray diffraction techniques. Results of this quantitative analyses, given in Table 2, confirm that mullite is a solid solution. The range of compositions of stable mullite solid solution, calculated using the data given in Table 2, is from 70.4 to 74.8 wt% Al_2O_3. Low-alumina and high-alumina ends of this range of compositions are in good agreement with the corresponding values obtained by Aksay and Pask[10] (70.5 wt% Al_2O_3) and Aramaki and Roy[11] (74.3 wt % Al_2O_3) respectively. The mullite rich in alumina, formed in mixtures containing 75.0 wt% Al_2O_3 exhibits a peculiar grain growth leading to a microstructure consisting of both elongated grains ($\approx$7 x 30 μm) and equiaxed grains ($\approx$ 5 x 5 μm). In contrast, the mullite of the same chemical composition formed in mixtures containing 77.3 and 80.0 wt% Al_2O_3 consists of only equiaxed grains. This difference is caused by α-alumina grains present at intergranular positions which retard the growth of mullite grains most probably by acting as grain boundary barriers. Despite the presence of a noncrystalline silicate phase ($\approx$ 5% in volume) , the mullite rich in silica, formed in mixtures containing 68.0 wt% Al_2O_3 consists of mainly equiaxed grains (a few slightly elongated grains are present). This shows that the presence of silica in excess of the solubility limit, which results in a noncrystalline (glassy) phase has almost negligible effect on mullite grain growth compared stoichiometric mullite formed in mixtures containing 71.8 wt% Al_2O_3 consists only of equiaxed grains ($\approx 5 \times 5\,\mu$m).

Final densification of powder compacts reaction sintered according to three different firing cycles are graphically shown in Figure 4. Powder compacts containing 71.8 wt% alumina, which convert into mullite (71.8 wt% Al_2O_3 and 28.2 wt% SiO_2) recorded the highest densification whereas those containing 75.0 wt% Al_2O_3, which convert into mixtures of mullite (74.7 wt% Al_2O_3 and 25.3 wt% SiO_2) and α-alumina recorded the poorest densification after each heat treatment. Better densification of powder compacts containing 77.3 and 80.0 wt% Al_2O_3, which also convert into mixtures of mullite(74.8 wt% Al_2O_3 and 25.2 wt% SiO_2) and α-alumina, is due to the presence of considerable amounts of free α-alumina after the completion of mullite formation.

Figure 5 shows the variation of flexural strength with chemical composition at 20(room temperature), 600, 900, and 1200 °C. The plot of room temperature flexural strength versus composition has a shape similar to that of the plot of densification versus composition (Figure 4). This close relationship between flexural strength and densification exists even at 600 °C. Therefore the low temperature mechanical strength depends more on densification than on composition. Flexural strength of products which do not contain a considerable amount of any phase other than mullite does not vary significantly with temperature up to 1200 °C. Presence of α-alumina precipitates leads to a 15 to 20 percent drop in strength at temperatures above 600 °C. Product containing 3 to 4 wt% free silica in the form of a vitreous phase shows the most remarkable dependency of flexural strength on temperature. Its strength increases with temperature from 210 MPa at 20 °C to a value in excess of 450 MPa at 1200 °C. This happens as a result of the presence of vitreous silica of high viscosity. Such a phase in a brittle polycrystalline material slows down the propagation of cracks at high temperatures as a result of the relaxation of stresses at crack tips when they enter the zones of vitreous phase of limited plasticity. Data given in Table 3 show that the fracture toughness at room tempera-ture is slightly improved by the presence of α-alumina precipitates.

Cell dimensions of mullite vary with its composition [12,13,14]. The changes in composition of mullite can be detected by measuring the cell parameter "a" which is the most sensitive one. It increases as the Al_2O_3 content in mullite increases. According to Camaron[12], the "a" versus composition plot does not deviate from linearity even when "a" values corrected for Fe and Ti in solid solution with mullite are incorporated(Figure 6).

Cell parameters of the mullite formed in three samples of the mixture containing 75.0 wt% Al_2O_3, sintered at 1550 °C for 1min and 1h and at 1600 °C for 10h were determined by powder X-ray diffractometry. $CuK\alpha_1$ peak positions were determined with an accuracy of $\pm$ 0.0025° relative to an α-alumina standard. Unit cell parameters of mullite calculated using a least-squares computer program and the relative amounts of other crystalline phases present determined by X-ray diffraction methods are given in Table 4. Gradual increase of "a" and "c" cell parameters with increasing sintering time and temperature indicates that the mullite formed at early stages of reaction enriches with Al_2O_3 on continued heating. The composition of mullite formed in each of the three samples was estimated using,

(1) quantitative analyses of crystalline phases present by X-ray diffraction methods (given under "x" in Table 4)

(2) values of cell parameter "a" and Figure 6(given under "y" in Table 4)

The significant difference between estimated "x" and "y" values for the composition of mullite formed in sample fired for 1min at 1550 $^{\circ}$C could be explained by the presence of noncrystalline SiO_2 owing to the insufficient time given for the transformation of amorphous silica into cristobalite. For other two samples "x" and "y" values are in good agreement. These data show that mullite begins to form with a composition close to the silica rich end of the stable mullite solid solution and it enriches with Al_2O_3 on continued heat treating.

CONCLUSION

The range of compositions of stable mullite is from 70.4 to 74.8 wt% Al_2O_3. Mullite begins to form with the composition close to the silica rich end of the range of solid solution irrespective of the Al_2O_3:SiO_2 ratio of the starting mixture. Densification, grain growth and mechanical properties of mullite ceramic are highly dependent on composition. $3Al_2O_3.2SiO_2$ stoichiometric mullite leads to the best densification and finest grain size. Nearly dense(97%) mullite ceramic possessing moderate mechanical properties up to about 1200 $^{\circ}$C can be produced by reaction sintering mixtures of α-alumina and amorphous silica at temperatures near 1600 $^{\circ}$C.

REFERENCES

1 Aksay, I.A. and Pask, J.A., Solid solution range and microstructure of melt-grown mullite, J. Am. Ceram. Soc., 66(1983)649.

2 Ghate, B.B., Hasselman, D.P.H. and Spriggs, R.M., Synthesis and characterization of high purity, fine grained mullite, Bull. Am. Ceram. Soc., 52(1973)670.

3. Mazdiyasni, K.S. and Brown, L.M., Synthesis and mechanical properties of stoichiometric aluminium silicate (mullite), J. Am. Ceram. Soc., 55(1972)548.

4 McGee, T.D. and Wirkus, C.D., Mullitization of alumino-silicate gels, Bull. Am. Ceram. Soc.,51(1972)577.

5 Kanzaki, S., Kumazawa, T. and Asaumi, T., Dependence of mechanical property of sintered mullite on chemical composition, Yoggo-Kyokai-Shi, 93(1985)407.

6 Metcalee, B.L. and Sant, J.H., The synthesis, microstructure and physical properties of high purity mullite, J. Brit. Ceram. Soc., 74(1975)193.

7 De Portu, G. and Henney, J.W., The microstructure and   mechanical   properties of mullite-zirconia composites, Trans. Brit. Ceram. Soc., 83(1984)69.

8 Rogeaux, B. and Boch, P., Influence of an ultrasonic assistance to the powder compaction on the Weibull modulus of sintered alumina, J. Mat. Sci. Lett., 4(1985)403.

9 Glandus, J.C., Meaning of the   biaxial   flexure   tests of discs for strength, Science of Ceramics13, Les Editions de Physique Publ., Paris (1985)C1-595.

10 Aksay, I.A. and Pask, J.A., Stable and metastable equilibria in the system SiO_2-Al_2O_3, J.

11 Aramaki, S. and Roy, R., Revised phase diagram for the system Al_2O_3-SiO_2, J. Am. Ceram. Soc.,45(1962)229.

12 Camaron, W.E., Composition and cell dimensions of mullite, Bull. Am. Ceram. Soc.,56(1977)1003.

13 Agrell, S.O. and Smith, J. V., Cell dimension, solid s olution, polymorphism, and identification of mullite and sillimanite, J. Am. Ceram. Soc.,443(1960)69.

14 Durovic, S.,Isomorphism between sillimanite and mullite, J. Am. Ceram. Soc.,45(1962)157.

Table 1.   Characteristics of starting materials

Material	Chemical analysis(wt%)					Particle size(μm) distribution		
	Al_2O_3	SiO_2	Fe_2O_3	Na_2O	CaO/ MgO	d_{10}	d_{50}	d_{90}
α-alumina	99.7	0.07	0.03	0.05	0.07	0.28	0.65	1.7
Amorphous silica	-	99.6	-	0.12	0.08	0.88	3.6	15

Table 2.   Quantitative analysis of phases present in mixtures of
            α-alumina and amorphous silica of different compositions
            reaction sintered at 1600°C for 10h.

Composition of the starting mixture (wt% Al_2O_3)	Amount (in wt%) of different phases present after reaction sintering at 1600°C for 10h.			Composition of mullite (wt% Al_2O_3)
	Al_2O_3	Amorphous silica	Mullite	
68.0	-	3.4	96.6	70.4
71.8	-	-	100.0	71.8
75.0	1.0	-	99.0	74.7
77.3	10.0	-	90.0	74.8
80.0	20.5	-	79.5	74.8

Table 3.   Fracture toughness of products obtained by reaction
            sintering at 1600°C for 10h.

Overall composition (wt% Al_2O_3)	68.0	71.8	75.0	77.3	80.0
Fracture toughness at 20°C($MNm^{-3/2}$)	2.0	2.1	1.8	2.3	2.4

Table 4.   Cell dimensions of mullite formed and relative amounts of
            crystalline phases present in mixtures of 75.0 wt% Al_2O_3
            composition reaction sintered under different conditions
            as indicated

Heat treatment (°C/min)	Cell parameters (Å)			Number of peaks considered for cell parameter estimation	Relative amounts of crystalline phases present		Composition of mullite (wt% Al_2O_3)	
	c	b	a		$\dfrac{A}{A+M}$	$\dfrac{C}{C+M}$	x	y
1550/1	2.8785	7.6758	7.5374	20	0.337	0.078	68.6	70.9
1550/60	2.8820	7.6858	7.5477	20	0.101	0.0	72.2	72.8
	2.8817	7.6832	7.5482	23				
	2.8821	7.6848	7.5480	25				
1600/600	2.8836	7.6824	7.5588	21	0.010	0.0	74.7	74.5
	2.8837	7.6824	7.5589	22				
	2.8835	7.6820	7.5591	23				

y: Estimated using values of cell parameter 'a' and Figure 6.
x: Estimated using [A/(A+M)] and [C/(C+M)] assuming that no
   vitreous phase is present. A, C and M are the quantities
   (by weight) of α-alumina, cristobalite and mullite respectively.

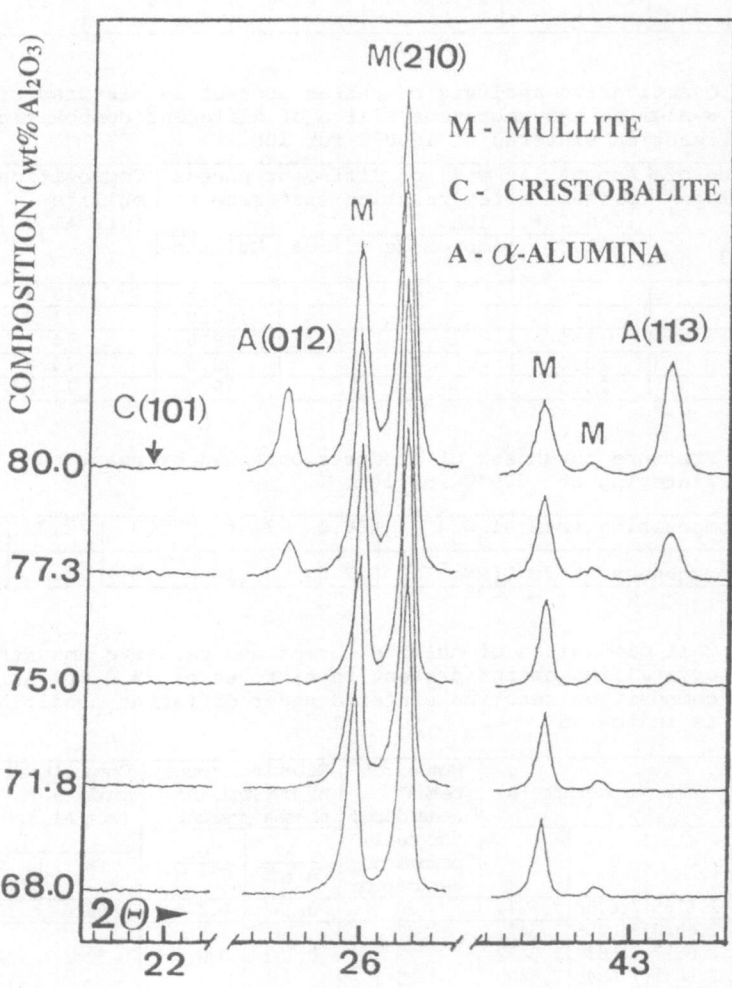

Figure 1. XRD patterns of mixtures of α-alumina and amorphous silica reaction sintered for 10h at 1600 °C.

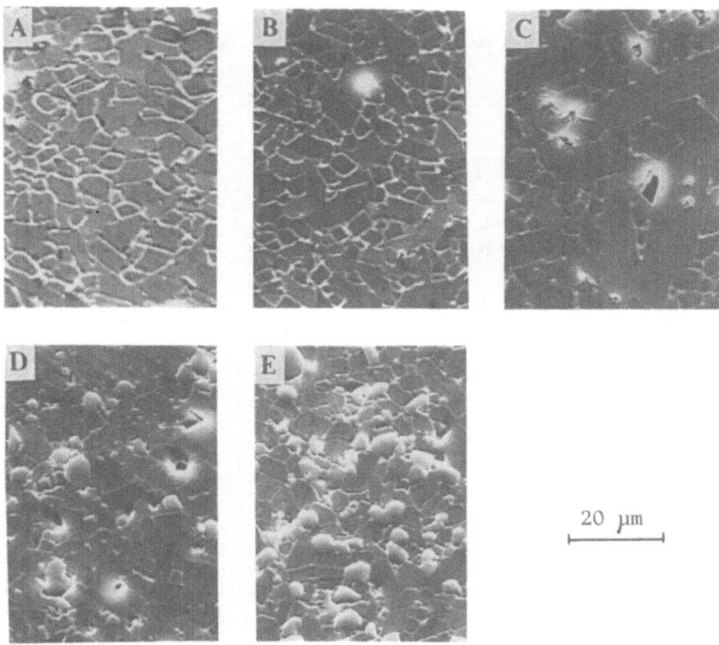

Figure 2. Microstructures of (A) 68.0, (B) 71.8, (C) 75.0, (D) 77.3, (E) 80.0 wt% Al_2O_3 compositions reaction sintered for 10h at 1600 $^\circ$C.

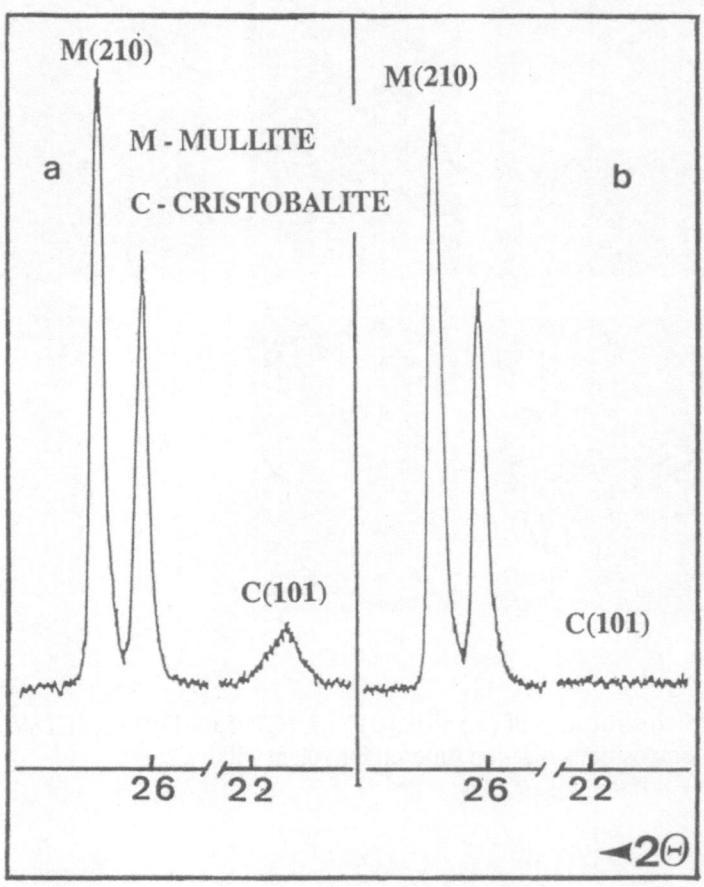

Figure 3. XRD patterns of the 68.0 wt% Al_2O_3 composition reaction sintered (1600 °C/10h) and reheated (1350 °C/2h) after (a) grinding, (b) grinding and leaching by 10% HF aqueous Solution.

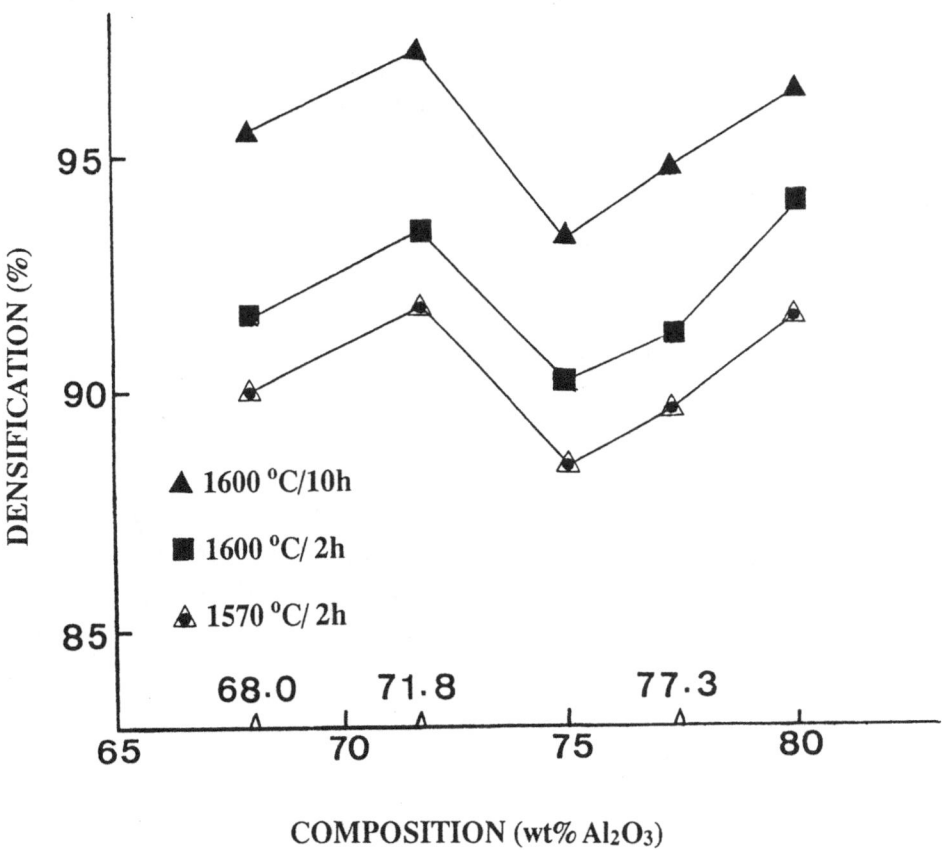

Figure 4.   Densification versus composition plots for powder com pacts of α-alumina and amorphous silica, reaction sin tered under different conditions as indicated.

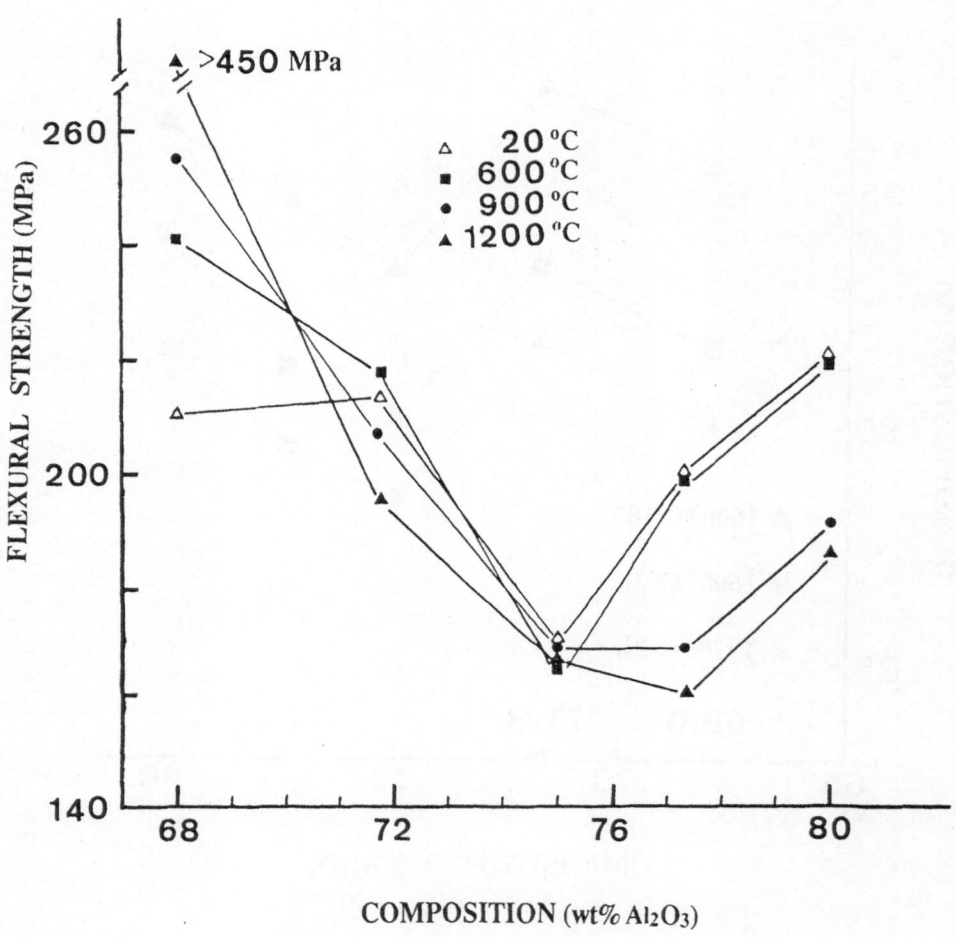

Figure 5. Flexural strength versus composition at 20, 600, 900 and 1200 °C for mullite based ceramics produced by reaction sintering for 10h at 1600 °C.

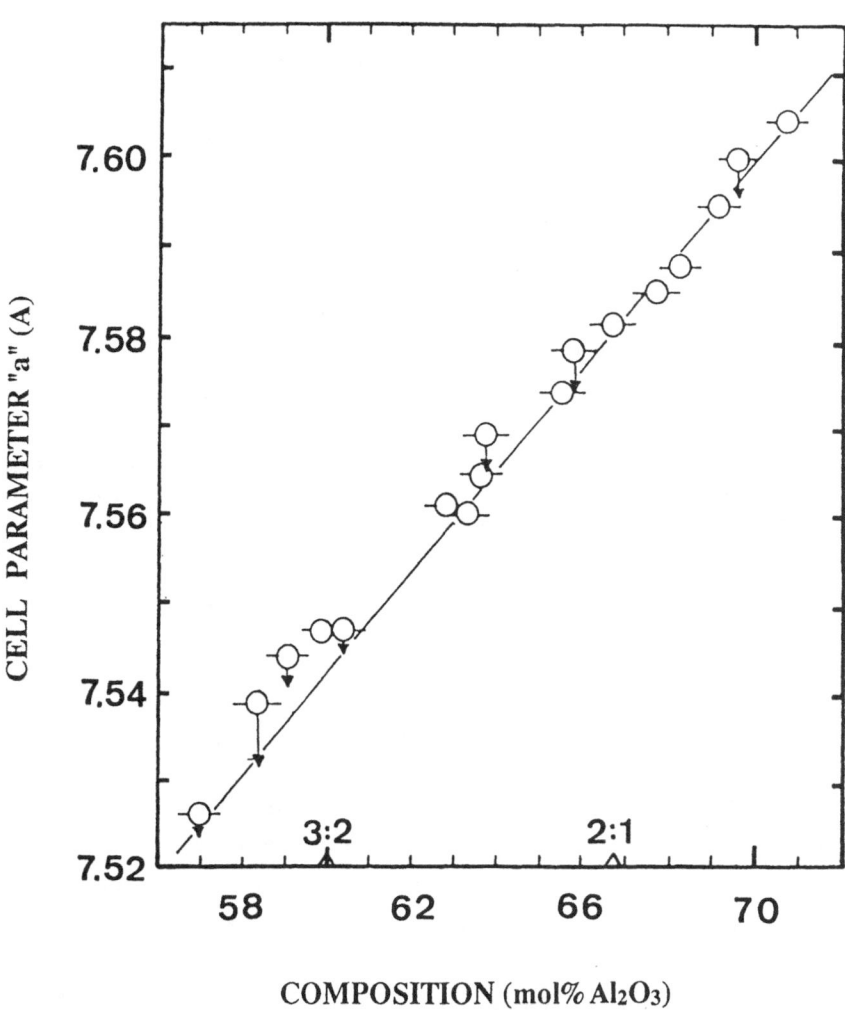

Figure 6. Cell parameter "a" versus composition for mullite (arrows indicate the values of "a" corrected for Fe and Ti in solid solution)(Ref 12).

DEVELOPMENT OF DIRECT OBSERVATION METHOD FOR INTERNAL STRUCTURE IN SILICON NITRIDE GRANULE AND GREEN BODY

MASAHIRO INOUE, JIN-YOUNG KIM, ZENJI KATO, NOZOMU UCHIDA,
KEIZO UEMATSU and KATSUICHI SAITO
Department of Chmistry, Nagaoka University of Technology,
Kamitomioka,Nagaoka,Niigata,Japan 940-21

ABSTRACT

Silicon nitride granule and green body were made transparent by immersion liquid which have a refractive index close to that of silicon nitride. Their internal structures were examined directly by means of optical microscope. Sharp void spaces existed even in the green body which was isostatically pressed at 600MPa. Present method is very useful for studying the processing of ceramics, because it gives us more detailed information than other methods.

INTRODUCTION

Defects and heterogeneities in ceramic green body are often persisted after sintering [1] and govern various properties of ceramics. It is essential that their size and concentration are both minimized for producing high performance ceramics. Therefore numerous forming processes, such as die pressing, extrusion molding, slip casting , injection molding, etc., have been developed depending on required shape and characteristics.

Requirements to produce highly reliable ceramics with high strength through powder process follows: (1) to characterize the microstructure in granule and green body precisely, (2) to feed back the information obtained from this characterization to forming process. Up to now, however, detailed chracterization of the internal structure in granule and green body has been difficult.

Recently, authors have developed a unique characterization method for internal structure of ceramics [2-4]; alumina granule and green body which were made transparent by immersion liquid were observed directly with traditional optical microscope. This method is based on a reduced internal reflection in the presence of immersion liquid which has a refractive index close to that of alumina. This method is superior to conventional methods; we can obtain only two dimensional information about fracture surface with SEM and can characterize only very large defects with scanning acoustic microscope. With the present method, we can examine microscopic internal structures over entire volume of specimen. Some additional advantages over other methods include simple experimental procedure and no requirement of any special equipments.

In principle, this method can be applied not only to alumina but also to other ceramics. Silicon nitride granule and green body were used as a model in the present study. Refractive index of silicon nitride determined on thin film was reported to be 2.05 [5]. Immersion liquids applied in this study are yellow phosphorus - methylene iodide and yellow phosphorus - sulfur - methylene iodide. They have refractive indexes as high as silicon nitride [6].

MATERIALS AND METHODS

Materials

Prototype silicon nitride granule(NKK, Ltd., Tokyo, Japan) was used as specimen in the present study. According to the supplier, the granule was produced by spray-dry method with high purity powder synthesized by thermal decomposition of silicon diimide(E-10, UBE Indust., Ltd., Ube, Japan) and normal sintering additives. TABLE 1 lists the composition and refractive indexes at room temperature of immersion liquids. Refractive indexes of the liquids were determined by the measurement of minimum deviation angles through a home-made hollow 30° prism with He-Ne gas laser(λ=632.8nm; GLG5040, NEC, Ltd., Tokyo, Japan).

Direct observation of internal structure in granule

The granule was immersed in immersion liquid after dewaxing, and its internal structure was observed directly with an optical microscope (OPTIPHOT, Nikon, Ltd., Tokyo, Japan). Subsequently, the granule was molded in resin and polished with diamond wheels(#260,#3000). Polished surface of the granule was also observed by means of SEM(JSM-T100, JEOL, Ltd., Tokyo, Japan).

Direct observation of internal structure in green body
The granule was uniaxially pressed into pellet(14mmϕ*5mm) at 20MPa and isostatically pressed(QIC12, NKK-ASEA, Tokyo, Japan) at 20, 100 or 600MPa. After dewaxing, specimens thinned (<0.2mm) with sandpaper were immersed in an immersion liquid(#4 in TABLE 1) and evacuated for 10-20min for direct observation with an optical microscope.

TABLE 1
Composition and refractive indexes (λ=632.8nm) of immersion liquids.

LIquid No.	Composition	Refractive index
#1	CH_2I_2	1.75
#2	P and CH_2I_2 (2:8 by weight)	1.82
#3	P and CH_2I_2 (4.5:5.5)	1.90
#4	P , S and CH_2I_2 (8:1:1)	2.02

RESULTS

Figure 1 shows optical micrographs of silicon nitride granule immersed in various immersion liquids. When immersed in liquid #1(n=1.75), granule was opaque and internal structure could not be examined. For immersion liquid #2(n=1.82), internal structure of only small granule could be seen. Large one(>40μm) was opaque. For immersion liquid #3(n=1.90), the internal structure of all granule could be clearly observed.

Figure 2 shows scanning electron micrograph of polished surface of granule. Defects in granule can be found with SEM. However, only two dimensional information can be obtained on polished surface.

Figure 3 shows a rare example of internal structure observed with the present method. There were small particles (10-15μm) in the pore of some large granules(>60μm). In principle, We may be able to find such a structure with SEM when it is accidentally exposed to polished surface. In practice, it is very difficult to find it with SEM, because it is rarely exposed to polished surface. The present method is very useful for observation of such a structure.

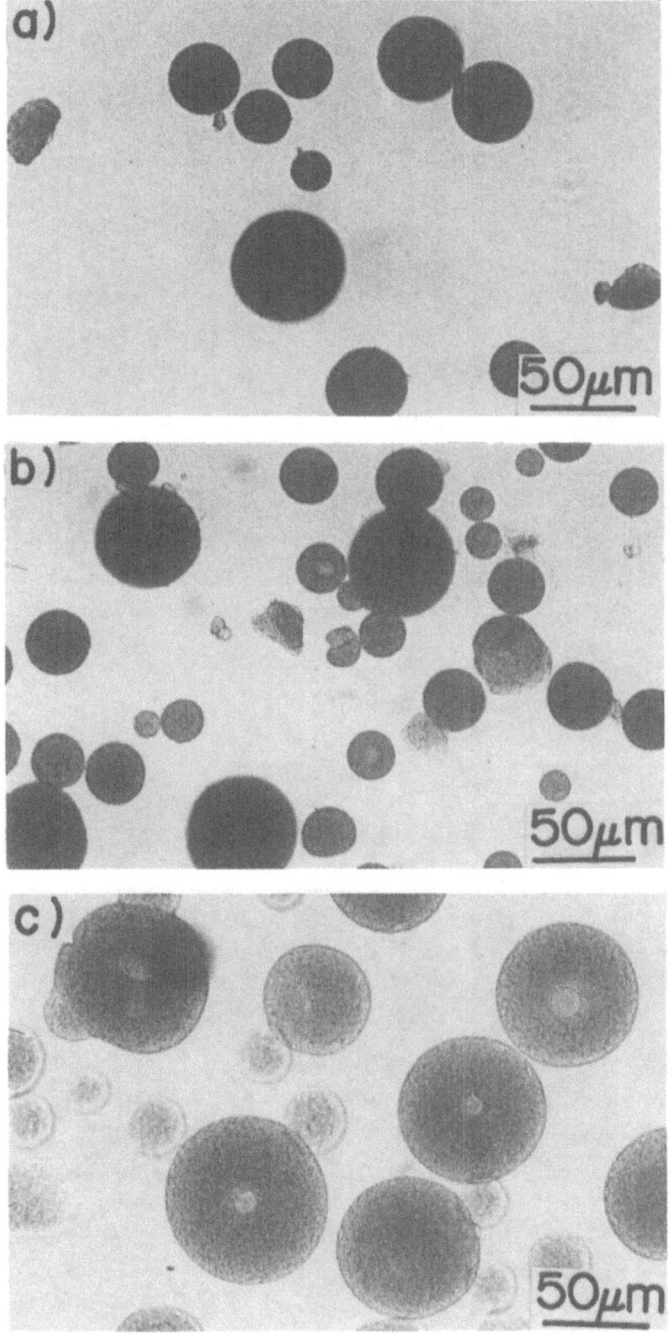

Figure 1. Optical micrographs of granule immersed in various immersion
liquids. a)#1 in TABLE 1, b)#2, c)#3.

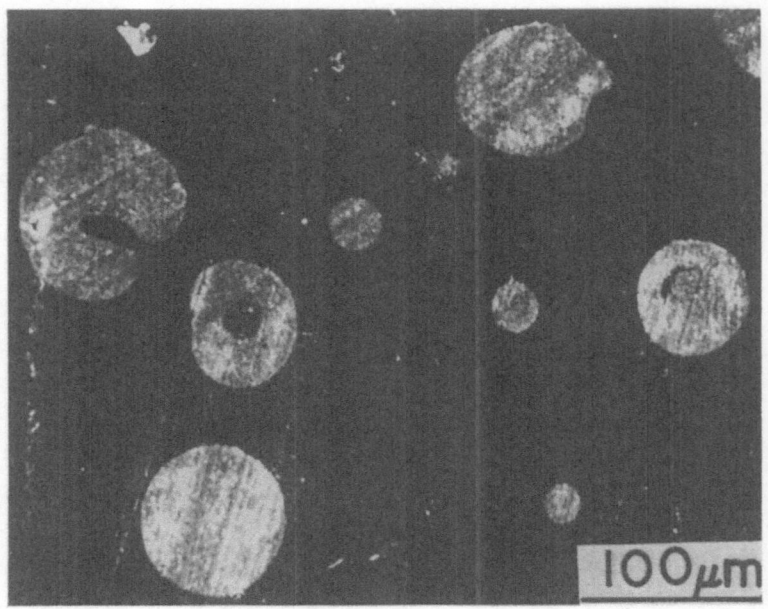

Figure 2. Scanning electron micrograph of polished surface of granule.

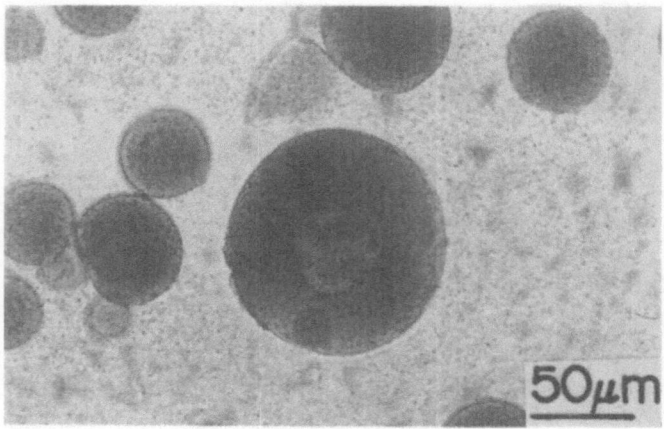

Figure 3. Optical micrograph of internal structure in large granule.

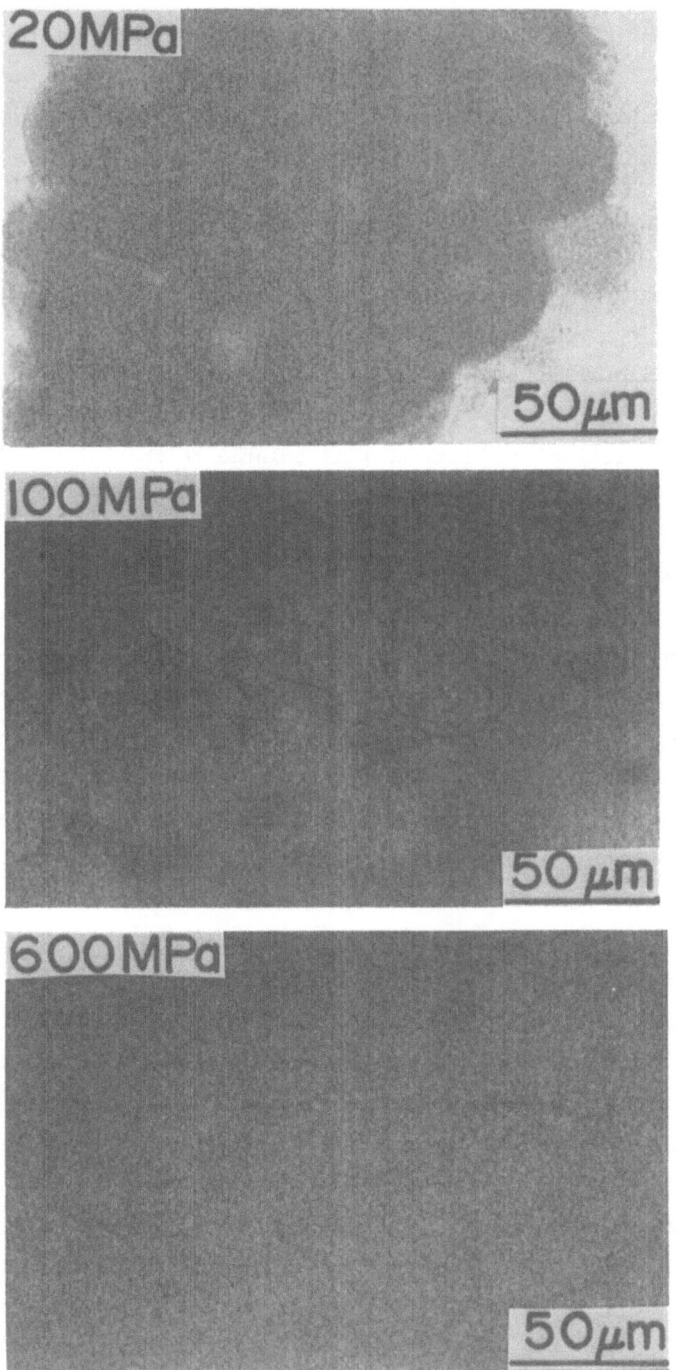

Figure 4. Optical micrographs of silicon nitride green bodies prepared at various pressures.

Figure 4 shows optical micrographs of typical internal structures in silicon nitride green bodies. Granule was not destroyed completely during powder compaction and sharp void spaces exist at interstices when isostatically pressed at 20MPa. Sharp crescent-like voids as large as 50-100 and 30-60μm were still found even at the applied pressure 100 and 600MPa, respectively.

DISCUSSION

The transparency of specimen ·in the immersion liquid is governed by scattering at the particle-liquid interface. Two governing factors for light scattering are; 1) Relative refractive index of the particle and the medium, 2) The ratio of the particle diameter to the wavelength of incident light [7]. The reflection of light R is related to the relative refractive index m by Eq.1 for normal incidence,

$$R=(m-1)^2/(m+1)^2 \quad (1)$$

The relative refractive indexes of silicon nitride and immersion liquid #1 and #2 are 1.17 and 1.13. For them, 0.61% and 0.37% of incidence light was reflected at each interface, respectively. Assuming that the granule is made transparent if 5% of incidence light were transmitted through specimen [3], number of interface allowed are 490 and 800, respectively. Recalling that the particle size of silicon nitride is 0.5μm and that light has to pass through two interfaces for each particle, the maximum diameter of granule allowed for transparency are estimated to be 120 and 200μm. This estimation does not agree to the experiment as shown in Fig.1. Granules could not be made transparent with immersion liquid #1 and only small one (<30μm) was made transparent with immersion liquid #2.

To explain the difference between estimation and experimental result, it is necesary to recall the second factor for light scattering. The maximum scattering occurs when the particle size is of the same magnitude as the radiation wavelength [7]. Particle size of silicon nitride (about 0.5μm) is close to the wavelength of white light used in optical microscope observation. The second factor is significant and reduces the maximum diameter of granule for which we can observe internal structure. We can similarly discuss the result on green body. The relative refractive index of silicon nitride and immersion liquid #4 is 1.01. Equation 1 suggests that reflection of light at each interface is 0.0025%. Assuming that direct transmission of 5% incident light corresponds to transparency, we can estimate the number of interfaces allowd to be $1.20*10^5$. The maximum thickness allowed for transparency is estimated to be about 30mm, which is much thicker than 0.2mm which was required for direct observation of internal structure.

CONCLUSIONS

The following conclusions are obtained:

(1) Silicon nitride granule was made transparent and its internal structure was successfully observed when the mixture of yellow phosphorus and methylene iodide, 4.5:5.5 by weight, were used as immersion liquid.

(2) Some interesting structures were observed in granule, e.g., small particles packed in a pore inside large granule.

(3) The internal structure in thinned(<0.2mm) silicon nitride green bodies could be observed with the ternary system of yellow phosphorus, sulfur and methylene iodide, 8:1:1 by weight, as immersion liquid.

(4) There were some narrow crescent-like void spaces in green bodies which were isostatically pressed at 100 and 600MPa. Their sizes were 50-100 and 30-60µm, respectively. They may cause heterogeneities in sintered bodies and deteriorate mechanical properties.

REFERENCES

1. Zheng, J., Reed, J.S., Effect of Particle Packing Characteristics on Solid-State Sintering. J. Am. Ceram. Soc., 1989, **72**, 810-17.
2. Uemastu, K., Kim, J.Y., Miyashita, M., Uchida, N., Saito, K., Direct observation of internal structure in ceramics:1. Spray-dried alumina granule. J. Am. Ceram. Soc., submitted for publication
3. Uematsu, K., Miyashita, M., Sekiguchi, M., Kim, J.Y., Uchida, N. and Saito, k., Direct observation of internal structure in ceramics:2. Alumina green bodies prepared by powder pressing, to be published
4. Uematsu, K., Kim, J.Y., Kato, Z., Uchida, N., Saito, K., Direct Observation Method for Internal Structure of Ceramic Green Body : Alumina Green Body as an Example, Seramikkusu Ronbunshi, submitted for publication
5. Tracy, C.E., Micromethod for Refractive Index Determination of Thin Films Using Liquid Standards, J. Electrochem. Soc., 1979, **126**, 103-6.
6. West, C.D., Immersion Liquid of High Refractive Index, Am. Mineral., 1936, **21**, 245-9.
7. Kingery, W.D., Bowen, H.K., Uhlmann, D.R., Chap.13 in Introduction to Ceramics, 2nd. Ed., Wiley, New York, 1976, pp646-706.

Structural, Electrical and Optical Characterization
of Sputtered ZnOx Thin Film.

T.Nishihara*, N.Fujimura and T.Ito

*Student: University of Osaka Prefecture
Metallurgical Engineering, College of Engineering
University of Osaka Prefecture
Mozu-Umemachi, Sakai, Osaka 591 Japan

1. Introduction

As Zinc oxide films have many available characteristics, a large number
of crystallographic studies have been reported on the films deposited by
sputtering.   However, most of those are about the influence of deposition
conditions for (002) (c-axis) oriented films(1)-(2). The growing process
of ZnOx films has not been necessarily manifested.

In this paper, the influence of deposition parameters on the (002) and
(110) preferred orientation were evaluated about ZnOx films deposited by a
rf magnetron sputtering. Moreover, the growing process of ZnOx films, the
electrical and optical properties were investigated.

2. Experimental Procedure

ZnOx films were deposited by a rf magnetron sputtering method.   A
pressed disk of ZnO(99.0% ,Inc.Kishidakagaku) was used as the target.
Corning 7059 was used as the substrates. The sputtering conditions were
rf power of 100W and gas pressure of 6m Torr. The gas composition (Ar:O_2)
was changed extensively(3)-(4). The structural change of prepared ZnOx
films was evaluated by using X-ray Diffraction (XRD)(Shimazu XD-3A type),
X-ray Photoelectron Spectroscopy (XPS)(Perkin-elmer PHI 5100).   The
electrical resistivity and the optical transmittance spectra in visible
radiation were   measured   by   using   a   conventional   four   probe
system(Kyowariken   K-705RD)   and   spectra photometer(Shimazu UU-265FS),
respectively.

3. Results and Discussion

ZnOx films deposited on Corning 7059 without heating substrates were
evaluated by X-ray diffraction. Figure 1 shows gas composition dependence
on preferred orientation of films. The (002) or (110) oriented films were

obtained by controlling gas compositions. Figure 2 shows the layered structure of ZnO, called wurtzite type. Only Zn or O atom exists in the each (002) plane, while, both Zn and O atoms exist in (110) plane alternately.

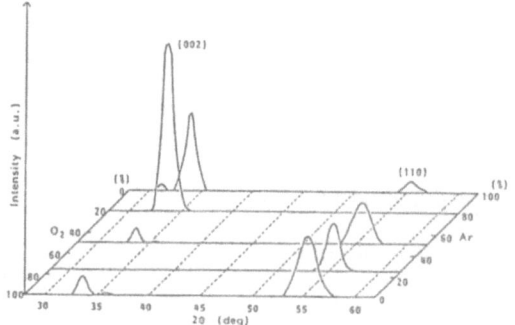

Fig.1 Gas composition dependence on the preferred orientation.

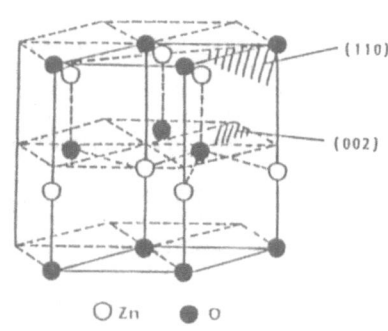

Fig.2 Crystal structure of ZnO.

The substrate temperature dependence on the orientation was studied under the sputtering condition which promoted the excellent (110) oriented films (Fig.3). The (110) oriented films can be formed under the low substrate temperature and the suitable O_2 partial pressure. As the substrate temperature(Ts) was raised, the (110) peak disappeared and the (002) peak became strong.

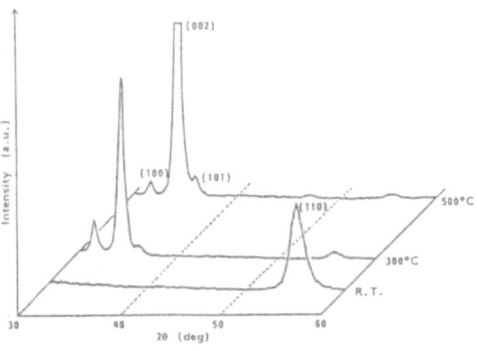

Fig.3 Effect of the substrate temperature on the preferred orientation.

But, when the excellent (110) oriented films deposited without heating substrates were post-annealed, the (110) preferred orientation was kept and became stronger with increasing annealing temperature. We suggest that as metallic binding can not exist at high temperature because of high vapor pressure of Zn, the structure necessarily become layered one, that is, (002) orientation at high Ts.

At the same time, the obvious peak shift was observed in X-ray diffraction angle, 2Θ. Those of (002) and (110) peaks of ZnOx films prepared without heating substrates were much lower than those of the bulk. These peak shifts to lower angles indicate the expanding of c-axis and a-axis. But these expansion decreased and the composition approached that of stoichiometric ZnO with increasing annealing temperature.

By the way, ZnOx films were darkened by X-ray irradiation(5). The optical transmittance at visible radiation decreased after X-ray irradiation, but the change in band gap was not observed. This darkened film returned by annealing, which is called photochromism. This may be caused by the F center. The degree of darkness by X-ray irradiation was not influenced by the difference in the orientation.

So, we evaluated as-deposited, X-ray irradiated and post-annealed (in vacuum) ZnOx films by XPS to investigate the composition and the change of valence electron state. Figure 4 shows spectra of Zn(auger electron, LMM) from ZnOx films with (110) orientation. The shoulder which is seen on the skirt of main peak at lower energy is considered to be Zn^+. But this shoulder did not change and remarkable peak shift was not observed by X-ray irradiation and annealing. Therefore, we suggest that

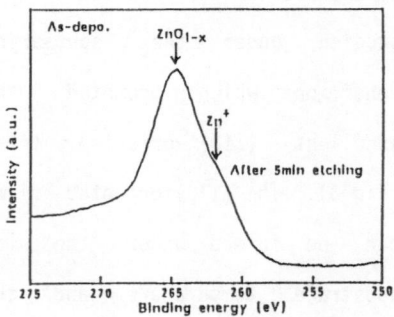

Fig.4 XPS spectrum of Zn
(auger electron,LMM)

this is because the volume of oxygen defects introduced by X-ray irradiation is not so much. While, the increase in the adsorbed oxygen is observed in the spectra of O 1s. The adsorbed surface oxygen would be pushed out from the inside of the film by X-ray irradiation. Comparing to

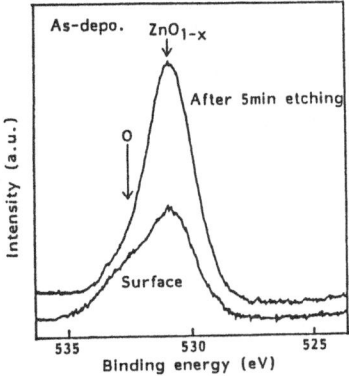
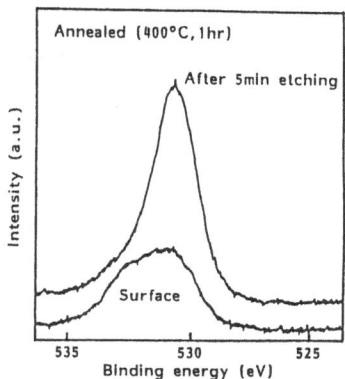

Fig.5 XPS spectra of O 1s.

the film after X-ray irradiation, the increase was more remarkable after annealing in vacuum(Fig.5). We suggest that the above-mentioned decrease of expansion by annealing in vacuum is mainly caused by interstitial oxygen atoms going up from the inside.

It is a preferable result for the application to a transparent conductive film that the resistivity decreased with increasing Zn concentration, but is not preferable one that the transmittance decreased at the same time. By annealing in vacuum, the shifts of fundamental absorption edge was investigated to estimate the band gap. I case of ZnO, it shifted to the longer wavelength, while in case of Al doped ZnO, it shifted to the shorter wavelength. These results are concerned with the change of carrier concentration(6).

References

(1) Characterization of ZnO piezoelectric films prepared by rf planar-magnetron sputtering
T.Yamamoto, T.Siosaki and A.Kawabata: J.Appl.Phys. 51(1980)P3113-3120.
(2) Reactive synthesis of well-oriented zinc-oxide films by means of the facing targets sputtering method
M.Matsuoka, Y.Hoshi and M.Naoe: J.Appl.Phys. 63(1988)P2098-2103
(3) The effect of O on reactively sputtered zinc oxide
C.R.Aita, A.J.Purdes, R.J.Lad and P.D.Funkenbusch: J.Appl.Phys. 51(1980)P5533-5536.
(4) Effects of oxygen partial pressure on the structure of ZnO thin films prepared by reactive RF sputtering
H.Kobayashi, Y.Igasaki and G.Shimaoka: Bull.Res.Inst.Electron. Shizuoka Univ. 19(1984)P47-53.

(5) Photochromism and anomalous crystallite orientation of ZnO films
    prepared by a sputtering-type electron resonance microwave plasma
    M.Matsuoka and K.Ono: Appl.Phys.Lett. 53(1988)P1393-1395.
(6) The interpretation of the properties of Indium Antimonide
    T.S.Moss: Proc.Phys.Soc.Lonson Sect. B67(1954)P775-782

Influence of external stress for non-equilibrium thin films.

N.Fujimura S.Tachibana* and T.Ito

Metallurgical Engineering, College of Engineering
University of Osaka Prefecture
Mozu-Umemachi, Sakai, Osaka 591 Japan
*Graduate school; University of Osaka Prefecture

INTRODUCTION

The structural controlling of bulk materials is widely industrialized. The grain size, the texture and the precipitation are perfectly controlled, which results in various preferable properties. While, in thin film materials, the non-equilibrium; that is to say, wide solubility limit; makes it complicated. For instance, in the case of Al-Si films used as a contact metal for Si devices, silicon dissolves in as-deposited aluminum more than solubility limit. During sintering, however, Si precipitates to the interface between silicon and aluminum, so-called Si precipitation. Moreover, we already reported that in Ti-rich TiNx/Si system, excess Ti reacted with Si substrate[1], and that in Si-rich WSix/Si system, excess Si precipitated at the film/substrate interface[2,3]. If there are no driving forces, excess atoms are to precipitate in the film (grain-boundary). What is the driving force? How is the structure of films controlled? Our final aims are to make these points clear. As one of the methods to control the structure of the films, we paid attention to the influence of stress. The structural changes were evaluated under external stresses.

EXPERIMENTS

The $WSi_{2.6}$ films were deposited by a plasma CVD system dilectly on P type (100)Si substrate. WF_6 and SiH_4 were used as reaction gases, and these flow rates were 2 and 120 cm^3/min, Helium was used as a dilution gas. The pressure was 40 Pa. The substrate temperature was 350 °C. The film compositon evaluated by Rutherford backscattering was $WSi_{2.6}$. The prepared samples were annealed in a vacuum of 10^{-7} Torr. The structural change and the depth profile of the film composition were evaluated by X-ray and electron diffraction, and X-ray photoelectron spectroscopy(XPS).

RESULTS AND DISCUSSION

The structure of as-deposited sample is amorphous, and it crystallizes to a hexagonal WSi_2 by annealing at 400 °C for 1 h in a vacuum. It is observed by XPS that the excess Si has precipitated at the WSi_2 /Si substrate interface(Fig.1). As the WSi film is deposited by CVD, it is considered that the internal film stress in as-deposited state is induced by the difference between the thermal expansion coefficient of the film and that of the substrate when the samples are cooled from 350 °C(substrate temperature on deposition) to room temperature, that is, there is little influence by the intrinsic stress. The structural change can be observed without the influence of the film stress by annealing at 350 °C. However, about 29 Å of precipitation of excess Si at the interface was observed (Fig.1). This fact suggests that another mechanism (maybe a relaxation of

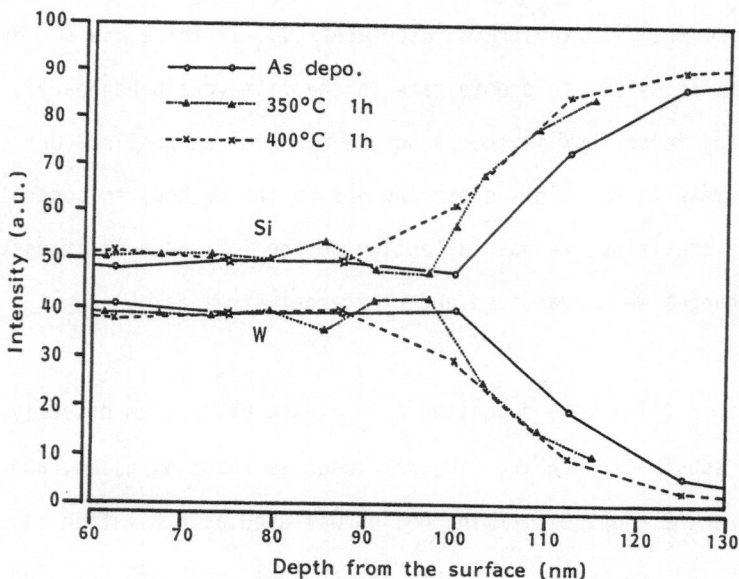

Fig.1 XPS depth profiles of the sample annealed at each temperature.

interfacial energy) is operating except film stress. If the film stress is concerned with the structural change, the process of the crystallization and the precipitation will be affected by applying larger elastic energy induced by the film stress than the interfacial energy.

To make clear the origin of the structural changes, the annealing samples under strong external film stress was performed. To discuss the results, it is necessary to describe about the difference between a concept of external stress and that of internal one.

To apply the external stress, the substrate, in a rectangular is clamped at the one side for the loading which is the constant displacement at another side (cantilever). In this case, the state of stress should be as in Table, by defining the elastic strain in the lattice of the film in contrast to the internal stress.

Table   A concept of the stress.

Since this external stress can be treated as the bending of cantilever, the stress is expressed as a linear function against the distance from the loading side, X,. Then, the external stress was obtained by calculating the radius of curvature at micro-distance ($R \Delta x$) and substituting it in Eq.(2).

$$\sigma = \frac{Ef}{6(1-\nu f)} \cdot \frac{ts^2}{tf} \cdot \frac{1}{R\Delta x} \qquad \text{Eq.(2)}$$

,where Ef, vf and tf are the Young's modulus, Poisson's ratio and thickness of the film, respectively. And, ts is the thickness of the substrate.

In this study, the value of Ef/(1-vf) was roughly estimated by assuming that the internal film stress in as-deposited state was induced by only the difference in thermal expansion coefficients.

Figure 2 shows the influence of external stress on the crystallization. By annealing at 350 °C for 1 h, samples with a external compressive stress (8.45x10^{11} dyne/cm^2) and without the external stress (maybe also without an internal stress at 350 °C) have not crystallized. While, those with an external tensile stress (1.13x10^{12} dyne/cm^2) have crystallized.

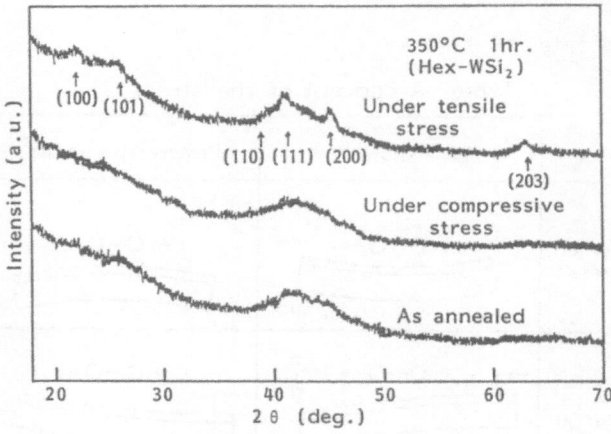

Fig. 2 X-ray diffraction patterns of each sample annealed at 350 °C.

The crystallization under the external tensile stress is caused by the enhanced diffusibility of atoms due to the expanding of the bonding. That is, it is considered that the external tensile stress is relieved by the crystallization(accommodation mechanism). Under the external compressive one, as the bonding is shrunk, the diffusion of atoms is suppressed and the crystallization is retarded.

Figure 3 shows the influence of external stress on the precipitation

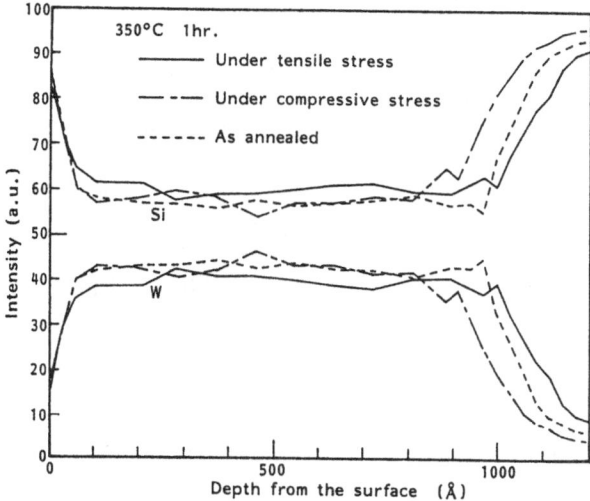

Fig.3 Depth profiles of XPS spectrum.

at the interface which is evaluated by XPS depth profile. The precipitation

at the film/substrate interface resulted in the volume change of the film.

The volume per unit of amorphous $WSi_{2.6}$ (initial state) was obtained by

calculating the film density. The film density was calculated using

Lorentz-Lorenz equation by measuring the refractive indices of amorphous

$WSi_{2.6}$ and crystallized WSi_2 using ellipsometry. The volume of WSi_2 +0.6Si

(precipitated in the film) and WSi_2 (precipitated at the interface) were

calculated using their lattice parameters. These calculations indicates

that if the precipitation of excess Si occurs in the film, the volume of

the film hardly changes. But, the precipitation at the interface makes the

film contract about 9.82 $\overset{\circ}{A}{}^3$/molecular.

The compressive stress can be relieved by precipitation at the

interface due to the volume decrement of the film. Therefore, the 88 $\overset{\circ}{A}$ of

precipitation at the interface was observed under the external compressive

stress(accommodation mechanism). The precipitation at the interface in the

sample without stresses is considered to relieve the interfacial energy as

mentioned above. On the other hand, under larger elastic energy induced by external tensile stress than the interfacial energy, the precipitation has not occurred, since the precipitation at the interface can make the elastic energy decrease in the film due to the volume decrement.

The obvious structural change of Si-rich $WSi_{2.6}$ was observed under the external film stress. When the interfacial energy between the film and the substrate is larger than the elastic energy induced by the film stress, the structure of the film changes to relieve the interfacial energy. In the contrary case, the structure changes to relieve the elastic energy induced by the film stress. In conclusion, the accommodation mechanisms on the structures are operating in the films to relieve a energy induced in the films.

REFERENCES

(1) N.Fujimura and T.Ito: Proc., Int. Conf. on Polycrystalline Semiconductors., Marente, W.Germany, Aug. (1988)

(2) N.Fujimura and T.Ito: Proc., Int. Conf. on Formation of Semiconductor interface., Takarazuka, Japan, Nov. (1988)

(3) N.Fujimura and T.Ito: Proc., Int. Conf. on Formation of Semiconductor interface., Takarazuka, Japan, Nov. (1988)

THIN FILMS OF YBaCuO PREPARED BY MULTILAYER EVAPORATION PROCESS

X.K. Wang, D.X. Li, J.Q. Zheng, J.B. Ketterson, and R.P.H. Chang

Materials Research Center and Science & Technology Center for
Superconductivity, Northwestern University, Evanston, IL 60208

ABSTRACT

Thin films of YBaCuO were prepared as a superlattice of three constituents from three electron guns using a computer-controlled evaporator. After annealing, the multilayer films are converted to the homogeneous superconducting phase. Highly epitaxial thin films with: (1) the a-axis perpendicular to (100) $SrTiO_3$; (2) the c-axis perpendicular to (100) $SrTiO_3$; and (3) the [110] axis perpendicular to (110) $SrTiO_3$ were confirmed by x-ray diffraction as well as scanning electron microscopy and high resolution electron microscopy. Both the a-axis oriented and the c-axis oriented films exhibit zero resistance at 91K. The [110] oriented film shows the sharpest transition with a transition width of 1K and zero resistance at 85K. The zero field critical current density, J_c, determined magnetically, is in excess of $10^7 A/cm^2$ at 4.4K and $1.04 \times 10^6 A/cm^2$ at 77K for the c-axis oriented film; for the a-axis oriented film we obtained $6.7 \times 10^6 A/cm^2$ at 4.4K and $1.2 \times 10^5 A/cm^2$ at 77K. The orientation dependence of the critical current density in the basal plane of the a-axis oriented film was studied. The largest J_c's occur along the in-plane <100> axes of the substrate.

INTRODUCTION

The preparation of the high quality superconducting YBaCuO thin films with high T_c and J_c is of great importance not only for technological applications but also for fundamental studies. A considerable activity has been focused on this area.[1-5] A primary goal of most high T_c thin film efforts is to increase the critical current.[6-9] It is known that critical current density is affected by stoichiometry, the atomic arrangement at grain boundaries, intergrain coupling, alignment of the crystal grains, and the critical current anisotropy.[10] Accordingly, it is expected that samples consisting of grains with the desired stoichiometry, with good intergrain contact and a high degree alignment will show high value of J_c. Therefore, the preparation and characterization of essentially epitaxial thin films with well defined but different orientations should be of considerable interest.

We have prepared epitaxial, oriented thin films of YBaCuO with: (1) the a-axis perpendicular to (100)$SrTiO_3$ (a-axis oriented); (2) the c-axis perpendicular to (100)$SrTiO_3$ (c-axis oriented); and (3) the [110] axis perpendicular to (110)$SrTiO_3$ ([110] axis oriented). A computer-controlled, three e-gun, multisubstrate evaporator was employed. Our films were characterized by x-ray diffraction (XRD), scanning electron microscopy (SEM), high resolution electron microscopy (HREM), conventional four probe resistivity, and magnetization measurements.

PREPARATION OF THE THIN FILMS

The most important requirement for high quality thin films is to achieve strict stoichiometry in order to avoid second phase formation and to minimize interdiffusion between the substrate and the film. The composition of a multilayer film can be controlled simply by adjusting the thickness of each sublayer. With our system, thin films with an artificial-superlattice structure were deposited from three e-guns containing Y, BaF_2, and Cu in an atmosphere of 5×10^{-5} Torr of O_2. The substrates were mounted on a substrate wheel and maintained at 450°C. The wheel was driven by a computer-controlled stepping motor. Any of the (up to 20) substrates could be positioned over any of the three e-guns. A second computer-controlled stepping motor drove a shutter wheel which allowed the flux from any of the three e-guns to reach the substrate directly above it. The flux from each of the three e-guns was monitored by individual quartz crystal sensors which controlled their respective fluxes via feedback to the e-gun power supply. In addition, when the accumulated thickness of the sensor associated with the e-gun depositing a given layer of the multilayer structure reached a preset thickness, the computer was activated to advance the substrate to the next e-gun (and the monitor was rest to zero thickness). A complete superlattice was deposited on a given substrate before commencing deposition on the next substrate.

A typical film was deposited under the following conditions: the deposition rates were 1.1 Å/sec., 1.3 Å/sec., and 2.9 Å/sec. for Cu, Y, and BaF_2 respectively. The sublayer deposition time was 40 seconds. The number of cycles in a complete deposition was 60. The run-to-run reproducibility of the composition was within the resolution of the energy dispersive x-ray spectroscopy (EDAX) which was of the order of two percent.

As-deposited films, with thickness approximately of 1 μm, were smooth, insulating and disordered. They were annealed in flowing O_2 saturated with H_2O at 860 °C to 900 °C for 1/2 hour and then maintained at this temperature for 1/2 hour in dry flowing oxygen; this was followed by a slow cool down at 2 °C/min.

CHARACTERIZATION OF THE FILMS

Figure 1 shows scanning electron micrographs of the microstructures of some highly oriented films. Fig. 1(a) shows that the a-axis oriented film (confirmed by XRD[6]) consists of an array of orthogonal, interconnecting, rectangular grains with mean in-plane dimensions of 0.3 μm by 3.0 μm. This aspect ratio of ~ 10 is consistent with that found for single crystals of YBaCuO.[11] The SAED pattern and the HREM image of this a-axis oriented film confirm that each grain is a single crystal of the 123 phase. The b-axis and c-axis are aligned with the long and short dimensions of the rectangular grains, respectively, and are also aligned with the two in-plane a-axes of the substrate; i.e., they are epitaxial. Furthermore, the 90° junctions between the grains are free of any second phase formation or other type of decoration. A HREM image of a typical a-axis oriented film is shown in Fig. 2. Cross-section results confirm that the rectangular grains, shown in Fig. 1(a), nucleate directly on the surface of the $(100)SrTiO_3$ substrate and grow through to the film surface. Further details regarding the microstructure studies are described in Ref. [12]. Fig. 1(b) shows a micrograph of a c-axis oriented film consisting of columns having their c-axis perpendicular to the substrate (again confirmed by x-ray data). Although it displays very well oriented c-axis crystallites, the x-ray diffraction pattern and a HREM micrograph show that small amounts of other phases are present in this c-axis oriented film.

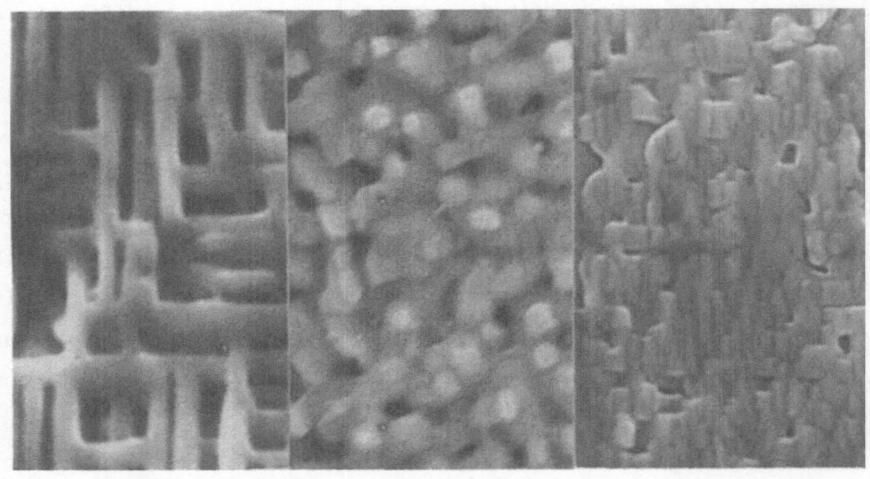

Figure 1.(a) The SEM micrograph of the a-axis oriented film showing a morphology consisting of an array of orthogonal, interconnecting, rectangular single crystal grains with well developed junctions, (b) The SEM micrograph of the c-axis oriented film, (c) The SEM micrograph of the [110] oriented film.

Figure 2. HREM image along [100] obtained from two interconnecting grains in the a-axis oriented film.

Fig. 1(c) shows a SEM micrograph, corresponding to a [110] oriented film (confirmed by XRD) in which only (hh0) reflections are observable. From the XRD data and SEM, as well as HREM micrographs, we conclude that the a-axis oriented film shows the best crystallinity with a high degree of grain alignment, both normal to and within the plane of the film. The crystallinity of the [110] oriented film is more perfect than that of the c-axis oriented film. We note in passing that films prepared in our lab, which were deposited on substrates such as MgO, ZrO_2, and Al_2O_3, where a considerable lattice constant mismatch exists, were never highly oriented. It is clear that lattice constant matching greatly affects the crystallographic orientations of the films.

Fig. 3 shows the resistive transition curves for the three oriented films. Transport measurements, performed by the conventional four probe technique, show that both the a-axis oriented and the c-axis oriented films exhibit zero resistance at 91K. The [110] oriented film shows the sharpest transition with a transition width of 1K, but achieves zero resistance only at 85K.

In addition to the characterizations mentioned above, we examined the films by performing extensive magnetic measurements using an SHE magnetometer equipped with a 50KG superconducting solenoid. The c-axis oriented film was measured with the field applied perpendicular to the plane of the film. The a-axis oriented film was measured with the magnetic field both perpendicular to and parallel to the plane of the film. In order to study the orientation dependence of the in-plane critical current, the film was positioned so that the magnetic field was at angles of 0°, 30°, and 45° to a substrate <100> axis, as shown in Fig. 4. The procedure described by Chaudari et al., [13] was used to determine J_c. On removing the applied field, a magnetic moment is observed that is associated with trapped flux and circulating currents in the film. The critical current density, J_c, can be deduced using Bean's or Kim's model.[14-16] The critical current density $J_{c\perp}^c$ of the c-axis oriented film is in excess of $10^7 A/cm^2$ at 4.4K and 1.04 x $10^6 A/cm^2$ at 77K. The critical current density $J_{c\perp}^a$ of the a-axis oriented film is 6.7 x $10^6 A/cm^2$ at 4.4K and 1.2 x $10^5 A/cm^2$ at 77K (here the symbol $\perp$ indicates that the magnetic field perpendicular to the plane of the film). The critical current density $J_{c\theta=0°}^a$ (H$\parallel$bars or equivalently the in-plane <100> substrate axis) is 1.6 time larger than $J_{c\theta=45°}^a$. We could not distinguish a difference between $J^a_{c\theta=45°}$ and $J^a_{c\theta=30°}$.

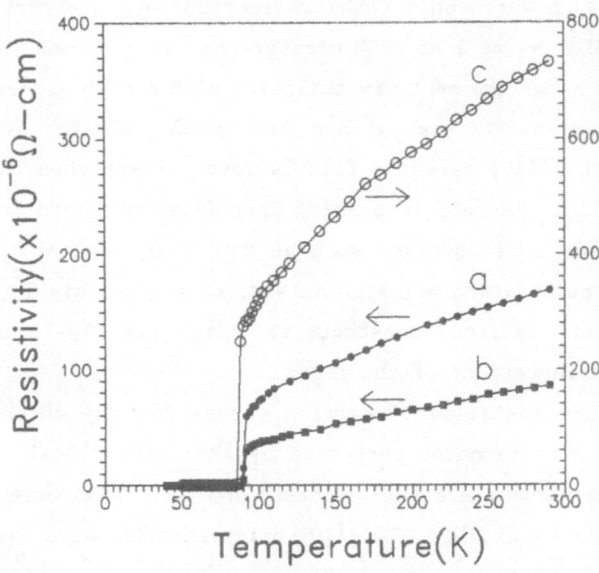

Figure 3. The resistive transition curves for: a) the a-axis, b) the c-axis, and c) the [110] oriented films.

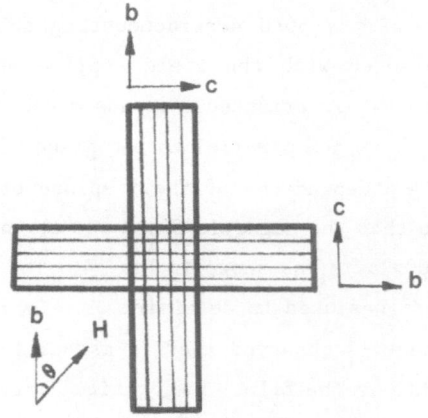

Figure 4. Schematic diagram showing the geometry of the crystal bars in the plane of the film and the reactive orientation of the magnetic field. Values of $\theta=0°$, 30°, and 45° were studied.

Our results argue that the degree of alignment of crystal grains is crucial for achieving high critical current densities. Furthermore, our a-axis oriented film (with its high degree of alignment both normal to and within the plane of the film) has a rather high critical current density.

CONCLUSION

We have prepared highly oriented films of YBaCuO by multilayer deposition and characterized them. We have shown that the superconducting critical current densities can be rather high, not only in the c-axis oriented film, but also in the epitaxial a-axis oriented films, which have excellent intergrain contact.

This work was supported by the NSF/S&T Center (DMR-89-117), the NSF/M (DMR-85-20280), and the Office of Naval Research (N00014-88-K-0100).

REFERENCES

1. B. Oh, M. Naito, S. Arnason, P. Rosenthal, R. Barton, M.R. Beasley, T.H. Geballe, R.H. Hammond, and A. Kapitulnik, Appl. Phys. Lett. 51, 852 (1987).

2. B.M. Clemens, C.W. Nieh, J.A. Kitt, W.L. Johnson, J.Y. Josefowicz, and A.T. Hunter, Appl. Phys. Lett. 53, 1871 (1988).

3. D.M. Hwang, T. Venkatesan, C.C. Chang, L. Nazar, X.D. Wu, A. Inam, and M.S. Hegde, Appl. Phys. Lett. 54, 1702 (1989).

4. C.H. Chen, J. Kwo, and M. Hong, Appl. Phys. Lett. 52, 841 (1988).

5. G. Linker, X.X. Xi, O. Meyer, Q. Li, and J. Greek, Solid State Communications 69, 249 (1989).

6. T.R. Dinger, T.K. Worthington, W.J. Gallagher, and R.L. Sandstron, Phys. Rev. Lett, 58 2687 (1987).

7. F.K. LeGoues, Philos, Mag. B57 167 (1988).

8. P. Chauhari, F.K. LeGoues, A. Segmüller, Science 238 324 (1987).

9. S. Nakahara, G.J. Fisanick, M.F. Yan, R.B. van Dover, T.Boone, and R. Moore, J. Crystal Growth 85 639 (1987).

10. K. Salama, V. Selvamanickam, L. Gao, and K. Sun, Appl. Phys. Lett. 54 5 (1989).

11. L.F. Scheneemeyer, J.V. Waszczak, T. Siegrist, R.B. van Dover, L.W. Rupp, B. Batlogg, R.J. Cava, and D.W. Murphy, Nature 328, 601 (1987).

12. D.X. Li, X.K. Wang, D.Q. Li, R.P.H. Chang, and J.B. Ketterson, J. Appl. Phys., 66, 5505 (1989).

13. P. Chaudhari, R.H. Koch, R.B. Laibowitz, T.R. McGuire, and R.J. Gambino, Phys. Rev. Lett. 58 2684 (1987).

14. C.P. Bean, Phys. Rev. Lett. 9 250 (1962).

15. Y.B. Kim, C.F. Hempstead, and A.R. Strand, Phys. Rev. Lett. 129 528 (1963).

16. X.K. Wang, D.X. Li, S.N. Song, J.Q. Zheng, R.P.H. Chang, and J.B. Ketterson, MRS Fall Meeting Proceedings, Boston 1989.

TOUGHENING MECHANISMS FOR MONOLITHIC CERAMICS

M.V. Swain

CSIRO Division of Materials Science and Technology
Locked Bag 33, Clatyon, Victoria 3168 Australia

ABSTRACT

The last decade has seen a dramatic improvement in the mechanical properties of structural ceramics. Strengths in excess of 2GPa and fracture toughness values higher than $1KJm^{-2}$ are now possible. These improved properties are possible because of the availability of highly sinterable powders coupled with hot isostatic pressing and the incorporation of new toughening mechanisms such as transformation toughening and fiber and whisker reinforcement. Means of combining more than one mechanism in a specific microstructure are also being developed. The primary role of the most effective troughening mechanisms is to reduce the crack tip stress intensity factor by 'crack tip shielding'. In this manuscript the various mechanisms for toughening are outlined along with examples of materials systems exhibiting this behaviour.

INTRODUCTION

The last decade has seen a tremendous upsurge of interest in ceramics. One of the worries of many old hands in the field is whether this is genuine commitment to this, the oldest of all manufacture commodities or whether it is cyclical market hype. No where has this enthusiasm been more evident than in Japan where "ceramic fever" was diagnosed as early as 1983 . Similar but less ambitious ceramic programs are underway in most technologically advanced nations. The market area with the greatest capacity for growth is considered to be in structural ceramics. Various estimates of the market for structural ceramics have been proposed from 0.1 to 3 x 10^9 US dollars in 1995, with the total ceramic market up to 15 to 20 x 10^9 US dollars [1].

What has lead to this upsurge of interest in ceramics? There have been a number of factors, such as: appreciation of new toughening mechanisms, availability of high purity sinter active powders, concern regarding supply and expense of strategic metallic materials such as cobalt, tungsten, etc, demand for more fuel efficient and lower cost maintenance engines. Interest in this area has tended to coincide with high oil prices. Of all these factors the major potential contributor to rapid growth for structural ceramic growth is the automotive industry, however as many commentators have warned it is somewhat speculative and depends upon future free market competitive forces and the price of energy. A significant increase in energy costs will greatly facilitate the improved demand for ceramics in heat engines, particularly insulating and gas turbine componentary. Other areas such as wear resistant components for valves, tappets, etc, will also find increasing demand when the production cost decreases and volume capacity increases. In the meantime demand for wear resistant ceramics is finding increasing application in the metal fabrication area (extrusion dies, cutting tools etc.) and chemical/mining/petroleum processing areas (valves, pumps, seals etc.)

In this paper the emphasis will be on the mechanical properties of structural ceramics particularly on the toughening mechanisms of such materials. Tremendous developments in the field of toughening ceramics have taken place over the last fifteen years. These developments have occured on the theoretical, experimental and technological levels in a manner that has been exceptional productive. The concept that conventionally processed polycrystalline ceramic materials could be substantially toughened was most convincingly demonstrated in partially stabilized zirconia (PSZ) ceramics by Garvie et al [2]. This conceptual breakthrough and the subsequent incorporation of a stress induced phase change into many matrices has taken place. More recently the notion of micro-fibre (on single crystal whisker) reinforcement of a brittle matrix has been appreciated. The combination of these mechanisms (transformation and whisker reinforcement) together with other mechanims leading to a hierarchy of toughening proceedures is leading to the development of exceptionally tough and damage tolerant ceramic materials. The basis of most of these toughening mechanisms is that the effective crack driving force experienced by the crack tip is reduced. Ritchie has termed this class of toughening mechanisms "Crack-tip shielding" [3].

The extension of a crack may be considered due to a **crack driving force** and opposed by the **resistance of the microstructure.** The driving force is generally defined by some parameter that quantifies the loading system, such as the stress intensity factor or path-independent integral J, which is meant to describe the stress and deformation fields about the crack tip. Crack extension is restrained by reducing the driving force or increasing the toughness of the microstructure. Two possible means are possible to increase the resistance or toughness namely increasing the **intrinsic toughness** of the matrix or by reducing the crack driving force by crack tip shielding, that is by a process termed **extrinsic toughening.** The latter is the more usual approach for toughened ceramics because of the inability to modify significantly the yield behaviour of the matrix or the intrinsic low toughness of ceramic materials.

The aim of this paper is to outline in more detail the concept of crack shielding and then to discuss various mechanisms that result in significant extrinsic toughening of ceramic materials. The emphasis will be placed on the physical basis, and limitations and consequences of toughening ceramic materials including the susceptibility of toughened materials, to cyclic fatigue crack extension. Examples of the various toughening mechanisms and combinations of mechanisms will also be presented.

2. BASIS OF ENHANCED MECHANICAL PERFORMANCE

The two major mechanical properties for engineering ceramics are strength and toughness. Other properties that influence strength, wear resistance and thermo-mechanical behaviour include creep behaviour, hardness, thermal expansion and thermal conductivity. These parameters will not receive attention here. For engineering design purposes the major consideration is the strength of the ceramic material. However unlike metals, ceramic materials do not have a well defined yield stress

or ultimate tensile strength but rather fail catastrophically and show a

wide scatter of strengths. This has lead to the development of extreme

value function statistical analysis, of which the Weibul analysis is one

variant, to provide a basis for interpretation of the observations and

means of predicting size and stress loading dependence of strength

[4]. It has been known for decades that the tensile strength of fibres

of glass, sapphire, etc, may have exceptionally high strength   (> 2-3

GPa) [5], but display considerable variability in strength as well as

being very vulnerable to damage. Contact with dust particles in the

atmosphere may result in 80% reduction of strength. This has

necessitated coating of high strength fibres with protective layers as

applied currently to telecommunication optical fibre links.

Similar approaches have been applied for polishing surfaces for the

improvement in strength of polycrystalline ceramics. Any such gains are

usually at great expense and considerable inconvenience and are readily

removed when the component enters into service. Other problems

associated with strength variability are the influence of pre-existing

or introduced flaws (<50μms) and the difficulty of NDE techniques to

detect such flaws [6]. The relationship between strength, toughness and

flaw or defect size in a ceramic is usually written,

$$\sigma = K_{1c} / Y\sqrt{c} \qquad\qquad (1)$$

where σ_f is the breaking stress, c the flaw size, Y a geometric

parameter dependent upon flaw shape and location. By improving

processing and or polishing the surface after grinding it is possible to reduce the flaw size and so increase the strength. However, as is well recognised ceramic materials have very low fracture toughness values and are prone to damage. Hence the only approach to improve the service reliability, is to increase the fracture toughness by mechanisms that impart crack tip shielding.

The concept of crack tip shielding, which is also the basis for the high toughness of metallic materials, is that the resistance to crack growth or measured K_{1c} consists of two or more components, the **intrinsic toughness** of the material plus some additional or **extrinsic toughness** component due to a range of possible options, that is

$$K_{1c} = K_1 + K_s \qquad\qquad (2)$$

where K_1 is the intrinsic matrix toughness and K_s is the contribution due to external shielding. This expression is similar to the expression proposed by for metals where K_s would be the equivalent plastic work component which is usually very much greater than the intrinsic toughness [7]. For polycrystalline materials it is difficult to decide upon a K_1 value because of the influence of grain boundaries etc on crack the path. Lawn and colleagues [8,9] are currently attempting to develop analytical approaches that encompass this feature.

An aspect of major significance associated with crack tip shielding concept is that events in the **wake** of the crack tip are the major contributors to increased toughening. Fig. 1 a. A logical consequence of such wake related toughening effects is that with the initiation of a crack, wake toughening mechanisms only develop with crack extension. That is, K_s is a function of crack extension and the material exhibits R-curve behaviour. Equation (2) should be then re-written

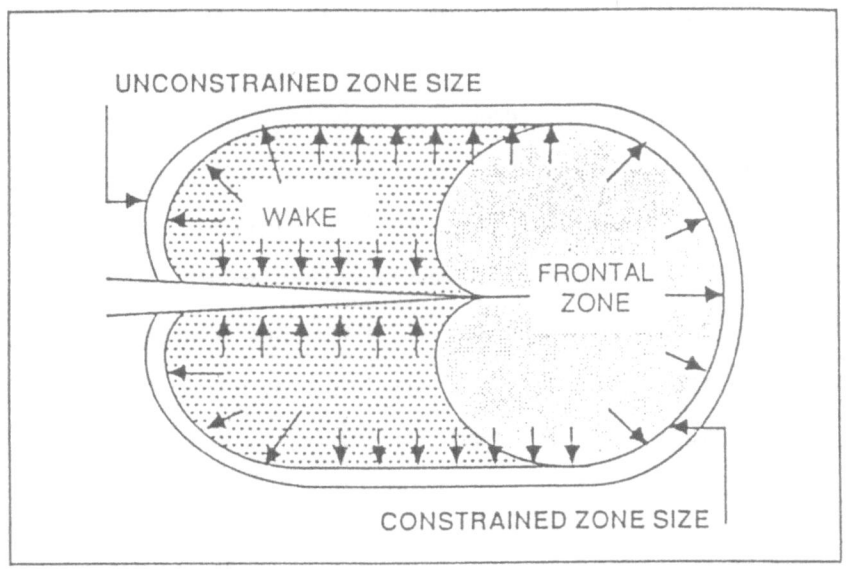

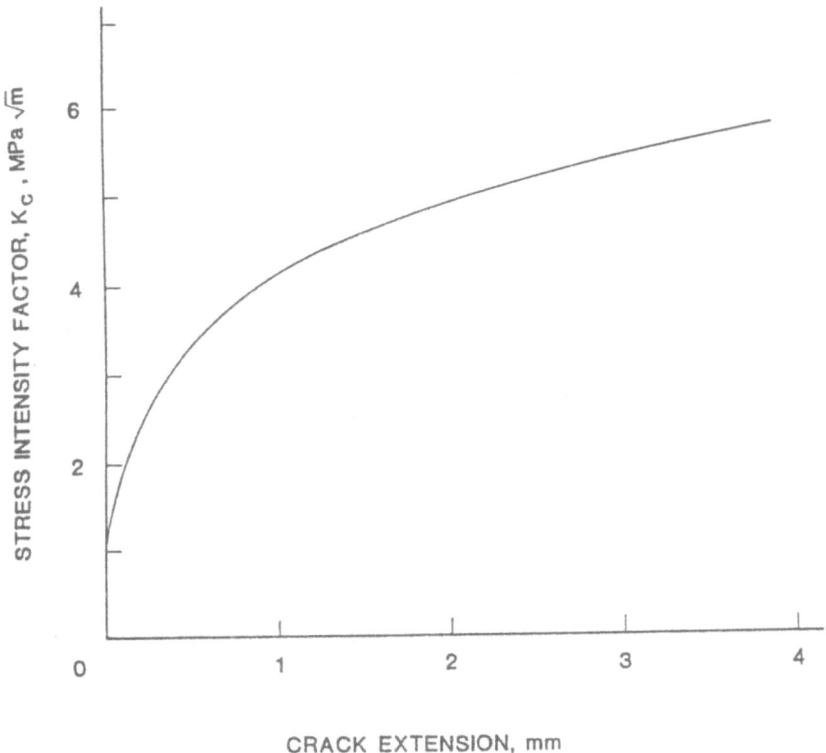

Figure 1. a, Wake region developed about an extending crack.
b, Influence of wake region on toughness.

$$K_{1c} = K_1 + K_s \text{ (c)} \qquad\qquad (3)$$

This consideration is shown schematically in Fig. 1b.

The simple relationship between strength and toughness is now more complex for materials exhibiting crack tip shielding or R-curve behaviour. This situation may be readily appreciated with the aid of Figure 2. A consequence of an R-curve is that the crack becomes unstable when the rate of change of strain energy G with crack length exceeds the rate of change of crack resistance,

$$\frac{dG}{da} \geq \frac{dR}{da} . \qquad\qquad (4)$$

Figure 2 [10,11]. Even this approach is somewhat simplistic as the form of the R-curve will be dependent upon the geometry of the wake, that is one might expect the R-curve for a surface crack to be different from an internal defact or through thickness crack. This feature will be taken up in the final section.

Whilst the consequences of R-curve behaviour are that lower strengths are experienced than anticipated by equation 1, less sensitivity of the strength on initial flaw size (i.e. higher Weibull modulus) is observed. This aspect was originally proposed by Kendall et al [12] and more recently extended by Cook and Clark [13]. Previous studies by Cook et al [14] had established that coarse grained alumina ceramics with significant R-curves were more damage tolerant to Vickers

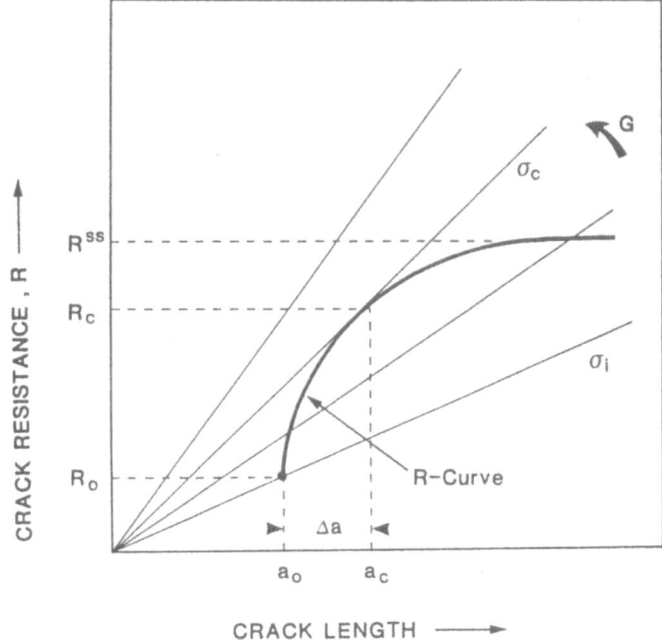

Figure 2. Schematic plot of crack stability due to R–curve response of
a material. Crack resistance (R) is plotted on the same
figure as the applied strain energy release rate G, equ
(3). Crack initiation commences at a stress σ_i when $G = R_C$
and catastrophic failure occurs at a stress σ_c when $G = R_C$
which is less than the steady state toughness R^{SS}.

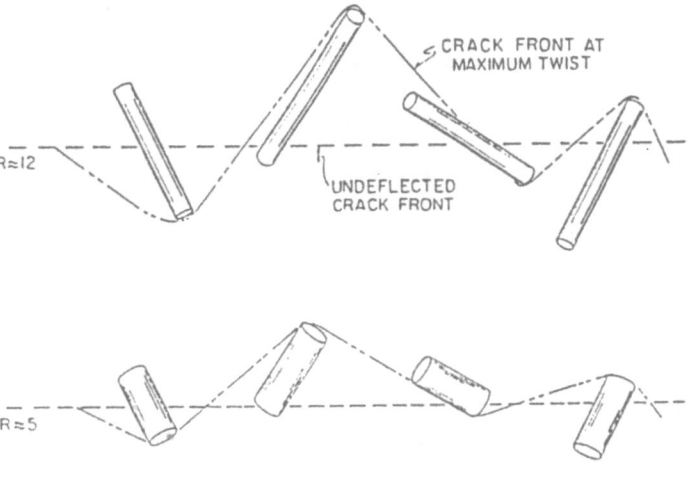

Figure 3. Schematic of crack front deflection and twisting about rods
in a matrix. R is the aspect ratio of the rods.

indentation than stronger finer grained alumina. It is also known that R-curve behaviour is beneficial to minimise thermal shock damage [15].

Various micro-mechanisms of crack tip shielding that impart increased toughness and associated R-curve behaviour are listed below characterised into several classes [3,10,16]. These include,

1. Crack deflection and meandering,

2. Zone shielding

   - Residual stress

   - Microcracking toughening

   - Transformation toughening,

3. Contact shielding

   - Crack bridging

   - Whisker and fibre reinforcement.

In the following section each of these classes will be dealt with in some detail along with the anticipated and where available examples illustrating the efficiency and range of there mechanisms. Certain systems have been developed that combine more that one specific shielding effect. This will most certainly be the direction for future research, namely the development of tailored microstructures that optimise the strengthening and toughening increments available via the above and other yet to be appreciated mechanisms.

3 TOUGHENING MECHANISMS

3.1 Crack Deflection and meandering

Cracks may be deflected from their planar path by a variety of means including; grain boundaries, fracture resistant second phase particles or residual stresses. A schematic diagram illustrating this mechanism is shown in Figure 3. The basis for the increased toughness associated with this mechanism is that reorientation of the crack plane away from normal to the applied tension causes a reduction in the crack driving force or crack tip stress intensity factor (K). Faber and Evans [17,18] have distingiushed two types of crack deflection: crack tilting about an axis parallel to the crack front, and twisting of the crack plane about on axis normal to the crack front. Fracture mechanics analysis suggests that the twisting component contributes most to the increased toughness.

The toughness increment possible with this mechanism is highly dependent upon the nature of the crack deflection and particulate phase that causes it. The toughening derived from randomly oriented, deflecting particles depends only on the volume fraction and shape of the particles. Rod shaped particles with large aspect ratio contribute maximum toughening because of their influence on the twist angle. The predicted toughening effect of rod shaped particles with different aspect ratio (R) is shown in Figure 4. The interesting feature of this graph is that toughening saturates with relatively low volume fractions of second phase particles. Another important feature about this mechanism is its independence of temperature and particles size.

Examples of materials with microstructures developed to enhance the toughness by this crack deflection mechanism are beginning to appear. Faber and Evans [18] showed this mechanism was relevant in SiC, Si_3N_4 and a glass ceramic materials. Lange [19] demonstrated the improved toughness of Si_3N_4 with increased accicular β phase present. A recent study by Hori et al [20] has clearly demonstrated the effect of volume

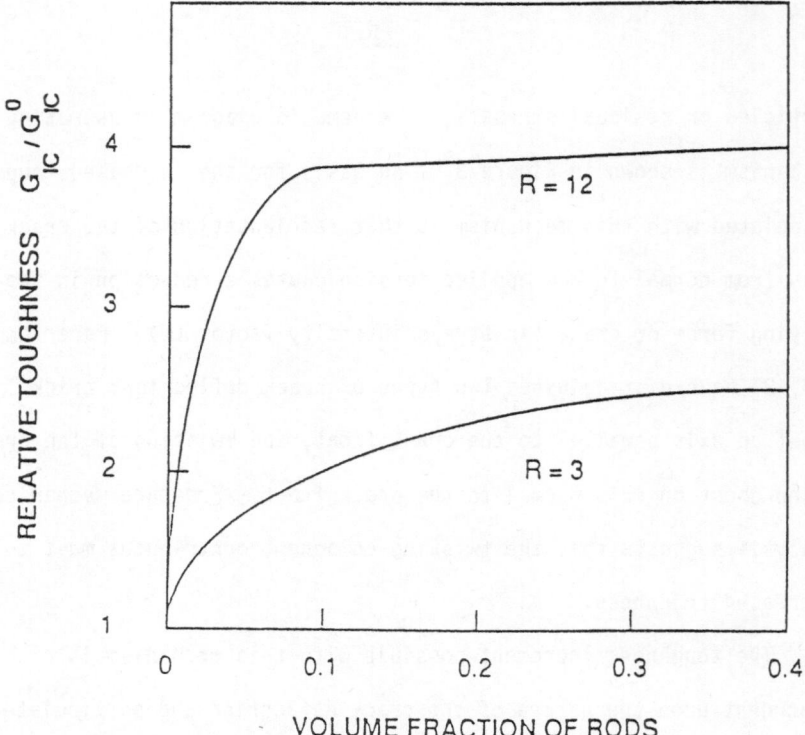

Figure 4. Toughening increment due to crack deflection as a function
of volume fraction and aspect ratio of crack deflecting
particles.

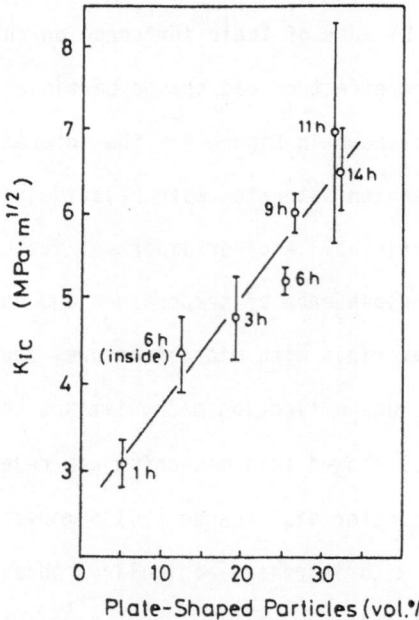

Figure 5. Observations of the toughening effect of plate like
particles in a rutile–corundum composite material [20].

fraction of plate like particles in a rutile-corundum composite material. Observations from Hori et al are shown in Figure 5. This mechanism of toughness enhancement will also be considered in the remaining two subsections.

3. 2 Zone Shielding

Mechanisms that provide zone shielding are usually extrinsic toughening mechanisms that reduce the apparent applied stress intensity at the crack tip to the intrinsic toughness of the material. These include: i) the presence of residual stress over a scale much larger than the strength controlling defect sizes, ii) microcrack toughening, and iii) transformation toughening. These three mechanisms will now be considered in greater detail.

3.2.1. Residual Stress

This approach has been recognised for many decades and technologically finds wide application in the glass industry. By thermal quenching glass heated above the softening temperature or chemical ion-exchange tempering, considerable improvements in strength and reduction in their sensitivity to surface damage. In this manner surface compressive stress and compensating internal tensile stresses are developed. Upon the application of a flexural stress these surface compressive tempering stresses must first be overcome before the surface flaws are placed in tension - this might be considered apparent toughening. One obvious disadvantage of this mechanism is that the internal tensile stresses may lead to the spontaneous fracture of glass plates due to inclusions or pre-existing flaws [21].

Need to include expression for apparent toughening based upon Lawn and Marshall's analysis. A detailed analysis of the "effective" toughening associated with a residual tempering stress has been made by Lawn and Marshall [22]. These authors used the scaling porometers of the relative tempering depth d and a modifying factor that scaled with crack depth to tempering zone depth, namely

$$K_s(c) = M \ (c/d) \ \sigma_R \ (\pi d^{\frac{1}{2}}) \tag{5}$$

where M (cd) is the dimensionless modifying factor that is determined by the tempering stress profile, σ_R is the surface compressive stress and d is the depth of the tempered zone.

Compressive surface stresses can be developed in ceramic materials by several mechanisms. These include thermal tempering as shown by Kirchener et al [23] for polycrystalline alumina, usually containing a glass phase. Volume changes of silicon nitride surfaces by oxidation have also been found to produce surface compressive stresses [24]. However if this process is taken too far the surface oxidised layer may spall off. In transformation toughened zirconia containing ceramics surface compressive stresses may be developed by grinding induced transformation of a surface layer, the depth and effectiveness of which increases with grinding severity [25]. It is also possible to destabilise the surface layer by chemically removing stabilisers and thereby enabling the volume expanding tetragonal to monoclinic phase change to take place on cooling leading to surface compressive stresses and increased strength [26].

The magnitude of such surface compressive stresses are given by the following simple relationship, namely

$$\sigma_R = \frac{\Delta V \ V_f \ E}{3 \ (1-\nu)} \tag{6}$$

where ΔV is the volume dilation, V_f the volume fraction of transformed phase, E Young's modulus and ν Poission's ratio.

3.2.2. Microcrack Toughening

Stress induced microcracking may take place in a range of single and polyphase ceramic microstructures that contain localised residual stresses. These residual stresses arise from thermal expansion anisotropy of non cubic polycrystalline materials, thermal expansion or elastic mismatch in poly phase materials and stresses arising from second phase inclusions that undergo a volume change due to a phase transformation upon cooling. Crack tip shielding results due to the instability of residually-strained regions in the near vicinity of the crack tip. This instability within a process zone results in micro cracking either circumferentially or radially (depending upon the internal stress within the second phase inclusion) or at a grain boundary in a single phase material. The critical size of grains or second phase particles for spontaneous cracking of such materials have been addressed in detail [27] and here we shall confine ourselves to subcritical sized materials that are triggered by the high tensile stresses about the crack tip. A schematic diagram of this mechanism is shown in Figure 6.

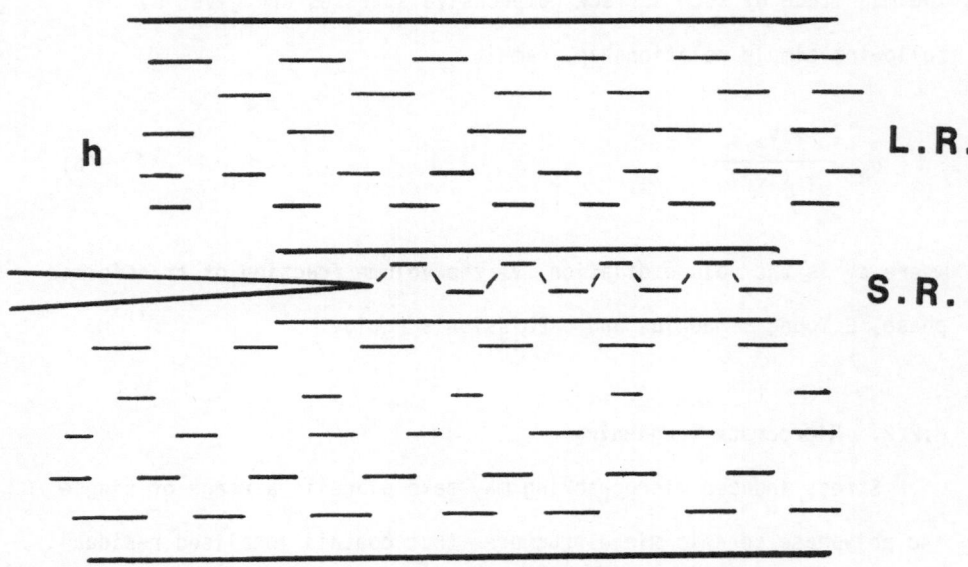

Figure 6. Schematic diagram of a microcrackedzone about a macrocrack.

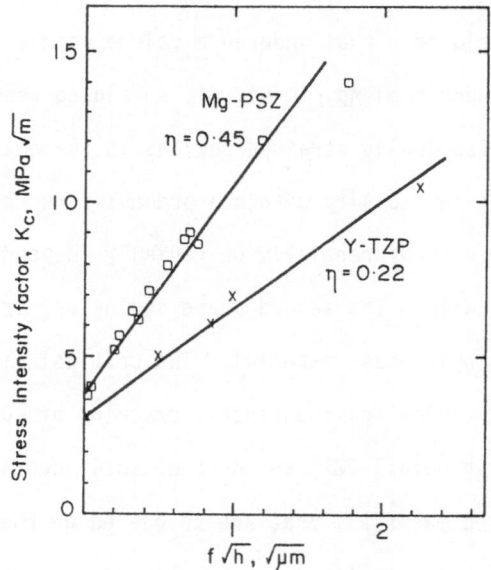

Figure 7. Dependence of K_c of PSZ and TZP ceramics on dimensions of
the transformation zone size and volume fraction of
transformed monoclinic zirconia.

Two features that are important in these materials is that upon the

initiation of microcracks in the process zone, permanent crack opening

occurs at each microcrack and associated with this micro-cracking is a

reduction of the elastic modulus within this zone.   There is still

considerable differences in the theoretical treatments of the toughening

increment due to stress induced micro-cracking.   Some authors have

treated the problem analogous to transformation toughening with the

additional feature of a reduction in elastic modulus [28].   Others have

approached the problem as an energy disipation zone within which a crack

linking region exists [29].   The former provides the simplist analysis

and will be presented here.   If we consider there is a critical stress

to initiate microcracking and results in a volume strain of ϵ^M, the

toughening increment maybe written

$$\Delta K^m = 0.21 \ E \ \epsilon^M \ h^{\frac{1}{2}} / \ (1-\nu) \qquad\qquad (7)$$

where E is the Young's modulus, h the micro-crack zone heighe and ν is

Poisson's ratio.   The value of ϵ^M depends upon the conditions for micro

cracking.   For spheres under a residual tensile stress of σ_R, it has

been shown that

$$\epsilon^M = 16 \ (1-\nu^2) \ \sigma_R V / \ (3 \ E) \qquad\qquad (8)$$

where V is the volume fraction of microcracking spheres.   This leads to

the expression

$$\Delta K^m = 1.12 \ V \ \sigma_R \ h^{\frac{1}{2}} \ (1 + \nu) \qquad\qquad (9)$$

Whereas for spheres under internal pressure, leading to radial micro cracking

$$\epsilon^M = 2 \ \epsilon^T (V + [3V/4\pi]^{1/3}) \ /3 \qquad\qquad (10)$$

where ϵ^T is the dilational strain due to transformation strains or thermal expansion mismatch strains ($\Delta\alpha \ \Delta T$). Substitution of this expression into equation (6) provides an estimate of the toughening increment.

Unfortunately there is very little definitive experimental data on the toughening increment due to micro cracking zone width on the internal stress or size of the second phase particles. This situation also exists for coarse grained polycrystalline materials like alumina for which it has been proposed that micro cracking is the predominant mechanism. Data by Ruhle et al [30] for zirconia toughened alumina (ZTA) with a K_{1c} value of 6 MPa $\sqrt{m}$. indicates that the toughening contribution due to microcrack shielding is only 1 MPa $\sqrt{m}$. This estimate is in good agreement with both a theretical [31] and finits element analysis [32] of microcrack toughening. Very recent experiments by Lutz et al [33] with ceramic materials consisting of a matrix containing second phase spheres of known size and internal stress have provided definitive evidence of micro cracking and long range R-curve behaviour.

3.2.3 Transformation Tougheneing

This mechansims for improving the toughness of ceramics relies upon a stress assisted martensitic volume expanding phase change that takes place about the crack tip. It was first appreciated by Garvie et al [2]

in 1975 in zirconia partially stabilized (PSZ) ceramics. Zirconia
(ZrO_2) exists in three phases depending upon temperature, namely

    ~2300°C   ~1200°C

  ZrO_2 ZrO_2 ZrO_2

 (cubic)   (tetragonal)   (monoclinic)

ρ = 6.09gm/cc ρ = 6.10 ρ = 5.83

The massive transformation strains, both dilational and shear, of the
tetragonal $\rightarrow$ monoclinic transformation cause destruction of articles
fabricated of pure zirconia. This feature is overcome by the addition
of various stabilizers such as MgO, CaO and various rare earth oxides.
Judicious selection of stabiliser content and sintering – heat treatment
conditions leads to retention of the tetragonal zirconia phase at room
temperature. Then the application of stress may lead to the following
reaction.

                 stress

ZrO_2 (tetragonal) ZrO_2 (monoclinic)

              heat

A wide range of microstructures containing tetragonal zirconia may
be fabricated varying in grain sizes from $0.2\mu m$ completely tetragonal
zirconia polycrystals (TZP) to $50\mu m$ grain size with lenticular
tetragonal precipitates in a cubic zirconia matrix in PSZ materials. A
third group containing tetragonal zirconia dispersed ceramics (ZDC) in
another matrix e.g. alumina, mullite etc [34].

A schematic drawing of the development of a transformed zone about
a crack is shown in Figure 1 . Only in this instance the zone is under
internal pressure resulting from the transformation. The toughening
increments achieveable in transformation toughened materials have been
discussed from the basis of energetics or mechanics perspectives
[35,36]. To date the theoretical arguments have taken the materials as
essentially isotropic with a smeared out zone of transformation about
the crack tip area. More detailed models are required to fully
appreciate the toughening mechanisms. Both approaches predict
comparable relationships, and are of the form [37]

$$K_c = K_o + \Delta K^T$$

$$(11)$$

and $\Delta K^T = \eta \, E^* e^T \, V_f \sqrt{h}$

where ΔK^T is the transformation toughening increment, K_o the toughness
of the matrix material, h the size of the transformation zone normal to
the crack, e^T the dilational strain, E^* the effective modulus of the
material and η a constant determined by the transformation zone
shape. The effective modulus E^* as pointed out by McMeeking [38] plays
a very important role in determining the effectiveness of the dilational
strain of the zirconia phase (E ~210GPa, ν = 0.3) on that the matrix in
ZDC materials. The influence of the dilation in an alumina matrix
(E ~380GPa, ν = 0.2) is only one third as effective as in a PSZ
material.

Substantial experimental support for the simple relationship in
equation (11) has been obtained for a range of transformation toughened

ceramics both PSZ and TZP materials [39]. This is shown in Figure 7
which plots the toughness of 3 different systems, Mg-PSZ, Y-TZP and Ce-
TZP against volume fraction and width of the transformed zone. The
latter have been measured with a Raman microprobe system which enables
volume fraction and zone width to be measured very accurately. The
observations show that a five fold increase in toughness is possible in
PSZ materials. The difference in slope (of value) of the Ce-TZP
material from the other materials is suggestive of a different zone
shape for the Ce-TZP material.

This is indeed found to be the case as shown in Figure 8 which compares
the transformation zone about a crack in Mg-PSZ and Ce-TZP material
[40]. The former is almost as predicted on the basis of a "small scale"
transformed zone modified by anisotropy in various grains along the
crack front. Whereas the zone shape in Ce-TZP material is closer to a
craze zone in a polymer or Dugdale stretch zone in a metal. This
similarity has been further explored to elucidate the basic
transformation mechanisms in TZP materials [41].

Another parameter that significantly influences the toughness of
transformation materials is temperature. With increasing temperature
the tetragonal phase becomes more stable making transformation
energetically more difficult and hence the zone size and toughness
decreases. Recently Becher et al [42] have shown that the critical
normalising parameter is the M_s temperature at which the tetragonal
phase spontaneously transforms to the monoclinic phase. The toughening
increment ΔK^T (equation 1) may then be modified to the following
relation [42]

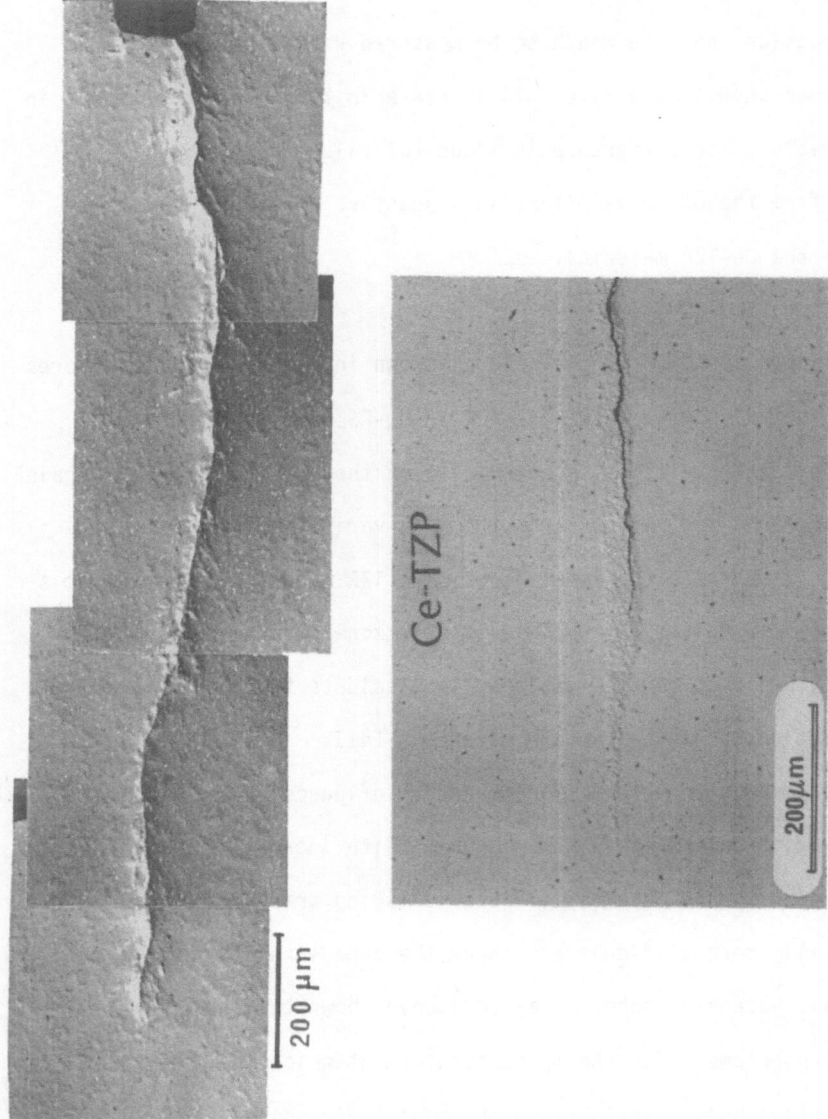

Figure 8. Comparison of the transformation zones about the crack tip in Mg–PSZ and Ce–TZP materials.

$$\Delta K^T = \frac{\Omega \, \eta (e^T)^2 \, V_f \, E \, K_o}{\Delta S \, (T-M_s)} \qquad (12)$$

where Ω and η are constants, ΔS is the entropy difference between tetragonal and monoclinic zirconia.

Experimental evidence in support of this expression is shown in Figures 9 which plot K_c versus temperature for Mg-PSZ materials [42], and a range of PSZ and TZP materials. The maxima in toughness corresponds with the M_s temperature and thereafter decreases linearly with temperature to the matrix K_o value. The realisation of the temperature sensitivity of transformation toughened ceramics has lead researchers to search for other less temperature sensitive toughening mechanisms to enable materials to have high toughness at temperature.

Recent studies have shown that it is possible to extend the analysis on which equation 2 is derived and predict the grain size dependence of the toughness of PSZ and TZP materials [43]. The approach combines the end point thermodynamic arguments developed by Garvie and Swain [44] and combines them with the fracture mechanics arguments of Evans and McMeeking [37]. This leads to the expression

$$\Delta K^T = \frac{\Omega \, \eta (e^T)^2 \, V_f \, K_o}{[< \Gamma.\tau + \sigma \, e > + \tau \, kd^{-\frac{1}{2}}]} \qquad (13)$$

where Γ_o is the single crystal transformation shear stress, k is a

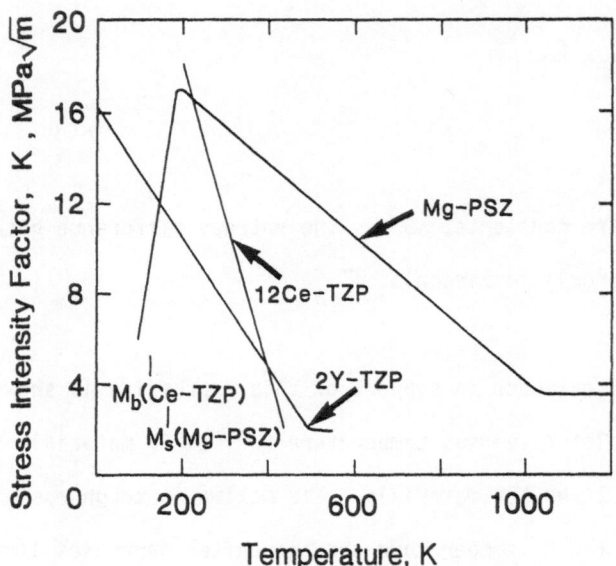

Figure 9. Temperature dependenced of the toughness of a range of
transformation toughened ceramics.

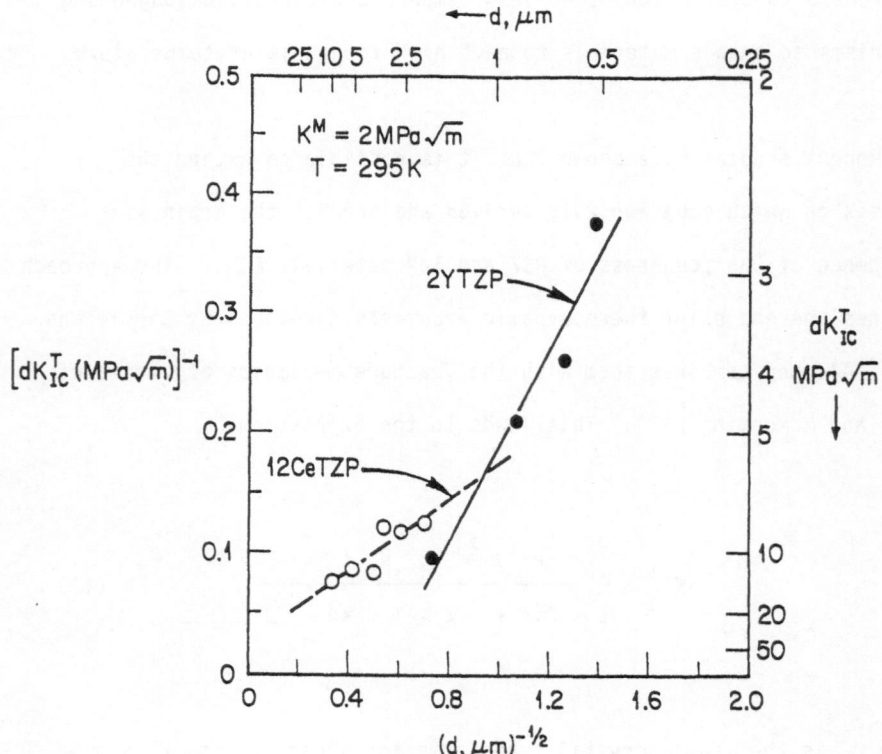

Figure 10. Grain size dependence of the toughness of Ce-TZP and Y-TZP
materials.

constant, d is the tetragonal grain size, σ the applied stress and e the resultant strain. Evidence in support of this relationship is shown in Figure 10.

The fracture toughness as observed and predicted theoretically for transformation toughened materials exhibit R-curve behaviour [35, 45]. This occurs because the volume dilation provides closure forces behind the crack tip, the observed R-curves in PSZ materials are shown in Figure 11. The rise in toughness occurs over approximately 5 times the transformed zone height. This feature only becomes significant for materials with well developed transformed zones or materials with high K_c values. R-curves have been observed for Mg-PSZ and Ce-TZP materials directly with standard fracture mechanics tests, e.g. DCB and SENB, as well as optically on tensile surfaces of flexure bars [46,47,48]. Associated with such cracks are surrounding zones of transformed zirconia which are readily observed using interference microscopy. It is observed that the onset of crack growth takes place at lower values of K_{1c} for surface cracks than through thickness cracks [48]. For transformation toughened ceramics where the toughness maybe modified by heat treatment the consequences of R-curve predict a maxima in a plot of the strength versus toughness (steady state) [46,48].

Another feature of transformation toughened materials is their observed non-linear stress-strain relationships in tension and compression [49]. The inelastic deformation to failure is much greater in compression than tension. The onset of ductility occurs because of the metastability of the tetragonal phase and the dilation is associated with the transformation to monoclinic due to transformational

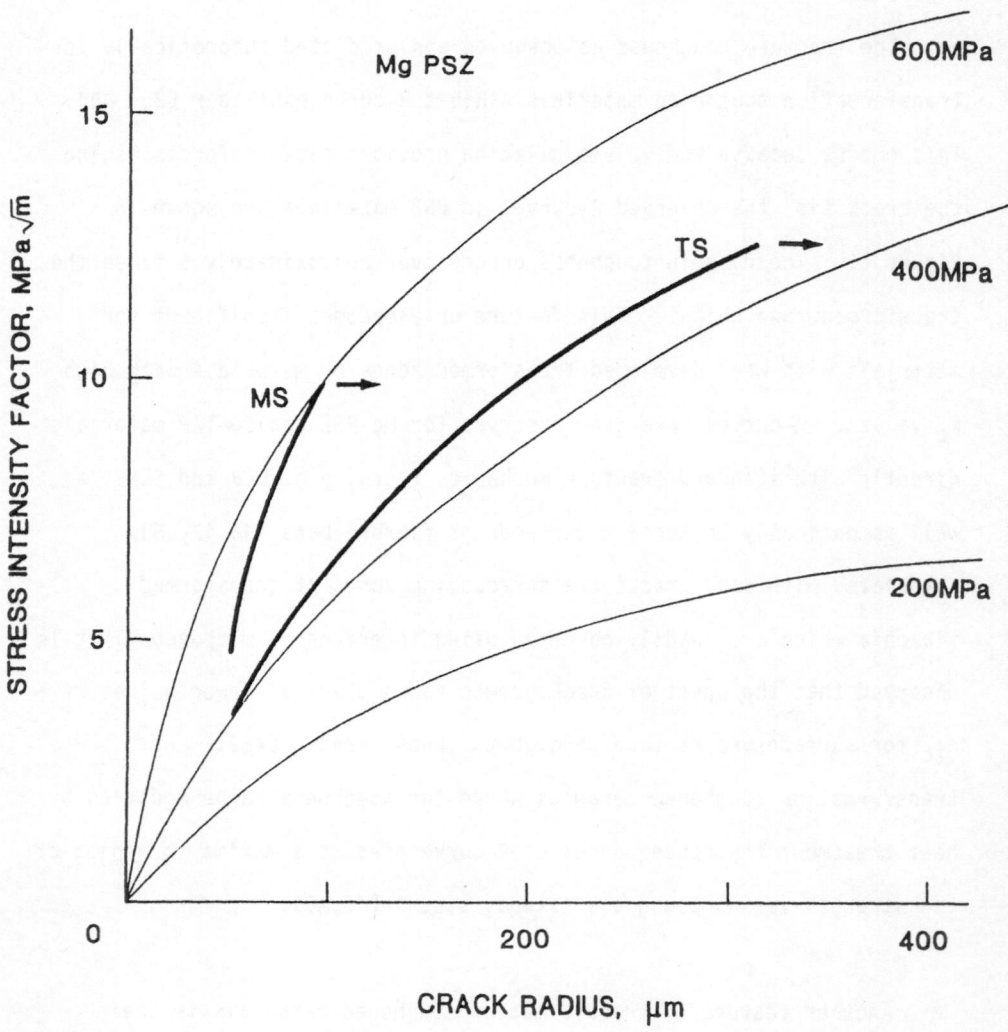

Figure 11. Observed R-curves in transformation toughened ceramics.

plasticity. This is determined by the M_s temperature which maybe modified by heat treatment, composition and grain size. In the PSZ materials the ductility takes place by means of a collaborative transformation of tetragonal precipitates often leading to microcracking at grain boundaries [50]. In TZP materials luders like bands are observed which occurs because of monoclinic laths are formed within tetragonal grains initiating adjacent grains [51]. The failure locus of a Ce-TZP material at room temperature due to stress induced transformation [52] is shown in figure 12. These observations are in excellent agreement with predictions of Chen and co-workers [53,54].

A consequence of the ductility of PSZ and TZP materials lead the author to propose an inverse relationship between strength and toughness [49]. Tougher materials such as Ce-TZP and also tough Mg-PSZ, "yield" prior to fracture. Also Y-TZP materials increase in strength with decreasing toughness initially. The inverse relationship proposed is shown in Figure 13. The highest strengths of 2.4 GPa are observed in hot isostatically pressed (HIP) specimens of Y-TZP containing 20-40 vol% Al_2O_3. The slope of the line through the origin indicates the critical flaw size to achieve such high strengths at a specific K_c value. For instance damage of the very high strength materials with a 10N Vickers indentation introduces flaws of approx. $200\mu m$ and the strength plummets to only 200 MPa. Whereas for Mg-PSZ material with a K_c value of 15 MPa√m, indents with loads as high as 500N causes no reduction in strength. More complete reviews of the Science and Technology of Zirconia Ceramics are available in three recent conferences devoted to this topic [55].

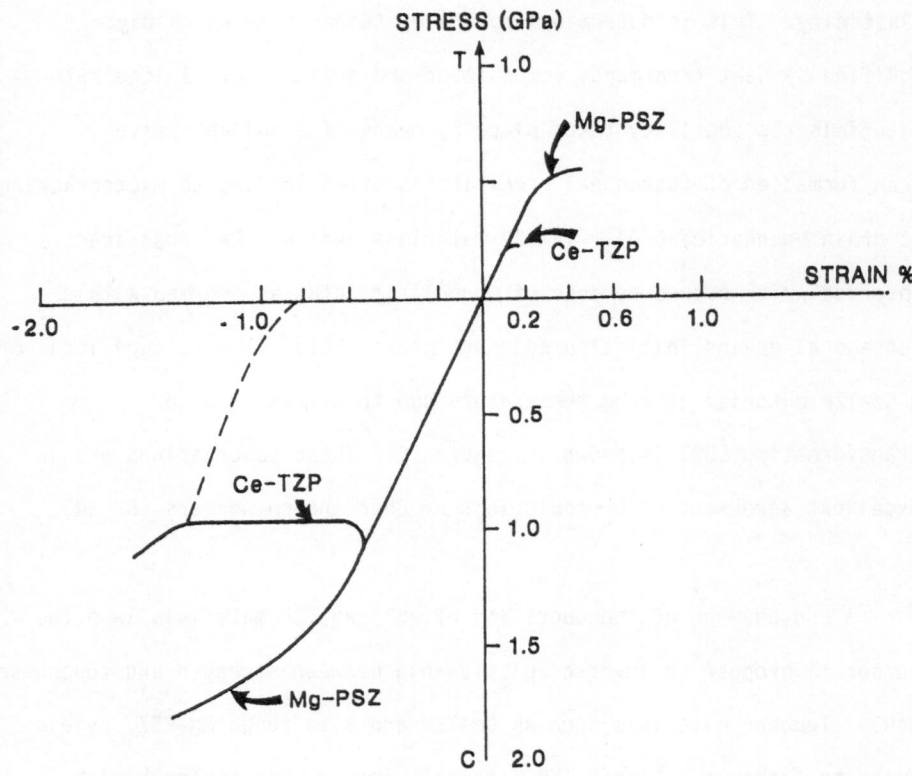

Figure 12. Stress Strain curves in tension/compression space for a 12
mol% Ce-TZP material and Mg-PSZ material.

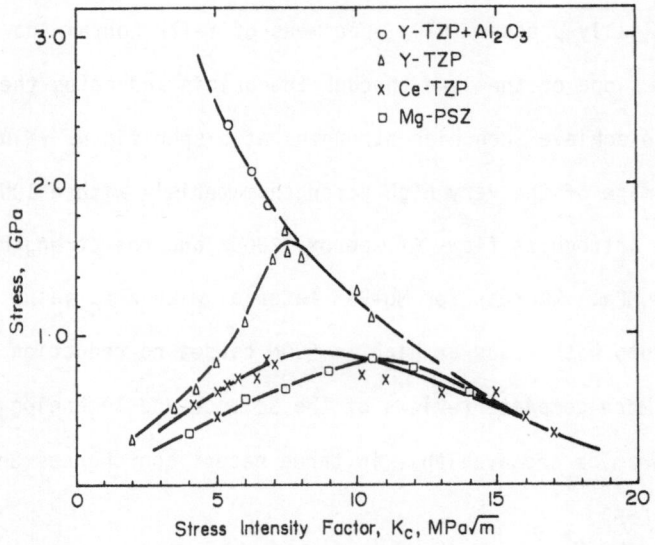

Figure 13. Strength-toughness relationships for several PSZ and TZP
transformation toughened materials.

The most recent developments in the field of zirconia toughened ceramics include fabrication of composite zirconia – non oxide systems such as $TiB_2-ZrO_2(Y_2O_3)$ and similar types of materials [56]. These materials under specific atmospheres maybe sintered to near theoretical density and HIP'ed to achieve high strengths 1 to 1.5 GPa. Examples of the toughness hardness and strength dependence of such composites is shown in Figure 14. These materials are electrically conducting and maybe shaped using conventional electrical discharge machining (EDM) techniques. Two other interesting new developments in ZTC materials are the observation of superplasticity in TZP materials between 1100–1300°C enabling novel forming–forging techniques [57]. This behaviour plus the observation of shape–memory behaviour in PSZ and TZP materials [58,59] confirms the original description of "ceramic steel" for these materials.

3.3 Contact Shielding

Contact shielding as proposed by Ritchie [3] involves physical contact between mating crack surfaces. This situation occurs through the presence of fracture surface asperities, and crack tip bridging via fibres, metallic inclusions or frictional intenlocks caused by major crack deviation. The most effective mechanism of contact shielding is via fibre or whisker bridging of the crack tip. The latter is particularly attractive because of its similarity to conventional processing whereas continuous fibre reinforcement introduces complex ceramic fabrication routes that are currently very expersive.

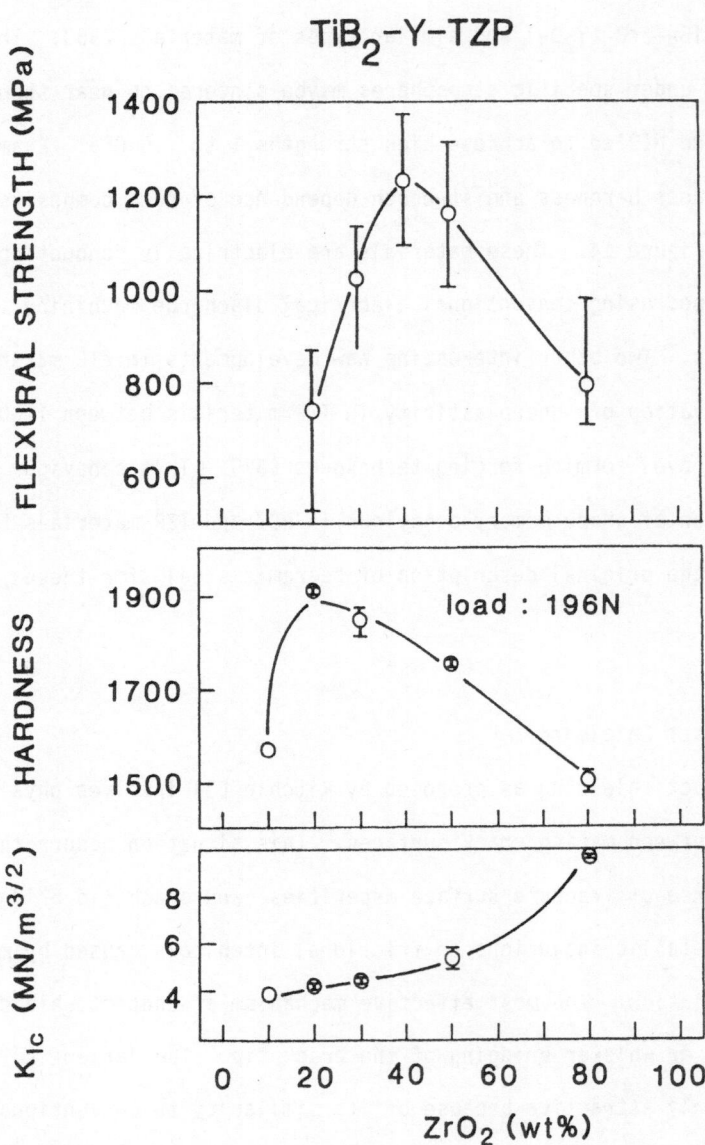

Figure 14. Dependence of strength hardness and toughness of TiB$_2$-Y-TZP
composites with composition.

3.31 **Whisker Reinforcment**

Almost two decades ago considerable interest was focussed on the preparation and properties of single crystal fibers or whiskers [29]. These materials are usually formed by vapour phase transport and preferential growth of certain crystal orientations. Because of their near perfect single crystal nature and fine diameters (μms) they exhibited in many instances near theoretical strength (> E/100). Interest in such materials as reinforcement for monolithic ceramics has resurged in the last three years. Most interest to date has centered on silicon carbide whiskers typically 0.5 - 1.0μm diameter and 100-200μm in length. Such materials are available from at least three suppliers although they tend to be very expensive (\$200-500/kg).

A number of studies have shown that substantial improvements in strength, toughness and creep resistance maybe developed in polycrystalline ceramics by incorporating up to 20-30 volume % of whiskers [60,61,62]. The attractiveness of the types of materials is that they allow more conventional powder processing techniques to be used in fabrication. The initial approach was to fabricate by hot pressing of milled and blended compositions particularly alumina-silicon carbide. More recently emphasis has shifted towards being able to sinter/HIP such materials and obtain comparable mechanical properties [63].

Whisker reinforcement may involve a number of toughening mechanisms, such as, fibre pullout, crack bridging and crack deflection due to the high aspect ratio fibres. At this stage of the development and understanding of such materials it is difficult to quantify which mechanisms are most effective. To date more attention has been placed upon the critical role of silicous layers on the whiskers leading to glassy films at the whisker—matrix interface. Such layers degrade properties above 800°C and lower the fracture toughness. For many whisker composites large thermal expansion coefficient (TEC) differences exist between whisker and matrix material with usually the TEC of the whisker < matrix. This leads to a clamping of whisker by the matrix making pullout more likely at higher temperatures, such behaviour is particularly, evident in mullite - silicon carbide whisker composites [64].

Becher et al [65] have recently proposed an analysis of the toughening contribution due to whisker reinforcement. The basis of their analysis is the development of the whisker bridging zone immediately behind the crack tip as a result of debonding of the whisker—matrix interface. Their analysis indicates the importance of the toughening increment upon composition and matrix, the interface and whisker properties. The approach adopted by these authors is to assume a Dugdale like crack tip zone where the increased toughness is associated with a closure stress on the crack by the bridging zone and a length of the zone. On the assumption of a uniform closure stress over the bridging zone the toughening increment is given by [65],

$$\Delta k^W = \sigma_f^W \left[\frac{V_f \ r}{3 \ (1-\nu^2)} \ \cdot \ \frac{E^C}{E^W} \ \cdot \ \frac{G^m}{G^i} \right]^{\frac{1}{2}} \qquad (14)$$

Where σ_f^W is the strength of the whiskers, E^C and E^W are the elastic modulus of the composite and matrix respectively, and G^m and G^i are the interfacial strain energy release rate of the matrix and interface respectively. This relationship highlights the importance of the strength, volume fraction, radius and elastic modulus of the fibres (whiskers). Evidence in support of the predictions of equation is shown in Figures 15 and 16. Figure is shows the dependence of the toughening increment on the square root of the volume fraction of whiskers, it also indicates the matrix modulus dependence as E mullite ~ 0.5 E alumina. The influence of whisker radius and interfacial toughness is shown in Figure 16. The higher toughness materials were treated prior to fabrication to remove the silica rich layer from the surface.

The strength of whisker reinforced alumina illustrating both the volume dependence and temperature dependence is shown in Figure 17. The strength shown a significant increase with volume fraction of whiskers in a manner similar to the toughness whereas only about 1000°C is their a significant decrease in strength.

The increase in strength with volume fraction of whiskers is partially due to a reduction in grain size. The creep properties of alumina–silicon carbide whisker materials are much improved over pure alumina [62].

One of the problems of whisker reinforcement of ceramics has been to obtain homogeneous distributions of the whiskers throughout the material. As fabricated whiskers usually are highly matted materials with clumps sometimes difficult to disperse. Such regions usually result in subsequent failure origins and may have remnant porosity about

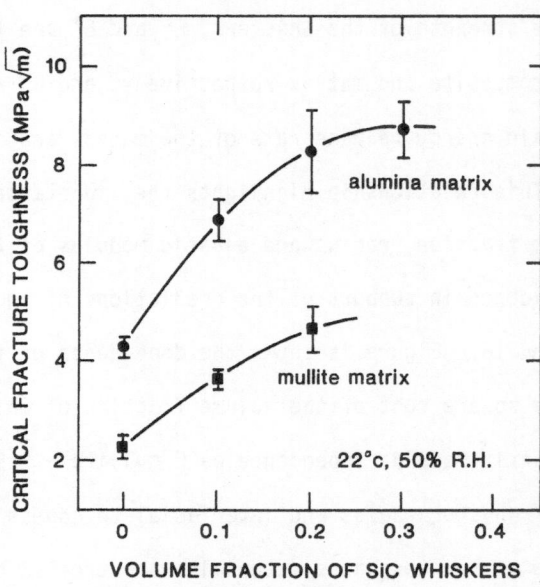

Figure 15.   Variation of the toughness of alumina and mullite with
volume fraction of SiC whiskers.

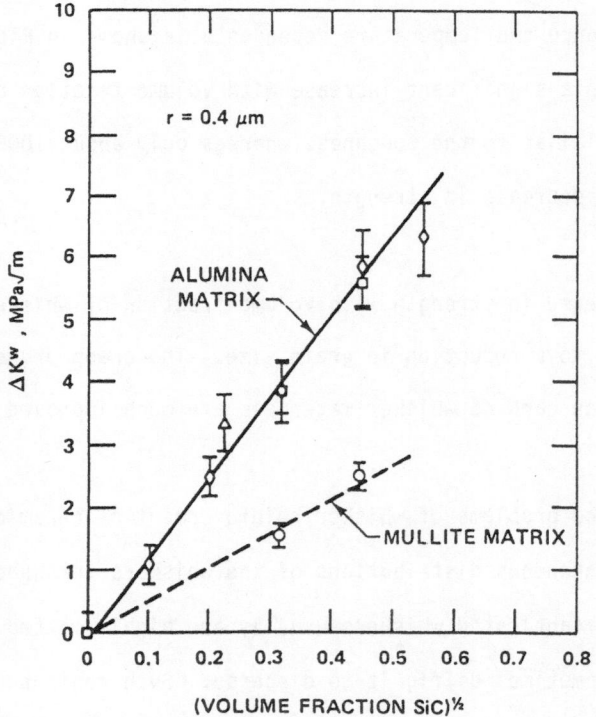

Figure 16.   Influence of whisker content on the strength of alumina –
silicon carbide whisker materials.

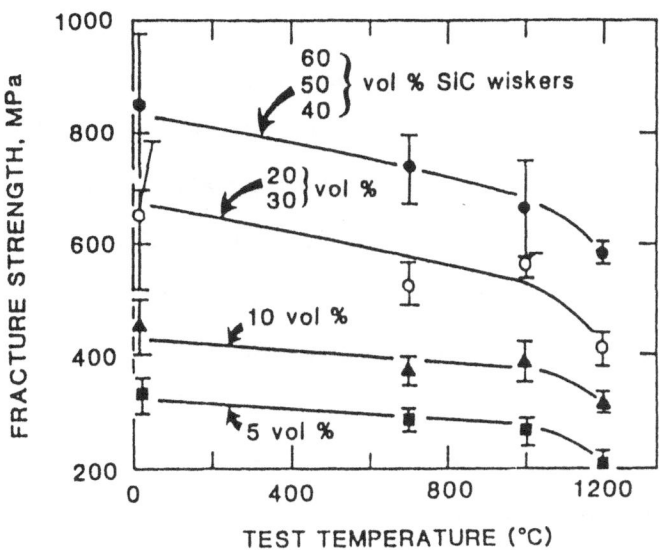

Figure 17. Strength of whisker reiforced alumina as a function of
temperature and volume fraction of whiskers.

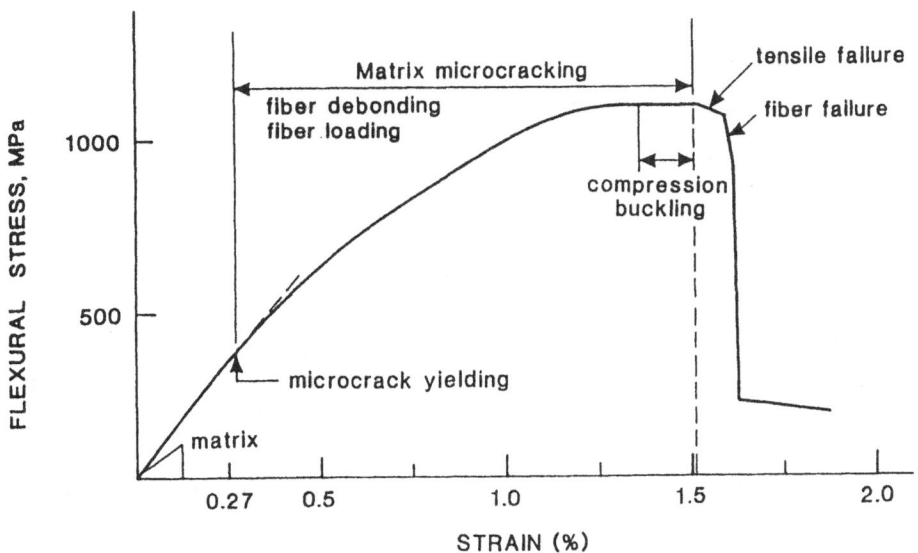

Figure 18. Flexural stress – strain behaviour of a SiC fibre reinforced
glass ceramic indicating deformation mechanisms that lead to
changes in slope of the curve.

them. Alternate approaches for achievement of whisker like micro-structural reinforcement has been developed by other workers. For instance Tani et al [66] have found by sintering silicon nitride with small additions of alumina and yttria or ceria at up to 2000°C and an overpressure of nitrogen that large high aspect ratio β-silicon nitride grains develop. Such materials have K_c values double that of conventional equi-axed silicon nitride materials.

3.3.2 Fibre Reinforcement

The recent availability of strong high quality fibres of various ceramics has lead to a renewed interest in fibre reinforced ceramics. Again this area achieved considerable attention well over a decade ago [67] however at that time only good quality carbon and glass fibres were available. Such materials were not suitable for high temperature composites in oxidising atmospheres. A major breakthrough occured about 10 years ago when Yajima [68] fabricated continuous silicon carbide fibres with diameters 15-20μms . These materials begin to degrade at temperatures above 1000°C. A detailed discussion of the properties and oxidation behaviour at elevated temperature is given by Mah et al [69]. These authors point out that behaviour of isolated fibers in various atmospheres does not necessarily determine their behaviour in a composite material. The important parameters tend to be, refractoriness, compatability between fibre and matrix and composite fabricability.

A range of potential matrix materials includes various glasses and glass ceramics, crystalline oxides, carbides, borides, nitrides, etc. The most widely used matrices have been glass and glass ceramics because of the relative ease of composite fabrication generally by low pressure hot pressing techniques. Attempts to fabricate ceramic matrix – continuous fibres have usually been unsuccessful because of the difficulties of densification about the fibres. Such problems also exaggerate thermal expansion mismatch cracking between matrix and fibre. More recently, following procedures developed for carbon–carbon fibre composites, infiltration by chemical vapour techniques as well as by polymer precursors have been used for carbide, nitride and oxycarbonitride matrices [70,71]. Chemical vapour infiltration can take place above 1000°C leading to dense composites with closed porosity and typically 80–85% theoretical density. A key parameter that maybe taken advantage of in such a processing route is that a thin interface coating maybe deposited on the woven structure prior to deposition of the matrix. In this manner the interface properties of the matrix to fibre maybe controlled and so determine whether genuine reinforcement occurs resulting in stable fracture fibre fracture and pullout or catastrophic fracture through the composite. Crack stability requires that the interface is strong enough to transfer load from the matrix to the fibres yet weak enough to fail preferentially prior to fibre failure.

The mechanical properties of fibre reinforced composites are impressive. The stress – strain behaviour of these materials has a closer resemblance to a metal than a ceramic. An more detailed example of a flexural stress–strain curve in a composite of SiC fibres in a matrix of barium osumilite are shown in Figure 18. The changes in the stress–strain behaviour indicate processes taking place in the material namely matrix microcracking followed by compression buckling just prior to fibre failure and reduction in load bearing capacity. Examples of the temperature dependence of strength and toughness of glass – ceramic – SiC fibres composite are shown in Figure 19. These materials are critically dependent upon the atmosphere and strain rate dependence at temperatures above 600°C, this is because of oxidation of thin layers of NbC at the interface changes the interface bonding between matrix and fibre [69,72].

More extreme values of toughness have been measured for SiC–SiC fibre composites fabricated by chemical vapour infiltration. Values of K_c as high as 39–40 MPa√m have been measured which are temperature insensitive to 1200°C [73]. An extremely steep R–curve has also been found. A consequence of such extreme toughness is that the material exhibits no reduction in strength upon thermal shock testing into water from temperatures as high as 1300°C. These materials response to crack extension are being addressed in a manner similar to that of tough metals which exhibit initial crack tip blunting prior to crack advance [74]. Work on such materials is currently limited by the high temperature degradation of fibres and the oxidation or creep response of the matrix. Attempts are underway to raise the operating temperature of such composites to 2000°C in air.

3.3.3 Crack bridging

As with all contact shielding toughening occurs due to events taking place behind the crack tip. As hinted at in section 3.2.2 (micro–crack toughening) many processes that were considered to generate a zone of micro cracks ahead of the crack tip. This situation was

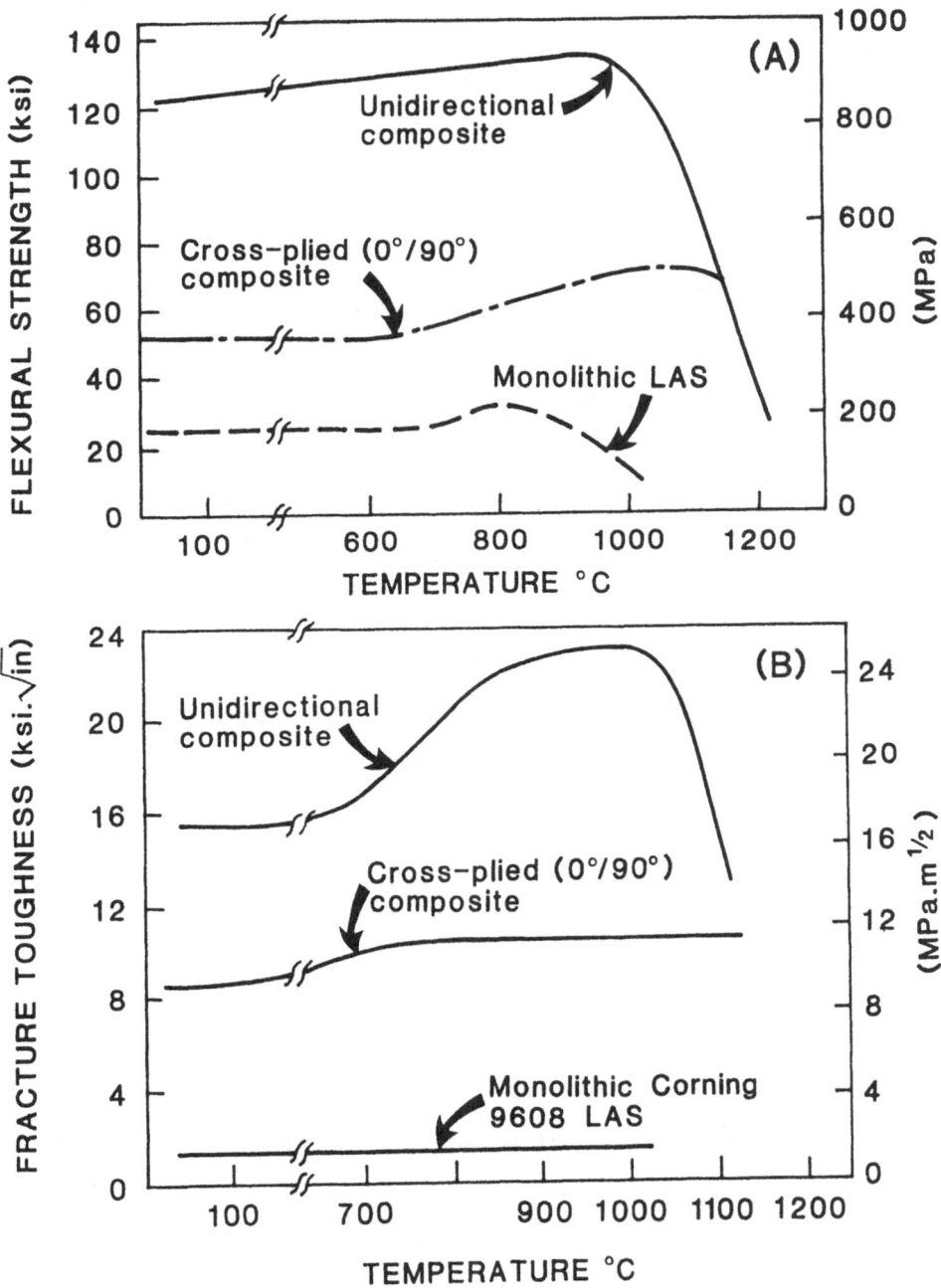

Figure 19. Temperature dependence of strength and toughness of SiC reinforced glass ceramic.

thought to occur in coarse grained alumina and other high thermally anisostropic materials. These materials were known to exhibit high steady state fracture toughness with considerable R-curve behaviour. However recent observations by Knehans and Steinbrech [75] and Swanson et al [76] indicate that events taking place at large distances behind the crack tip are responsible for a major part of the toughening. Knehans and Steinbrech [75] were able to definitively show that events behind the crack tip were responsible for the toughening by re-notching to just behind the crack tip, this lead to a significant reduction in the fracture toughness and re-development of the R-curve with further crack extension. Optical and scanning electron microscope observations confirm that frictional (mechanical) interlocking events generate closure forces behind the crack tip and are responsible for the increased toughness. Analysis of the problem has been very difficult as it is virtually impossible to estimate the force-displacement (stiffness) of these crack bridges and so estimate the closure forces.

Examples of the frictional interlock events taking place along a crack in a three phase TiB_2 based aluminium evaporator boat material [76] are shown in Figure 20. This material exhibits a significant R-curve behaviour, the range of which scales with the extent of crack bridging events behind the crack tip. Observations of the crack opening displacement with distance behind the crack tip for the material in Figure 20 were found to exhibit an inflection at the end of the bridging region.

Other mechanisms of crack bridging in ceramics is by the incorporation of a ductile phase into the matrix. These materials, one example of which has been the WC base hardmetals,

Figure 20. Observations of regions of frictional interlock in a TiB_2 based aluminium evaporator boat.

4 Strength in Service

The maintenance of strength, or understanding of factors controlling its degradation of in service, is one of the major considerations of current ceramic materials research. Considerable attention over the last decade has been given to the parameters that are important for the nucleation of cracks through contact with sharp or blunt indenters and in abrasion and machining operations [78,79]. This has entailed a detailed understanding of deformation and contact fracture mechanics and has lead to simple indentation techniques being routinely used to characterise K_c values of materials. Other studies have explored the influence of defect shape, modulus and thermal expansion mismatch of inclusions on the resultant strength. The other key issue has been the influence of environment on slow crack growth resulting in failure of a component sometime after loading. These areas are reasonably well understood and procedures exist to enable design of ceramic structures within well defined safety limits or probabilities [80].

4.1 Fatigue

A number of key areas in the field of mechanical behaviour of ceramics still require considerable attention. They include the influence of cyclic loading on the lifetime of a structure and measurement of simple S-N curves. This becomes particularly important for crack tip shielded toughened materials or those that exhibit non-linear stress-strain behaviour prior to fracture. Ritchie [3] has provided guidelines for the anticipated behaviour of crack tip shield materials response to cyclic loading. These features are shown schematically in Figure 21. For contact shielding the anticipated response is that the da/dN versus ΔK curveis shifted to higher ΔK values. With zone shielding, such as transformation or microcrack toughening the threshold value is barely affected but a less steep shape is expected which is exactly opposite to the influence expected for contact shielding. Evidence in support of these mechanisms is well accepted in the metals fraternity and is beginning to be more appreciated for ceramics. For instance in Mg-PSZ materials the lifetime predictions based upon static loading greatly overestimate the observations of lifetime under cyclic loading [81]. This is shown in Figure 22. The reason for this difference has recently been shown to be due to a genuine fatigue crack growth behaviour in PSZ materials with a slope much less than seen for static fatigue loading [82], Figure 23. Similar behaviour might be anticipated for fibre composite materials for stresses exceeding the onset of matrix-microcracking [6].

The other area where serious problems arise is in the area of high temperature application. At elevated temperatures ceramic materials are no longer chemically or physically stable. Chemical reactions between ceramics and the environment lead to the formation of new populations of flaws and or result in chemical or physical modification of the ceramic microstructure. All these processes may result in strength degradation

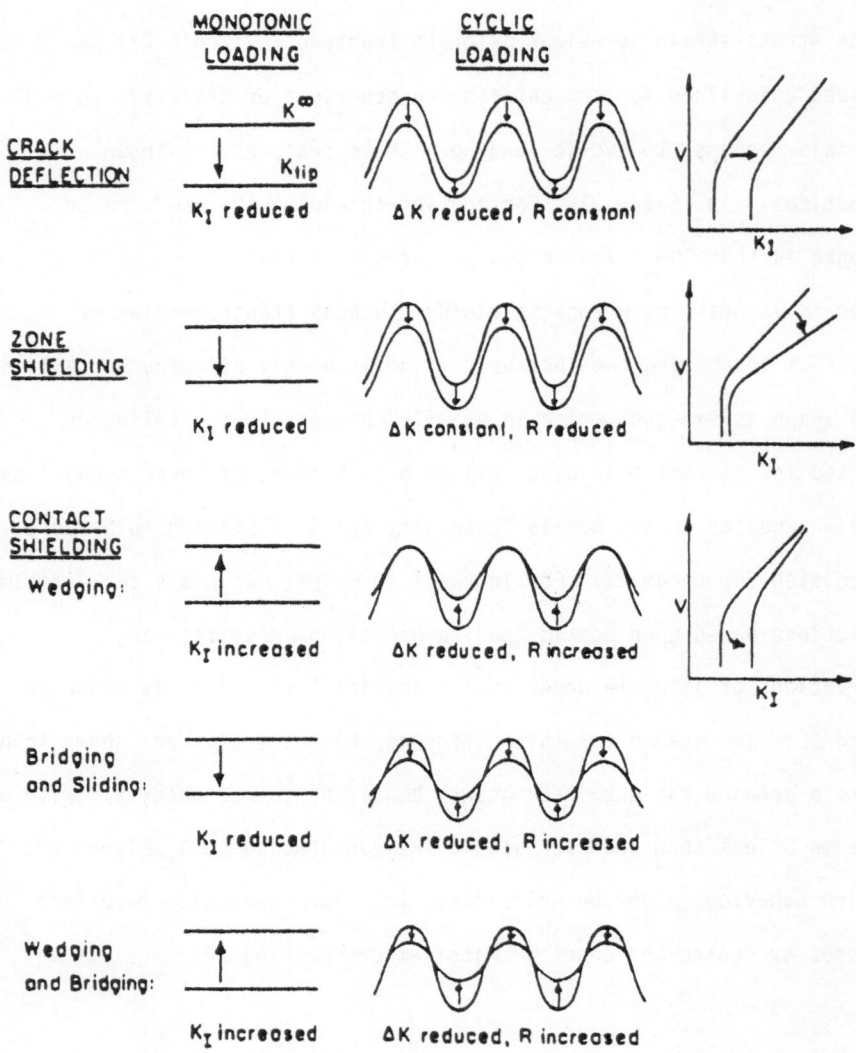

Figure 21. Influence of specific crack tip shielding mechanisms on the driving force, ΔK and crack velocity on cyclic loading.

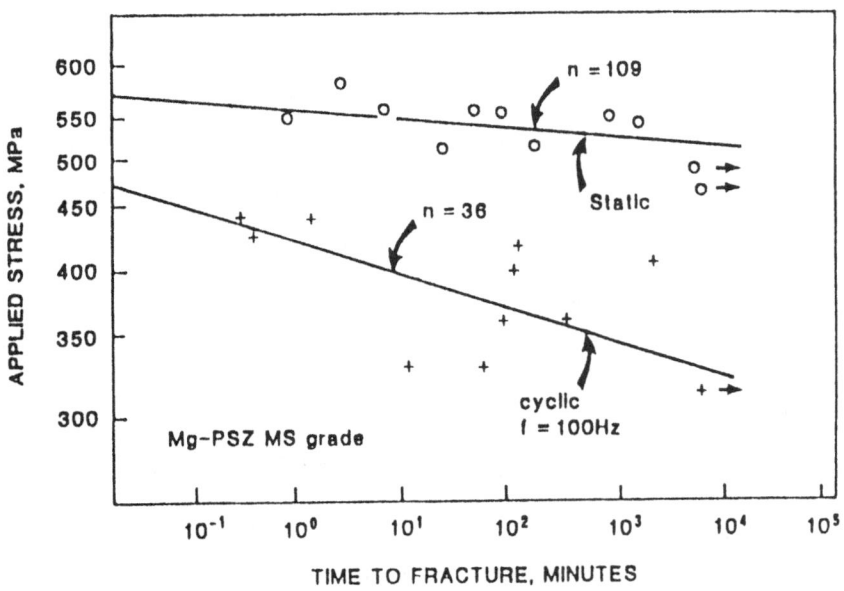

Figure 22. Comparison of lifetimes of Mg–PSZ tested under static and
rotational cyclic flexural loading with applied stress.

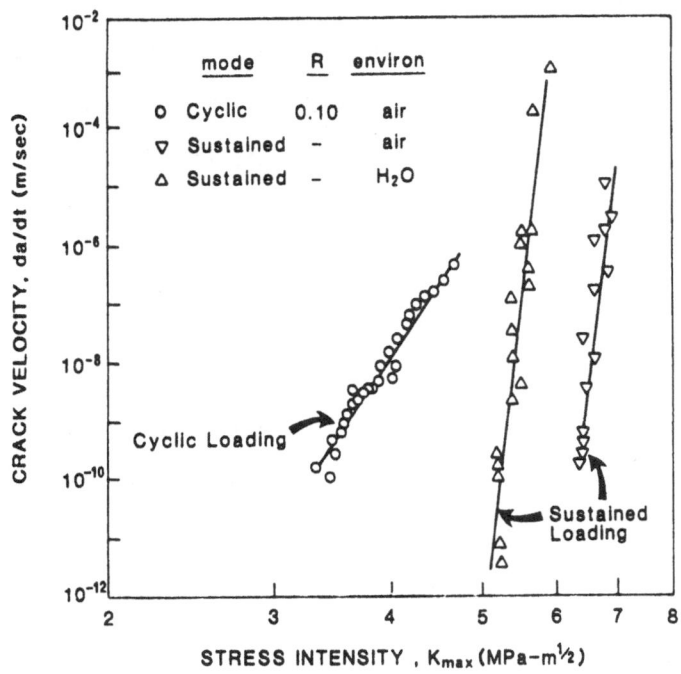

Figure 23. Comparison of the crack propogation rates with applied
stress intensity factor for Mg–PSZ under static and cyclic
testing conditions.

in service. Transport processes such as bulk diffusion or viscous flow of a grain boundary phase are highly temperature dependent. Hence crack nucleation and growth are observed to occur at elevated temperatures in oxide and non-oxide ceramics as a result of creep. In this manner flaw populations are generated in the material with time, leading to mechanical failure of a component [83].

Attempts to understand behaviour of materials at elevated temperatures are still very much at the experimental stage. Quinn [84] has generated a fracture map for silicon nitride at elevated temperatures, this is shown in Figure 24. These observations are for the lifetime of static loaded silicon nitride (MgO doped) that been given a 16N Knoop indentation prior to testing in order to reduce scatter of the results. These observations and others in alumina [85], and silicon carbide [86], suggest that, for short periods of time, failure occurs by crack growth from pre-existing flaws.
At lower stresses and higher temperatures failure is controlled by creep fracture processes. These results have been found to fit empirically derived creep fracture relationships for metals, the Monkman-Grant equation (namely $t_f \epsilon^\alpha = \beta$, where ϵ is the minimum creep rate, β and α constants), or modified such relationships. Hence for a particular material data must be obtained to determine the mechansims of failure. Parametric relationships may then be obtained which provide the basis for structural design.

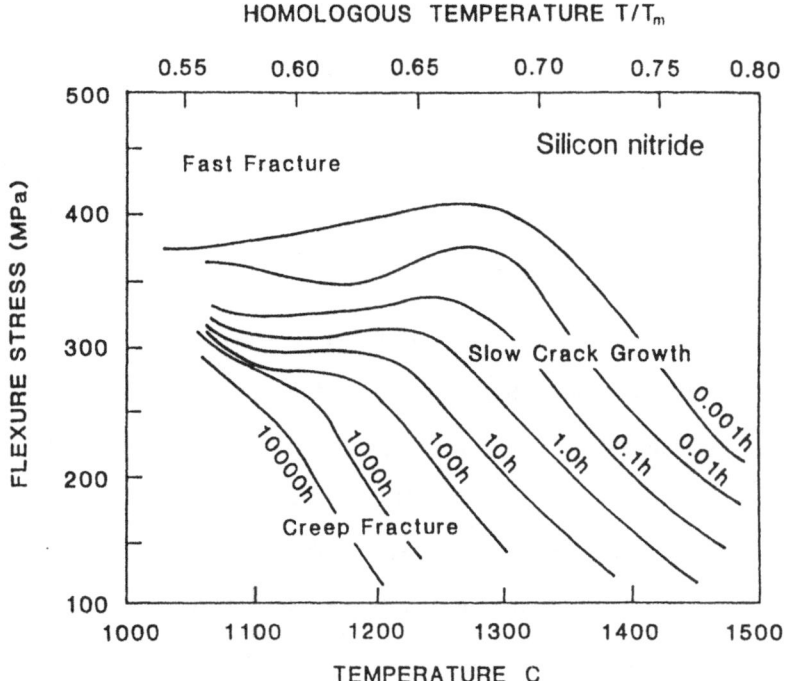

Figure 24.   Fracture – mechanism – lifetime map of silicon nitride at
temperatures where creep becomes significant.

5. FABRICATION DEVELOPMENTS

Associated with recent advances in the toughening mechanisms available to ceramics has been associated developments in the area of ceramic powder production and consolidation. Many of the processes are not novel but equipment and techniques are readily available or sufficiently well understood to be utilized on semi-production/commerical basis.

Until recently most ceramic processing consisted of familiar milling of powders to achieve particle sizes that may be sintered with or without the assistance of sintering aids. Such processing even with the advent of more rapid attrition milling systems was time consuming and introduced many steps where contamination was inadvertantly introduced into powder. A number of processes are now available that completely eliminate such milling/mixing procedures and result in greatly improved purity of powders and performance of products. A listing of such processes (not meant to be exhaustive) includes,

Sol-gel (aero-gel)

Polymer precursors

Melt processing - rapid quenching

Chemicl reaction deposition (plasma, laser)

Chemical vapour deposition/infiltration

Extensive reviews of the progress possible in these areas may be found in two recent MRS conference proceedings and elsewhere [87]. The usual problem with many of these techniques is that the cost of powders produced is much greater than by current production methods, and as mentioned this factor limits penetration of ceramics in the highly cost competitive automotive engine market. Where these alternate processes are making a significant impact is in the area of highly refractory materials such as carbides and borides. Plasma chemical reaction techniques have been used to fabricate highly sinter active powders of titanium diboride with incorporated sintering aid and grain growth inhibition (carbon) into powders typically $0.1 - 0.4\mu m$ diameter. These materials may then be sintered to near theoretical density at temperatures as low as 1700°C – a feat not possible by traditional techniques [88]. her advantages of such an approach to powder fabrication is the it enables homogenous solid solutions of various species and so develop complex precipitate laden highly refractory materials.

Although chemical vapour deposition (CVD) may appear as an expensive route to fabricate monolithic components it does offer some unique advantages. Materials maybe fabricated over a range of compositions at reasonably rapid deposition rates and variable crystallinity. Hirai and colleagues [89] have shown that CVD materials have exceptionally clean grain boundaries that improves the oxidation resistance of silicon nitride materials by at least 200°C over that of sintered materials. Other modifications in processing enable fibre like microstructures to be developed and the density of stacking faults to be modified, both means of increasing the fracture toughness [90].

The recent widescale availability of hot isostatic pressing (HIP) facilities with operating conditions exceeding 2000°C and 200 MPa has been another great means for improvement of the strength of ceramics.

The most spectacular examples of this have been in the area of zirconia toughened ceramics where strengths up to two times that achievable by conventional sintering are possible [91]. However HIP'ing may lead to the undesirable introduction of carbon into the ceramic. This is observed in zirconia ceramics when HIP'ed in argon atmospheres with carbon heating elements. This aspect is discussed elsewhere and at this conference [92]. The most recent vintage HIP facilities provide for oxygen containing atmosphere up to 1500-1600°C. Another advantage of such an approach is that a range of oxides may now be fabricated to near transparency.

With the appreciation that superplasticity is possible in fine grained ceramic materials the opportunity for a range of shaping techniques more usual considered for metallic materials such as forging becomes possible [57].

The future direction in processing and fabrication of ceramics is likely to follow in some instances along similar directions to metals processing. That is, greater emphasis placed upon solid solution/melt processing followed by heat-treatment precipitation processing as currently applied to PSZ materials. Already the literature on TTT curves for the development of various phases and the role of dopants in PSZ ceramic is available [93,94]. The other direction will be the greater emphasis placed upon coating technology to assist in overcoming the problems of high temperature oxidation/erosion/corrosion currently experienced by all non oxide materials. Already such approaches are being utilized in the fabrication of continuous fibre composite structures.

269

6. Conclusions

It is now well established at the research, technology, applications and production that a number of very effective toughening mechanisms are available for ceramics. Unfortunately mosts of these mechanisms are appropriate at lower temperatures as they are intrinsicelly temperature sensitive. The two exceptions would appear to be crack deflection toughening and whisker reinforcement, and even these have not been shown to be applicable in any materials at 1400°C or above.

The concept of crack tip shield provides a framewwork within which to analyse most of the effective toughening mechanisms. As a corrolary it also provides the basis upon which the more severe cyclic fatigue behaviour may be explained or possibly designed for. Current research attention is focussed on this area of fatigue crack growth and it is anticipated as with metallic materials that short cracks will initiate at much lower thershold cyclic stressing levels than longer cracks.

The major new ceramic materials initiatives would appear to be in the design of complex composite materials with a number of incorporated toughening mechanisms. These mechanisms would be anticipated to operate at different length scales from the nanometer to the millimeter range depending upon the size and application of the material. Increasingly novel and radical fabrication routes will be used to develop these complex microstructured composites.

Acknowledgements:

The author wishes to thank Professor S. Somiya for the invitation to Japan to present this paper.

REFERENCES

1. See for example Report by D.W. Lee of Arthur Little Inc. (Nov. 1986) similar estimates have been proposed by Kline, SRI and other market forecasting bodies.

2. R C Garvie and R H J Hannink and R T Pascoe: Nature, Vol 258, 703 (1975)

3. R.O. Ritchie Mat. Sci and Eng in press (1988)

4. Y. Matsuo and S. Kimura J. Ceram. Soc Japan, Int. Edt 96 C-125 (1988).

5. A Kelly: "Strong Solids" Clarendon Press (1966).

6. D Munz et al: Fracture Mechanics of Ceramics Vol 7, p 265 1986, Plenum Press N.Y. Edts. R C Bradt, A G Evans, D P H Hasselman and F F Lange.

7. E. Orowan Rept. Prog. Phys. 12, 185 (194a)

8. Y.W. Mai and B.R. Lawn Ann. Rev. Mater. Sci. 16, 415 (1986).

9. B. R. Lawn and S. Lathabai to be published.

10. M V Swain and L R F Rose: "Advances in Fracture Research" Vol 1, p 473 ICF 6. Edts. S R Valluri et al. Pergamon (1984).

11. M. Sakai and R.C. Bradt. J. Ceram. Soc. Japan 96, 779 (1988).

12. K. Kendall, N. Mc Nalford, S.T. Tan and J.D. Brichall J. Mater. Res. 1, 120 (1986).

13. R.F. Cook and D.R. Clarke Acta Met 36, in press (1988).

14. R.F. Cook, B.R. Lawn and C.J. Fairbanks J. Am. Ceram. Soc. <u>68</u>, 604 (1985).

15. D B Marshall and J E Ritter: Bull. Am. Ceram. Soc. Vol 66, [2] 309 (1987).

17. K.T. Faber and A.G. Evans Acta Metal <u>31</u>, 565 (1983).

18. K.T. Faber and A.G. Evans Acta Metal. <u>31</u>, 577 (1983).

19. F.F. Lange, Bull Am. Ceram. Soc. <u>62</u>, 1369 (1983).

20. S. Hori, H. Kaji, M. Yoshimura and S. Somiya MRS proceedings Vol. 78, 283 (1987).

21. M.V. Swain J. Mater. Sci. <u>16</u>, 151 (1981).

22. B.R. Lawn and D.B. Marshall Phys. Chem. Glasses <u>18</u>, 7 (1977).

23. H. Kirchener, R.E. Walker and D.R. Platts J. Appl Phys. <u>42</u>, 3685 (1971).

24. F.F. Lange J. Am. Ceram. Soc. <u>63</u>, 38 (1980).

25. M.V. Swain J. Mater. Sci. <u>15</u>, 1577 (1988).

26. D. Green J. Am. Ceram. Soc. <u>66</u>, c-178 (1983).

27. D.J. Green J. Am. Ceram. Soc., <u>64</u>, 138 (1981).

28. A.G. Evans and K.T. Faber J. Am. Ceram. Soc. <u>67</u>, 255 (1984).

29. W. Pompe and W. Kreher Adv. in Ceramis Vol, <u>12</u>, 283 (1984).

30. M. Ruhle, A.G. Evans, R.M. McMeeking, P.G. Charalambides and

J.W. Hutchinson Acta Met. <u>35</u>, 2701 (1987).

31. J.W. Hutchinson Report No. Mech - 87 Division of Applied Sciences, Harvard University. Comb. Mass. (1986).

32. P.G. Charalambides and R.M. McMeeking J. Am. Ceram. Soc. <u>71</u>, 465 (1988).

33. E. Lutz, N. Claussen and M.V. Swain unpublished work (1988).

34. D.J. Green, R.H.J. Hannink and M.V. Swain. "Transformation Toughening of Ceramics" CRC in press 1988.

35. D.B. Marshall, A.G. Evans and M. Drory, Fract. Mechs of Ceramics Vol 6. p 289 Plenum N.Y. 1983.

36. B.·Budiansky, J. Hutchinson and J. Lambropulos. Int. J. Solids and Structures <u>19</u>, 337 (1985).

37. R.M. McMeeking and A.G. Evans. J. Am. Ceram. Soc. <u>65</u>, 242 (1982).

38. R.M. McMeeking. J. Am. Ceram. Soc. 69, C–301 (1986).

39. P.F. Becher and M.V. Swain unpublished work 1987.

40. M.V. Swain, R.H.J. Hannink and J. Drennan, Ceramic Microstructures – 86. Mater. Res. soc. Vol. 21, 819 (1987) Plenum Press N.Y.

41. L.R.F. Rose and M.V. Swain Acta Metal <u>36</u>, 955 (1988).

42. P.F. Becher, M.V. Swain and M.T. Ferber J. Mater. Sci, <u>22</u>, 63 (1987).

43. P.F. Becher and M.V. Swain in preparation.

44. R.C. Garvie and M.V. Swain, J. Mater Sci., <u>20</u>, 1193 (1985).

45. M.V. Swain and R.H.J. Hannink Adv. in Ceramic 12, 225 (1983).
46. M.V. Swain and L.R.F. Rose J. Am. Ceram. Soc. 69, 511 (1986).

47. D.B. Marshall J. Am. Ceram. Soc. 69, 173 (1986).

48. D.B. Marshall and M.V. Swain J. Am. Ceram. Soc. 71, 399 (1988).

49. M.V. Swain Acta Metal. 33, 2083 (1985).

50. R.H.J. Hannink and M.V. Swain J. Mater. Sci., 16, 1428 (1981).

51. R.H.J. Hannink and M.V. Swain J. Am. Ceram. Soc., in press 1988.

52. D. Shetty and K. Wu, J. Am. Ceram. Soc. in press.

53. I. Weh Chen and P.E. Reyes Morel J. Am. Ceram. Soc., 69, 181 (1986).

54. I. Weh Chen and P.E. Reyes Morel, Mat. Res. Soc Symp., Proc. Vol. 78, 75 (1987).

55. Adv. in Ceramics Vol. 3 (1981)
        ibid   Vol. 12 (1984)
        ibid   Vol. 24 (1988)
    Am. Ceram. Soc.

56. K. Shobu, J. Watanabe, J. Drennan, R.H.J. Hannink and M.V. Swain Adv. in Ceramics 24, in press 1988.

57. F. Wakai, S. Sakaguchi and Y. Matsuno Adv. in Ceramic Mater 1, 33 (1986).

58. M.V. Swain Nature (Lord) 322, 234 (1986).

59. A.P. Levitt "Whisker Technology" Wiley-Interscience (1970).

60. P.F. Becher and G.C. Wei, J. Am. Ceram. Soc. 67, C-267 (1984).

61. S.T. Buljan, J.G. Baldoni and M.L. Huckakeee, Bull Am. Ceram. So. 66, 347 (1987).

62. A.H. Chokski and J.R. Porter. J. Am. Ceram. Soc. 68, C-144 (1985).

63. T.N. Tiegs and P.F. Becher, Bull. Am. Ceram. Soc. 66, 339 (1987).

64. M.C. Shaw and K.T. Faber Materials Science Res. Vol 21, 929 (1987) Plenum Press N.Y.

65. P.F. Becher, C.H. Hseuh, P. Angelini and T.N. Tiegs, J. Am. Ceram. Soc 71, in press (1988).

66. E. Tani, J. Mat. Sci. Lett. 4, 1454 (1985)

67. R.A.J. Sambell, D.H. Bowen and D.C. Phillips J. Mater, Sci. 7, 663 (1972), 676 (1972).

68. S. Yajima, K. Okamura, J. Hayaski and M. Omoni J. Am. Ceram. Soc. 59, 324 (1976).

69. T. Mah, M.G. Mendiratta, A.P. Katz and K.S. Mazdiyasni Bull. Am. Ceram. Soc. 66, 304 (1987).

70. J.W. Warren, Ceram. Eng. and Sci. Proc. 6, 684 (1985).

71. A.J. Caputo Bull. Am. Ceram Soc. 66, 368 (1987).

72. K. Prewo and J. Brennan. J. Mater. Sci. 15, 463 (1980) 17, 1201 (1982).

73. P.J. Lamicq et al. Bull Am. Ceram. Soc. 65, 336 (1986).

74. M. Gomina Frac. Mech. Ceram. 7, 17 (1986) Plenum Press N.Y.

75. R. Knehans and R. Steinbrech J. Mater. Sci. $\underline{1}$, 327 (1982)

76. P.L. Swanson, C.J. Fairbanks, B.R. Lawn, Y.W. Mai and B.J. Hockey
     J. Am. Ceram. Soc. $\underline{70}$, 279 (1987).

77. M.K. Bannister and M.V. Swain, Materials Science Forum Vol. 34-36
     669 (1986).

78. B.R. Lawn, A.G. Evans and D.B. Marshall J. Am. Ceram. Soc., $\underline{63}$, 574
     (1980).

79. B.R. Lawn Fract. Mechs. Ceram. $\underline{5}$, 1 (1983) Plenum Press N.Y.

80. S.M. Weiderhorn, S.W. Freiman, E.R. Fuller and J. Richter ASTM
     Spec. Publ. 884, p 95 (1984).

81. M.V. Swain and V. Zelizko, Adv. in Ceramics $\underline{24}$, in press (1988).

82. R.H. Dauskardt, W. Yu, and R.O. Ritchie, J. Am. Ceram. Soc. $\underline{70}$, C-
     248 (1987).

83. S.M. Weiderhorn and E.R. Fuller. Mat. Sci. and Eng. $\underline{71}$, 169 (1985).

84. G.D. Quinn ASTM STP 884 p 177 (1984).

85. S.M. Johnson, B.J. Dalgleish and A.G. Evans J. Am. Ceram. Soc. $\underline{67}$,
     759 (1984).

86. S.M. Weiderhorn et al "Tailoring Multiphase and Composite
     Ceramics". p 755 Plenum Press N.Y. (1986).

87. "Better Ceramics Through Chemistry" MRS Symp. Proc. Vol. 32 (1984)
     Vol. 73 (1986).

88. H.R. Baumgartner and R.A. Steiger, J. Am. Ceram. Soc., $\underline{67}$, 207
     (1984).

89. T. Hirai Mat. Sci. Res. <u>17</u>, p 329 (1984) Plenum Press N.Y.

90. K.Niihara, A. Suda and T. Hirai, Ceramic Components for Engines p. 480 (1983) Edt S. Somiya et al KTK. Tokyo.

91. K. Tsukuma et al J. Am. Ceram. Soc. <u>68</u>, C-56 (1985).

92. C.L. Hogg, R.K. Stringer and M.V. Swain J. Am. Ceram. Soc., <u>69</u>, 248 (1986).

93. R.R. Hughan and R.H.J. Hannink, J. Am. Ceram. Soc., <u>69</u>, 529 (1986).

94. C.A. Leach. J. Mater. Sci. Lett. <u>6</u>, 303 (1987).

OPTIMISATION OF Y_2O_3 DOPED ZrO_2 POWDER (Y-TZP)

A.J. Hartshorn[A] and M. Baba[B]

[A] ICI Australia Research Group : Newsom St., Ascot Vale, Australia.

[B] Z-TECH Japan, Central P.O. 411, Tokyo 100-91.

Abstract

Fine sinteractive Y-TZP powder, produced by a conventional wet chemistry process, has been optimised to give a strong, dense and tough ceramic. Very high conductivity has also been achieved.

Introduction

The requirements for optimum mechanical properties[1] of 2 to 3 mole% Y_2O_3 doped Zirconia (Y-TZP) are now clear.

Green bodies made up of uniformly distributed submicron particles are required [2,3] if the necessary high sinteractivity is to be achieved.

For conductivity at elevated temperatures very high purity powders which have a low grain boundary resistivity are required[4].

There is a considerable technological challenge in producing powders which meet these very demanding requirements without recourse to the cost of such techniques as sol gel processing for example.

This paper describes the preparation of such powders which make up the Z-TECH SY-ULTRA range.

The powder is produced by calcination of a chemically coprecipitated intermediate. It is then milled and spray dried. The latter part of the process will be discussed.

The process produces a powder with a very uniform Yttrium distribution. A new method of measuring the uniformity of Yttrium distribution in Zirconia has been developed[5].

Milling

The calcined Y_2O_3 doped Zirconia Powder consists of agglomerates approximately 25 microns in size consisting of tetragonal crystallites. Comminution of this material was investigated initially by ball milling in laboratory test apparatus as follows:

```
Polypropylene Jar  Height 6 cm  Diameter 7 cm
290g 3 mm diameter  Y-TZP Balls
rotation speed 120  r.p.m.
slip volume 33 ml (eg. for 50% wt. slip, 28g powder + 28 ml water plus
dispersant)
milling time  24 hours (unless otherwise stated)
```

The first experiments were carried out without dispersant (see Table 1).

TABLE 1

Slip Concentration (wt.% ZrO_2)	Particle Size (microns) *		
	10%	50%	90%
4	0.26	0.70	2.23
8	0.19	0.43	1.44
12	0.18	0.42	1.33
20	0.21	0.54	1.49
50	0.27	1.66	4.38

* Measured with a Leeds and Northrup MICROTRAC

The smallest particle size was obtained with a 12% wt. slip. No further size reduction was observed after 24 hours.

Next the effect of a dispersant (Dispex A40 Allied Colloids) an ammonium acrylate polymer was investigated. (see Table 2)

TABLE 2

Slip Concentration (wt.% ZrO_2)	Dispex-A40 (g/100g ZrO_2)	Particle Size (microns)		
		10%	50%	90%
30	1.0	0.13	0.33	0.80
	2.0	0.13	0.33	0.82
40	1.0	0.13	0.35	0.84
	2.0	0.13	0.30	0.65
50	0.1	0.25	1.01	3.20
	0.5	0.13	0.39	0.96
	1.0	0.13	0.29	0.65
	2.0	0.13	0.29	0.69
60	1.0	0.13	0.41	1.01
	2.0	0.13	0.37	0.88
70	2.0	0.13	3.28	7.59

The smallest particle size was obtained with a 50% slip and 1% Dispex A40.

The effect of Powder Calcination temperature on milling rate and final particle size is given in Table 3. No significant differences were observed.

TABLE 3

Calcination Temperature (^{o}C)	Particle Size (8 hr milling)*		Particle Size (24 hr Milling)*	
	50%	90%	50%	90%
775	0.53	1.64	0.34	0.73
825	0.54	1.77	0.35	0.63
925	0.54	1.73	0.34	0.66

* Milling Conditions : 50% Slip, 1% Dispex A40.

These experiments were used as a basis for scaling up to a pilot plant.

In production it has been found that an increase in milling rate can be achieved using an attritor mill using similar conditions to the ball milling described above. The properties of typical slips produced by this method are given in Table 4.

TABLE 4

Slip Concentration	60 to 70%
Particle Size	50% < 0.3 , 90% < 0.6 microns
Viscosity	5 to 10 mPa.s (Newtonian)

Spray Drying

A slip dispersed with Dispex A40 was spray dried on a Niro Production Minor spray dryer using an inlet temperature of 300°C and an outlet temperature of 105°C. Two products were collected, the major product, Powder A, with an agglomerate size of 30 microns (see Figure 1, A) and material from the cyclone, Powder B, with a size of 14 microns (see Figure 1, B).

Product Evaluation

Powder A was uniaxially pressed at 20 MPa followed by isostatic pressing at 200 MPa. It was fired at 1500°C. A low sintered density and strength were obtained on the sintered product (Powder A, Table 5).

Figure 1A

Figure 1B

Figure 2A

Figure 2B

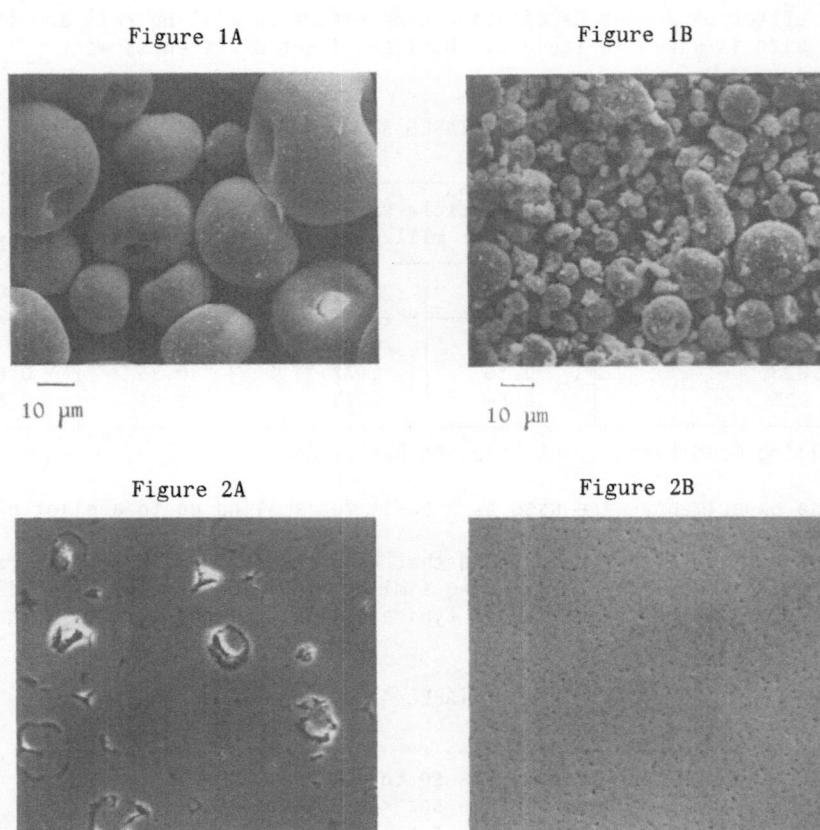

10 µm

10 µm

10 µm

10 µm

Figure 3

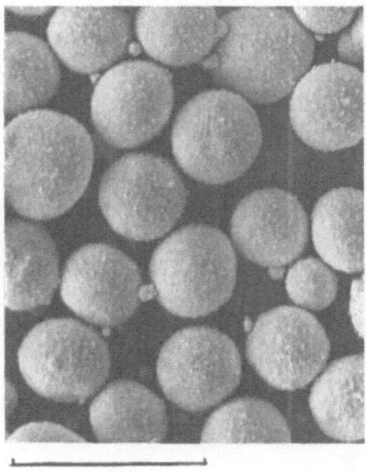

100 µm

TABLE 5

POWDER	MEDIAN SPRAY DRIED AGGLOMERATE SIZE (Microns)	TAP DENSITY (g/ml)	FLOW TIME** (Sec.)		GREEN DENSITY (g/ml)	SINTERED DENSITY (g/ml)	MOR + (MPa)
			3mm	5mm			
A	24	1.89	67	23	3.25	5.94	733±59
B	14	2.00	*	65	3.26	6.04	954±90
SY−ULTRA	40	1.40	80	30	3.10	6.07	1110±151

* No Flow
** Measured on a Hall flowmeter
+ 4-Point Bend

S.E.M. examination of a ground and polished surface of the sintered body clearly shows the toroidal shape of the original spray dried agglomerates (see Figure 2, Powder A).

The fine fraction (Powder B) was evaluated in a similar manner. Considerably higher strength and density was obtained (see Table 5). S.E.M. examination (Figure 2, Powder B) showed fairly uniform fine porosity. This result is consistent with the pores between the spray dried agglomerates in compacts made from Powder B being smaller than in compacts made from Powder A.

Unfortunately reducing the spray dried agglomerate size to that of Powder B is not an acceptable option because the bulk density is low and the flow properties are poor (see Table 5). Such a powder is very difficult to handle in dry pressing.

SY−ULTRA

The particle size, surface area and spray drying conditions for Powder B were optimized at the Pilot Plant scale. The poor flow properties however could not be overcome. Consequently a new product SY−ULTRA has been developed. It consists of spheroidal uniformly dense spray dried particles (see Figure 3). This weakly agglomerated powder deforms readily on pressing to give a uniform green body with high sinteractivity.

Mercury porosimetry shows a narrow pore size distribution in the green body (see Figure 4). A polished surface (see Figure 5) of the sintered body shows only an occasional small pore and excellent strength and density (see Table 5) are obtained.

FIGURE 4.

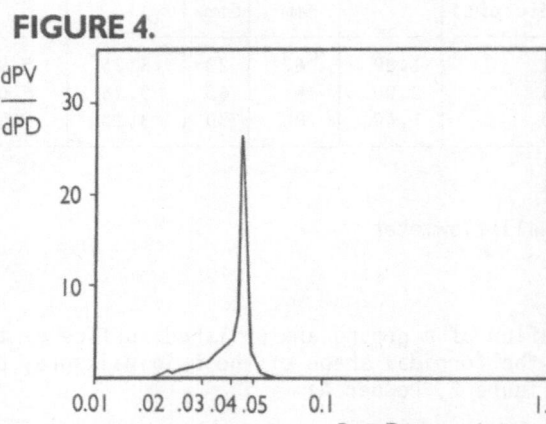

The pore size distribution of SY-ULTRA pressed and heated to 750°C.

Figure 5

10 μm

Conductivity Studies

Yttria doped Zirconia is a solid electrolyte. This property is being used in oxygen sensors and several other applications are under development including, solid oxide fuel cells (SOFC), electrochemical reactors and oxygen pumps.

The resistivity of zirconia can be separated into components due to the lattice (within grains) and across interfaces (eg. grain boundaries). Grain boundary resistivity is generally very high, consequently large grained fully stabilized cubic zirconia has been used to minimise this effect. Recently however it has been demonstrated [6] that very high purity Y-TZP can achieve better conductivity than fully stabilized yttria-zirconia because the grain boundary resistivity has been reduced dramatically.

SY-ULTRA performs particularly well in this application [7].

Figure 6 shows a TEM of sintered SY-ULTRA. No glassy phase can be detected in the grain boundaries.

Acknowledgements

We would like to thank Dr. J. Sellar for the TEM results on SY-ULTRA and other members of the research team for their contributions to this work.

Figure 6

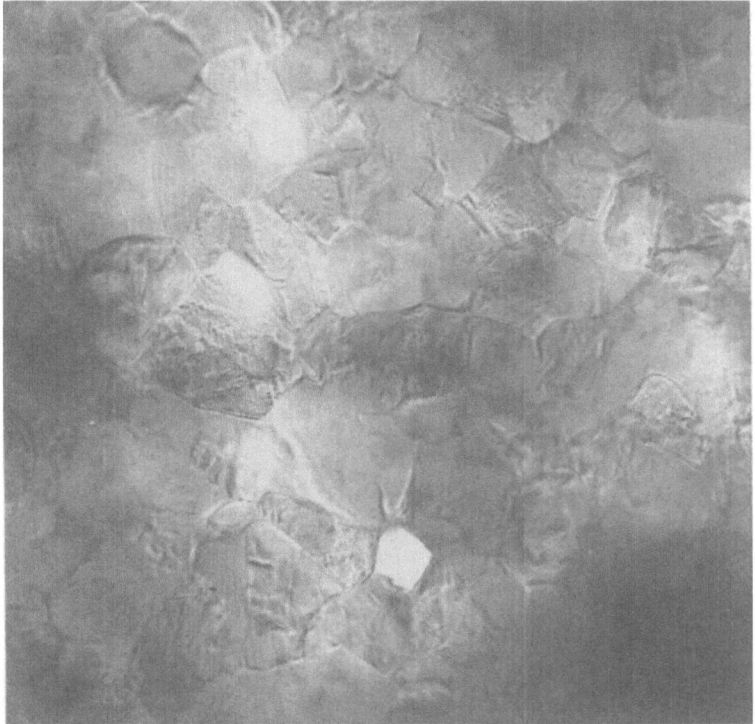

284

References

1   (a)   Nettleship, I. and Stevens, R. : Int.J.
High. Technol. Ceramics 1987,$\underline{3}$,1.

    (b)   Kendall, K. : Materials Forum, 1988,$\underline{11}$,61.

2   van de Graaf, M.A.C.G. and Burggraaf, A.J. : Advances in
Ceramics Vol.12, 744.

3   Pampuch, P. and Haberko, K. : Ceramic Powders (ed. P. Vincenzini)
1983 (Elsevier Amsterdam) 623.

4   Badwal, S.P.S. and Drennan, J. : J. Mater. Sci., 1987 $\underline{22}$, 3231.

5   Hartshorn, A.J., Hill, R.J. and Houchin, M.R. : Mater. Sci. Forum
1988 $\underline{34-36}$, 153.

6   Badwal, S.P.S. and Swain M.V., J. Mater. Sci. Lett. 1985 $\underline{4}$ 487.

7   Murray, M.J. and Badwal, S.P.S. : Mater. Sci. Forum 1988 $\underline{34-36}$,
213.

Thermal Shock Fracture of Zirconia Ceramics

Tsugio Sato, Masayuki Ishitsuka, Tadashi Endo and Masahiko
Shimada
Department of Molecular Chemistry and Engineering, Faculty of
Engineering, Tohoku University, Aoba, Sendai 980, Japan
Haruo Arashi
Research Institute for Scientific Measurements, Tohoku
University, Katahira, Sendai 980, Japan

Abstract

Thermal shock fracture behaviour of various kinds of
zirconia ceramics such as magnesia partially stabilized zirconia
(Mg-PSZ), yttria and ceria-doped tetragonal zirconia polycrystals
(Y-TZP and Ce-TZP), Y-TZP/Al_2O_3 composites and yttria-doped cubic
stabilized zirconia (Y-CSZ) was evaluated together with that of
alumina, mullite, silicon nitride and silicon carbide by
quenching method using water, methyl alcohol and glycerin as
quenching media. Thermal shock fracture of all ceramics was
proceeded by the thermal stress due to convective heat transfer
accompanied by boiling of solvents under the present experimental
conditions. Thermal shock resistance of zirconia based ceramics
increased with increasing the fracture strength, but that of Y-
TZP and Y-TZP/Al_2O_3 composites was anormalously lower than the

predicted value, since the toughening mechanism of zirconia by the stress-induced phase tranformation did not sufficiently function against the thermal stress fracture of Y-TZP based ceramics.

1. Introduction

Since ceramic materials show excellent high temperature fracture stength and chemical stability, they have been leading candidates for high temperature structural applications. Thermal shock resistance of ceramic materials is one of the most important properties, because brittle ceramics are susceptible to catastrophic failure under conditions of thermal stress introduced by the temperature difference (ΔT). Many studies have been carried out to elucidate the basic principles governing the thermal stress fracture of brittle ceramics. The quenching test into liquid media such as water, silicon oil, etc. has been used extensively for characterizing the thermal shock resistance of ceramics [1-3]. It was reported [4] that the observed thermal shock resistance was generally in good agreement to the predicted value by the thermal shock resistance parameter expressed by equation (1).

$$R_O = \sigma_f(1-\nu)/\alpha E \tag{1}$$

where σ_f is the fracture strength, ν is Poisson's ratio, α is the linear thermal expansion coefficient and E is Young's modulus. Since zirconia based ceramics such as Y-TZP show extensively high fracture strength such as 800-2500 MPa [5,6] due to the transformation toughening mechanism, they are expected to show excellent thermal shock resistance. However, thermal shock resistance of zirconia based ceramics were modest [7-11] and the details have not been clarified yet. In the present paper, the thermal stress fracture behaviour of zirconia based ceramics such as Y-TZP, Y-TZP/Al$_2$O$_3$ etc. was evaluated by quenching method in detail.

2. Experimental

Yttria doped tetragonal zirconia powders containing 2 and 3 mol% Y$_2$O$_3$ (2Y-TZP and 3Y-TZP), yttria doped cubic stabilized zirconia powder containing 6 mol% Y$_2$O$_3$ (6Y-CSZ), ceria doped tetragonal ziconia powders containg 8, 12 and 16 mol% CeO$_2$ (8Ce-TZP, 12Ce-TZP and 16Ce-TZP), and 2Y-TZP/Al$_2$O$_3$ powders containing 10, 20 and 40 wt% Al$_2$O$_3$ supplied by Toso Co. were used as starting materials. These powders were isostatically pressed at 200 MPa to form plates (5x30x50 mm) and sintered at 1500°C for 3-10 hr in air. The sintered bodies of 2Y-TZP/Al$_2$O$_3$ were hot isostatically pressed at 1450°C and 150 MPa for 1 hr in Ar gas atmosphere. The sintered body of mullite was fabricated by the procedures described in the previous paper [12]. The sintered bodies of Mg-PSZ, SiC and Si$_3$N$_4$ were supplied by Nilcra Ceramics PTY Ltd., NGK Spark Plug Co., Ltd. and Toshiba Co., respectively. The characteristics of the samples used are summarized in Table 1 [10]. The samples were cut into bars (5x2x15 mm) and polished to parallel mirror like plane. The thermal shock resistance of each specimen was determined by quenching test using water, methyl alcohol and glycerin at 0°C as quenching media. The bending strength of the specimen was determined by 3-point bending test with a cross head speed of 0.5 mm/min and span length of 10 mm. The Raman spectra around the cracks introduced by the thermal stress were recorded using a double monochrometer and photon

Table 1. Characteristics of the Sintered Bodies [10]

Material	α $(\times 10^{-6}/K)$	ν	E (GPa)	k (W/m·K)	σ_{3b} (MPa)	R_O
Al_2O_3	7.4	0.27	393	18.5	300	75
SiC	3.2	0.25	330	91	330	234
Si_3N_4	3.2	0.25	330	25	500	355
Mg-PSZ	10.1	0.23	205	1.8	460	171
6Y-CSZ	9.0	0.25	200	3.5	240	100
3Y-TZP	9.0	0.25	200	3.5	900	292
2Y-TZP	9.0	0.25	200	3.5	1300	542
2Y-TZP/10 vol% Al_2O_3	8.8	0.25	229	3.5*	1720	640
2Y-TZP/20 vol% Al_2O_3	8.6	0.26	254	5.7*	2060	698
2Y-TZP/40 vol% Al_2O_3	8.2	0.26	298	7.8*	2070	627
8Ce-TZP	5.2	0.25	200	3.5*	600	433
12Ce-TZP	10.9	0.25	200	3.5*	425	146
16Ce-TZP	5.8	0.25	200	3.5*	160	103

$R_O = \sigma_{3b}(1-\nu)/\alpha E$ *:Estimated

counting system using 514.5 nm lines of Ar-ion laser as exciting

lighl for the following optical conditions: probe diameter ca. 4

um, objective lens x40, spectrum scan rate 6 cm^{-1}/min and

resolution power 2.5 cm^{-1}.

3. Results and Discussion

    The thermal shock fracture tests of 3Y-TZP and Al_2O_3 were

carried out using water, methyl alcohol and glycerin at $0^{\circ}C$ as

quenching media. The relationship between the 3-point bending

strength of quenched samples and the temperature difference

between the samples and quenching media are shown in Figs. 1 and

2 [10]. A variety of the critical quenching temperature

difference (ΔT_c) above which the fracture strength of the

queched sample degrades such as 275-475$^{\circ}C$ for 3Y-TZP and 200-

350$^{\circ}C$ for Al_2O_3 were observed by using different quenching media.

These results indicated that the magnitude of the thermal stress

greatly depended on the characteristics of the quenching media.

    The thermal stress, S_t, introduced into the ceramic

materials can be described by following equation.

$$S_t = \sigma^* \alpha E \Delta T/(1-\nu) \tag{2}$$

where S_t is the tensilestress introdued into the sample at the

temperature difference of ΔT and σ^* is a nodimentional maximum

thermal stress depending on the quenching conditions. Since S_t

equals the tensile strength, σ_t, of the materials at the critical

quenching temperature difference, ΔT_c, equation (3) can be

deribed from eqation (2).

$$\Delta T_c = \sigma_t(1-\nu)/\sigma^* \alpha E \tag{3}$$

Two kinds of heat transfer mechanism, i.e., conductive heat

transfer and convective heat transfer, have been considered to

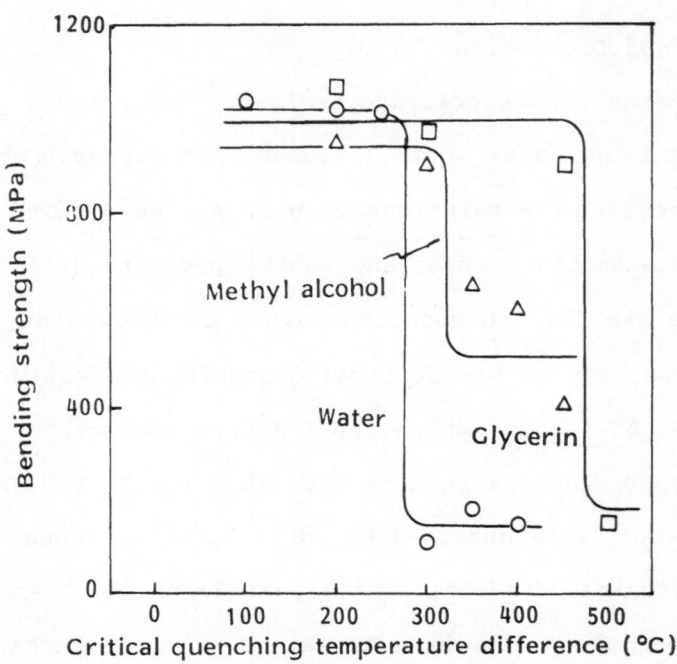

Fig. 1 Relation between 3-point bending strength of 3Y-TZP and quenching temperature difference for various quenching media [10].

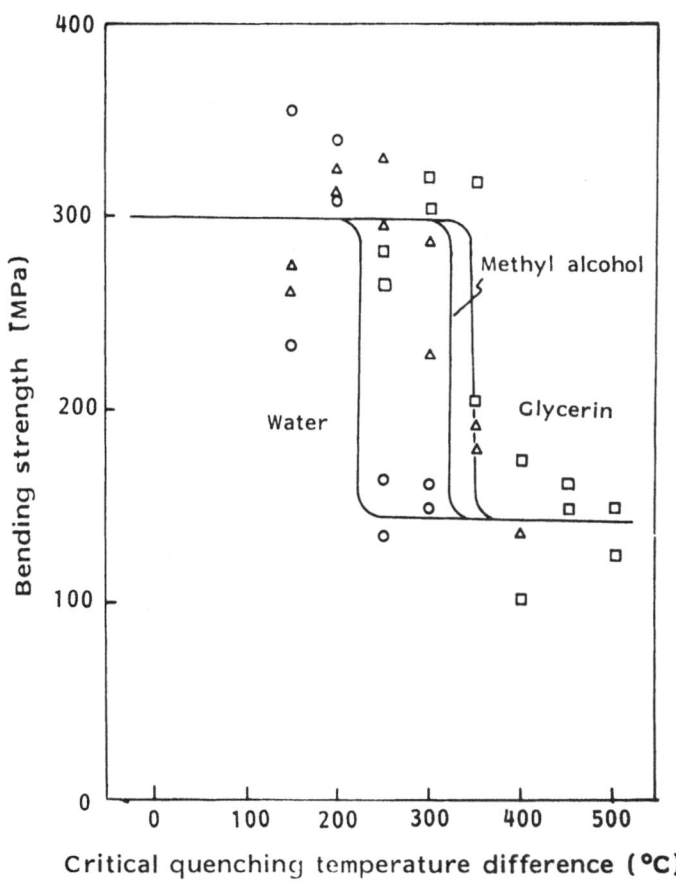

Fig. 2   Relation between 3-point bending strength of Al_2O_3 and quenching temperature difference for various quenching media [10].

interpret the experimental data for thermal shock resistance of ceramics obtained by the quenching test. σ^* for convective heat transfer and conductive heat transfer mechanism can be expressed by equations (4) and (5), respectively [13,14].

$$1/ \sigma_v^* = 1.451(1 + 3.42k_1/rh) \tag{4}$$

$$1/ \sigma_d^* = (k_1\rho_1c_1/k_2\rho_2c_2)^{1/2} + 1 \tag{5}$$

where r is radius for cylinder sample and half thickness for plate sample, h is heat transfer coefficient, k is thermal conductivity, C is specific heat, ρ is density and subscripts 1 and 2 refer to the specimen and quenching medium, respectively.

The nondimentional thermal stresses, σ_v^* and σ_d^* at the quenching test of 3Y-TZP and Al_2O_3 were calculated by equations (4) and (5) and listed in Table 2 [10] together with the thermophysical properties of each solvent, where h was calculated by Holman's equation (6) [15] by assuming natural convection.

$$h = 0.53(Gr \cdot Pr)^{1/4}(k_2/2r) \tag{6}$$

$$Gr = gB_2(T_1-T_2)(2r)^3 \rho_2^2/\mu_2^2, \quad Pr = c_2\mu_2/k_2$$

where Gr and Pr are Grashof number and Prandtl number, respectively, g is the gravitational constant, B is the volumetric thermal expansion coefficient and μ is the viscosity. The quantity (T_1-T_2) was taken to be 300^oC for all calculations. As seen in Table 2, for all quenching media, σ_d^* was greater than σ_v^*. Therefore, it was suspected that the conductive heat transfer caused greater thermal stress than convective heat transfer caused by natural convection. The relationship between ΔT_c and $1/ \sigma_d^*$ for 3Y-TZP and Al_2O_3 is shown in Fig. 3 [10]. The straight lines were calculated by equations (2) and (5), where the value of σ_t was calculated by equation (7) [16] by using the value of Weibull modulus, m, of 10.

Table 2   Charactaristic of the solvents, $\sigma_v{}^*$, $\sigma_d{}^*$ and ΔT_c of 3Y-TZP and Al_2O_3 in various quenching media [10]

	Methyl alcohol	Glycerin	Water
$k(W/m\cdot K)$	0.216	0.285	0.574
$cx10^{-3}(J/kg\ K)$	2.51	2.39	4.20
$\rho \times 10^3 (kg/m^3)$	0.792	1.26	1.00
$\mu \times 10^6 (kg/m\cdot sec)$	0.59	1500	1.79
$Bx10^6 (1/K)$	1700	610	53
$\sigma_v{}^*(3Y-TZP)$	0.167	0.035	0.189
$\sigma_d{}^*(3Y-TZP)$	0.180	0.236	0.341
$\Delta T_c(3Y-TZP)$	350	475	275
$\sigma_v{}^*(Al_2O_3)$	0.017	0.003	0.020
$\sigma_d{}^*(Al_2O_3)$	0.069	0.095	0.150
$\Delta T_c(Al_2O_3)$	350	325	200

r_m = 2 mm

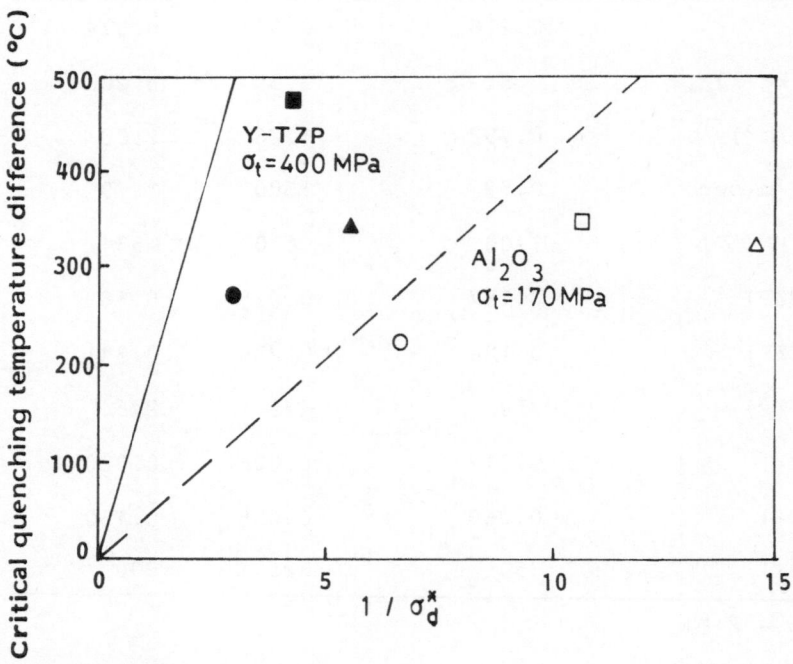

Fig. 3 Relationship between the critical temperature difference
and the reciprocal of the nondimentional maximum stress for
conductive heat transfer [10]. Quenching media- o:Al_2O_3 into
water,   △ :Al_2O_3 into methyl alcohol, □:Al_2O_3 into glycerin,
●:3Y-TZP into water, ▲:3Y-TZP into methyl alcohol, ■:3Y-TZP intc
glycerin.

$$\sigma_t / \sigma_{3b} = [1/2(m+1)^2]^{1/m} \tag{7}$$

As seen in Fig. 3, the experimental values of ΔT_c were significantly smaller than the calculated ones. These results indicated that thermal shock fracture of the samples was not initiated by the thermal stress due to conductive heat transfer, but the convective heat transfer accompanied by boiling of the solvents played an important role for the thermal stress fracture under the present experimental conditions.

ΔT_c of various zirconia based ceramics, Al_2O_3, mullite, SiC and Si_3N_4 determined by quenching test into water at $0°C$ are listed in Table 3 [10] together with 3-point bending strength, σ_{3b}, of the samples. Since σ_{3b} of zirconia ceramics significantly decreased with increasing temperature, the values of σ_{3b} both at room temperature and at $300°C$ were listed in Table 3. From equations (2) and (3), it can be expected that the thermal shock resistance of ceramic materials is improved by increasing the fracture sterngth and decreasing the thermal stress. The relationship between ΔT_c and σ_{3b} at $300°C$ for zirconia ceramics is shown in Fig. 4 [10]. As expected, ΔT_c linearly increased with increasing σ_{3b}, but these plots were divided into two groups. The slope of the straight line for 2Y-TZP, 3Y-TZP and $2Y-TZP/Al_2O_3$ composites was noticeably smaller than that for other zirconia ceramics.

Since k/rh is positive, equation (8) can be derived from

$$\Delta T_c > 1.451 \ \sigma_t(1-\nu)/ \alpha E \tag{8}$$

equations (3) and (4). Therefore, the value of $1.451 \ \sigma_t(1-\nu)/\alpha E$ can be considered as a parameter for thermal shock resistance. The plot of observed ΔT_c versus $1.451 \ \sigma_t(1-\nu)/\alpha E$ is shown in

Table 3 Bending strength and critical temperature difference of various ceramics quenched into water at $0^{\circ}C$ [10]

Material	$\sigma_{3b}(MPa)$ at $25^{\circ}C$	$\sigma_{3b}(MPa)$ at $300^{\circ}C$	$\Delta T_c(^{\circ}C)$
Al_2O_3	300		225
Si_3N_4	500		750
SiC	330		425
Mg-PSZ	460	353	300
6Y-CSZ	240		200
3Y-TZP	900	700	275
2Y-TZP	1300	900	250
2Y-TZP/10vol%Al_2O_3	1720	1200	250
2Y-TZP/20vol%Al_2O_3	2060	1300	300
2Y-TZP/40vol%Al_2O_3	2070	1450	325
8Ce-TZP	600	439*	360
12Ce-TZP	425	311*	290
16Ce-TZP	160	117*	260

* Estimated

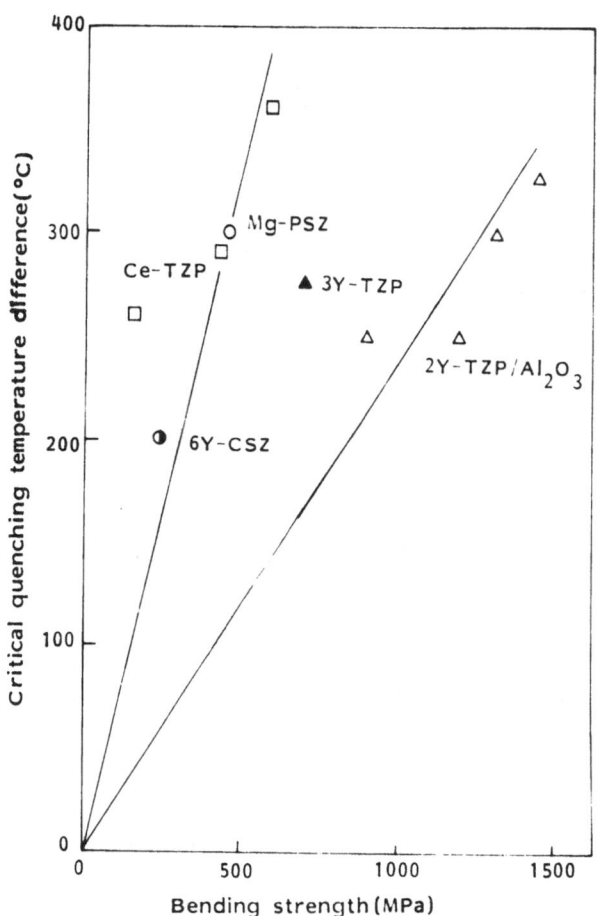

Fig. 4 Relationship between the fracture strength and the critical water quenching temperature difference of 6Y-CSZ, Mg-PSZ, Ce-TZP, 3Y-TZP and 2Y-TZP/Al$_2$O$_3$ composites [10].

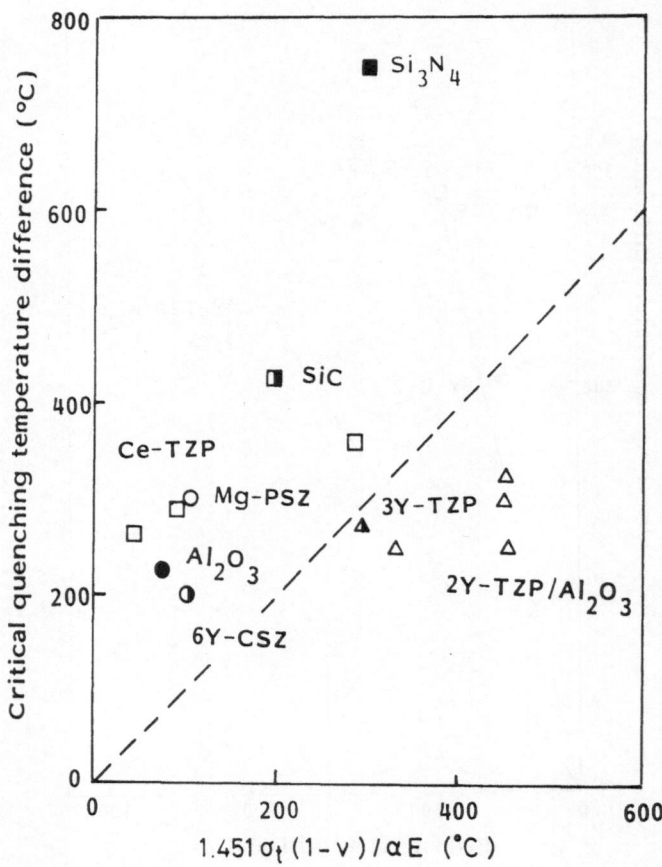

Fig. 5 Relationship between T_c and thermal shock resistance parameter, $1.451 \sigma_t(1-\nu)/\alpha E$ [10].

Fig. 5 [10]. The slope of the dashed line is 1 which coressponds to the case h=∞. The plots of ΔT_c in Ce-TZP, Mg-PSZ, 6Y-CSZ, Al_2O_3, mullite, SiC and Si_3N_4 located above the dashed line as expected by equation (8), but those of Y-TZP based ceramics such as 2Y-TZP, 3Y-TZP and 2Y-TZP/Al_2O_3 composites were below the dashed line. These peculiar results indicated that the cracks which cause strength degradation in Y-TZP based ceramics were propagated by the thermal stress lower than the tensile stress calculated by equation (7) using the value of σ_{3b}.

The heat transfer coefficient, h, calculated by equations (3) and (4) using the results listed in Table 3 for Al_2O_3, mullite, SiC, Si_3N_4, Mg-PSZ, 6Y-CSZ and Ce-TZP is shown in Fig. 6 as a function of ΔT_c. The values of h shown in Fig. 6 were significantly larger than that calculated for natural convection, h=1910 $W/m^2 \cdot K$ (log h = 3.28) and increased with increasing ΔT_c. These results indicated that the heat transfer between the sample and solvent were promoted by boiling of the solvent. From the results shown in Fig. 6, it might be possible to assume the value of h as about 10^4 $W/m^2 \cdot K$ when the thermal shock fracture of Y-TZP and Y-TZP/Al_2O_3 ocurred under the conditions of ΔT_c = 250-350°C. Using the value of h=10^4 $W/m^2 \cdot K$, the thermal fracture stresses of 2Y-TZP, 3Y-TZP, 2Y-TZP/Al_2O_3 composites were calculated as 235-285 MPa by equations (3) and (4). These values were significantly smaller than σ_t=520-1200 MPa determined by equation (7) using the value of σ_{3b}. These peculiar results indicated that the thermal shock fracture of Y-TZP based ceramics proceeded by the different mechanism with the mechanical fracture.

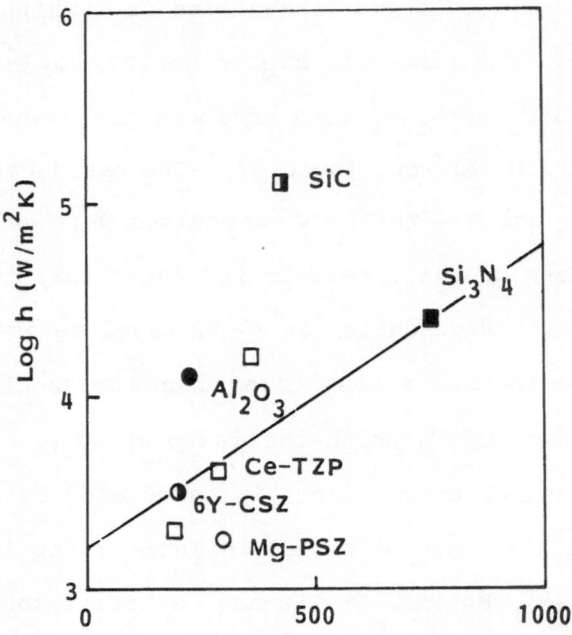

Fig. 6   Calculated values of heat transfer coefficient of variou

ceramics at critical temperature difference [10].

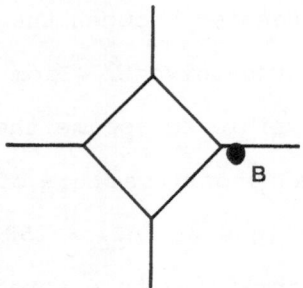

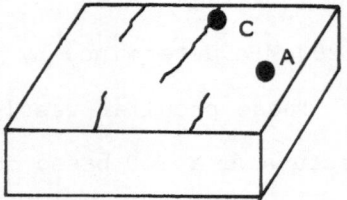

A: Without crack

B: Around crack induced by the
   Vickers indentation

C: Around crack induced by the
   thermal stress

Fig. 7   Schmatic diagram illustrating the crack introduced by

Vickers indentation and thermal stress and the positions of the

laser beam.

The tetragonal to monoclinic phase transformation behavior of Mg-PSZ and 2Y-TZP/40 wt% Al_2O_3 ceramics around cracks introduced by both Vickers indentation and by thermal stress was investigated by using Raman microprobe spectroscopy. The Ar-ion laser beam was focussed arounf the crack. The beam position of the Raman microprobe around the cracks is schematically shown in Fig. 7. Raman microprobe spectra of as-polished surface (A), around the crack introduced by Vickers indentation (B) and by thermal stress (C) in Mg-PSZ and 2Y-TZP/40 wt% Al_2O_3 are shown in Figs. 8 [17] and 9 [17], respectively. Only the peaks at 148 and 264 cm^{-1} corresponding to the tetragonal bands [18] were observed in the Ramn spectra of as-polished surface of both samples. On the other hand, the monoclinic doublet at 181 and 192 cm^{-1} and additional monoclinic band at 224 cm^{-1} were observed together with the tetragonal bands in the Raman spectra around the indentation cracks of both samples. The amount of the monoclinic phase formed by the Vickers indentation evaluated from the Raman intensity ratio of $(I_m^{181}+I_m^{192})/(I_m^{181}+I_m^{192}+I_t^{148}+I_t^{264})$ [19] were in the order of Mg-PSZ > 2Y-TZP/40 wt% Al_2O_3. These results agreed with the facts that Mg-PSZ shows the large transformation zone and excellent fracture toughness about 12 $MPa \cdot m^{1/2}$ [20] and 2Y-TZP/40 wt% Al_2O_3 possesses high fracture strength of 2000 MPa, but moderate transformation zone and fracture toughness about 7 $MPa \cdot m^{1/2}$ [6]. The Raman microprobe apectrum around the crack introduced by thermal stress in Mg-PSZ was almost similar to that around the indentation crack. On the other hand, the Raman spectra around the cracks introduced by thermal stress in 2Y-TZP/40 wt% Al_2O_3 showed much smaller peaks of the monoclinic phase than those around the Vickers indentation. These results

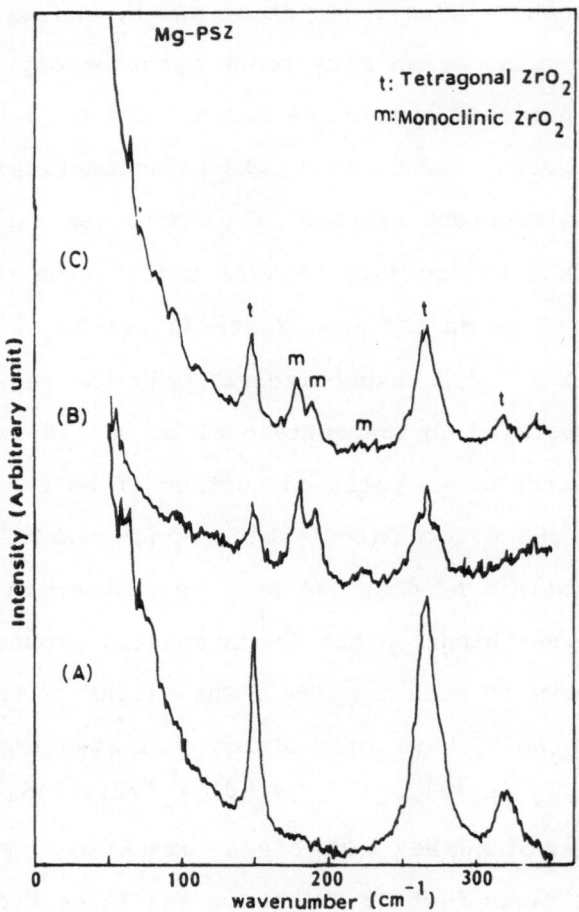

Fig. 8 Raman spectra of Mg-PSZ: (A) As-polished, (B) Around the crack introduced by Vickers indentation, (C) Around the crack introduced by the thermal stress [17].

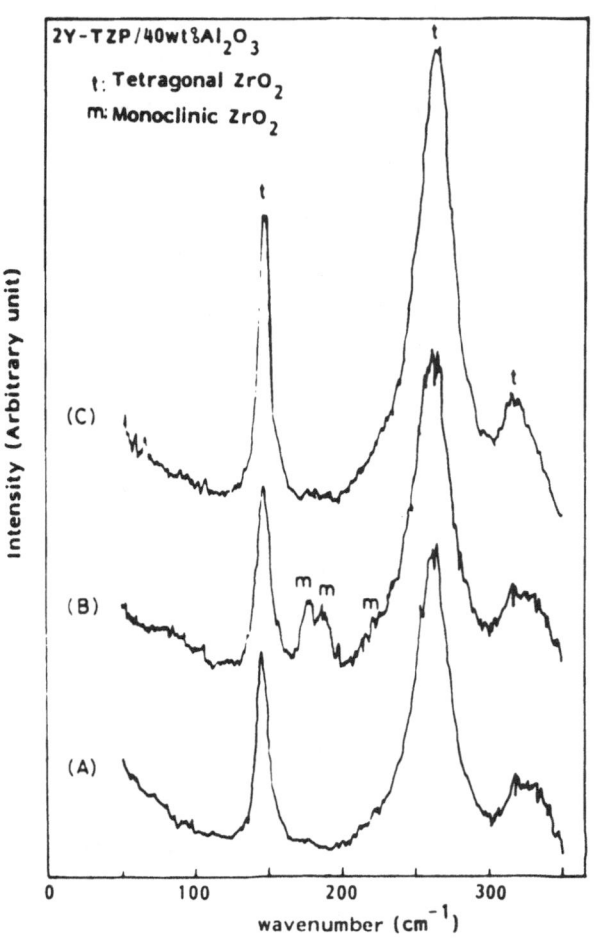

Fig. 9 Raman spectra of 2Y-TZP/40 wt% Al$_2$O$_3$: (A) As-polished, (B) Around the crack introduced by Vickers indentation, (C) Around the crack introduced by the thermal stress [17].

indicated that the tetragonal to monoclinic phase transformation was caused similarly by both the mechanical stress and thermal stress in Mg-PSZ, but the toughening mechanism by the stress-induced phase transformation did not function well against the thermal stress in Y-TZP based ceramics. Consequently the thermal shock fracture of Y-TZP based ceramics is caused by the thermal stress significantly smaller than the original fracture stress.

Acknowlegement

This work was supported in part by a grant-in-aid for Scientific Research of the Ministry of Education and a grant-in-aid for Developmental Scientific Research of the Ministry of Education.

Reference

1. D.P.H. Hasselman, "Strength Behavior of Polycrystalline Alumina Subjected to Thermal Shock," J. Am. Ceram. Soc., 53 490-495 (1970).

2. J.P. Singh, Y. Tree and, D.P.H. Hasselman, "Effect of Bath and Specimen Temperature on the Thermal Stress Resistance of Brittle Ceramics Subjected to Thermal Quenching, " J. Mater. Sci., 16 2109-2118 (1981).

3. M. Oguma. C.J. Fairbanks and, D.P.H. Hasselman, "Thermal Stress Fracture oof Brittle Ceramics by Conductive Heat Transfer in a Liquid Metal Quenching Medium," J. Am. Ceram. Soc., 69 C87-C88 (1986).

4. D. Lewis, "Comparison of Critical ΔT_c Values in Thermal Shock with the R Parameter," J. Am. Ceram. Soc., 63 713-714 (1980).

5. K. Tsukuma and M. Shimada, "Hot Isostatic Pressing of Y_2O_3 Partially Stabilized Zirconia," Am. Ceram. Soc. Bull.,$\underline{64}$ 310-313 (1985).

6. K. Tsukuma, K. Ueda, K. Matsushita and, M. Shimada, "Strength and Fracture Toughness of Isostatically Hot-Pressed Composites of Al_2O_3 and Y_2O_3-Partially-Stabilized ZrO_2," J. Am. Ceram. Soc., $\underline{68}$ C4-C5 (1985).

7. M. V. Swain, "The Effect of Decoompoosition on the Thermal Shock Behavior of Mg-CSZ," J. Mater. Sci. Lett.,$\underline{2}$ 279-282 (1983).

8. T. Sato, T. Fukushima, T. Endo and, M. Shimada, "Thermal Shock Resistance of Yttria-Doped Tetragonal Zirconia Polycrystals: Effect of Solvent in Quenching Test," J. Mater. Sci. Lett.,$\underline{6}$ 1287-1290 (1987).

9. A.H. Heuer and L.H. Schoenlein, "Thermal Shock Resistance of Mg-PSZ," J. Mater. Sci.,$\underline{20}$ 3421-3427 (1985).

10. M. Ishitsuka, T. Sato, T. Endo, and M. Shimada, "Thermal Shock Fracture Behavior of ZrO_2 Based Ceramics," Thermal Shock Fracture Behavior of ZrO_2 Based Ceramics," J. Mater. Sci., in press.

11. M. Anzai, Y. Kimura, H. Fujii, K. Abe and, Y. Kubota, "Thermal Shock Behavior of Y_2O_3-Partially Stabilized Zirconia,"Yogyo-Kyokai-Shi, $\underline{94}$ 577-582 (1986).

12. M. Ishitsuka, T. Sato, T. Endo and M. Shimada, "Sintering and Mechanical Properties of Yttria-Doped Tetragonal ZrO_2 Polycrystal/Mullite Composites," J. Am. Ceram. Soc., $\underline{70}$ C342-C346 (1987).

13. J.P. Singh, J.R. Thomas and D.P.H. Hasselman, "Analysis of Effect of Heat-transfer Variables on Thermal Stress Resistance of Brittle Ceramics Measured by Quenching Experiments," J. Am.

Ceram. Soc.,63 140-144 (1980).

14. H. Hencke, J.R. Thomas, and D.P.H. Hasselman, "Role of
Material Properties in the Thermal-Stress Fracture of Brittle
Ceramics Subjected to Conductive Heat Transfer," J. Am. Cream.
Soc.,67 393-398 (1984).

15. J.P. Holman, Heat Transfer, 3rd. ed., McGraw-Hill, New York
(1981).

16. D.G.S. Davis, "The Statistical Approach to Engineering
Design in Ceramics," Proc. Brit. Ceram. Soc., 22 429-452 (1973).

17. M. Ishitsuka, T. Sato, T. Endo, M. Shimada and H. Arashi,
"Raman Microprobe Spectroscopic Studies on Thermal Shock Fracture
of ZrO_2 Based Ceramics," J. Mater. Sci. Lett., in press.

18. C.M. Phillippi and K.S. Mazdiyasni, "Infrared and Raman
Spectra of Zirconia Polymorphs," J. Am. Ceram. Soc., 54 254-258
(1971).

19. D.R. Clarke and F. Adar, "Measurement of the
Crystallographically Transformed Zone Produced by Fracture in
Ceramics Containing Tetragonal Zirconia," J. Am. Ceram. Soc., 65
284-288 (1982).

20. D.B. Marshall, "Strength Chracteristics of Tranformation-
Toughened Zirconia," J. Am. Ceram. Soc., 69 173-180 (1986).

PHASE TRANSFORMATION CHARACTERISTICS OF NANO-SIZED ZIRCONIA-ALUMINA COMPOSITE POWDER

J.L.Shi, Z.X.Lin and D.S.Yan
Shanghai Inst. of Ceramics, Academia Sinica
865 Chang-ning Rd., Shanghai 200050, P.R.China

ABSTRACT

A series of highly homogeneous ZrO_2-Al_2O_3 composite powder was prepared by the spray-drying and calcination of mixed oxalate solution with any desired component ratio. The homogeneity extent(the separation range between ZrO_2 and Al_2O_3) is below 10Å as determined experimentally. The homogeneous mixing between ZrO_2-Al_2O_3 leads to a postponed crystallization of the tetragonal zirconia phase from amorphous phase at elevated temperatures. The grain growth of zirconia is greatly restrained by Al_2O_3 present in the composite powder as soon as ZrO_2 particles appear as an isolated phase surrounded by Al_2O_3 matrix. For pure zirconia, 50mol%alumina doped and 80mol%alumina doped composite powders calcined at 1000°C-25min the average grain sizes are 100nm, 30nm and 13nm(by XRD-Line Broadening) respectively. In the mean time, with grain growth restraining, the phase transformation of zirconia particles from tetragonal to monoclinic is obviously repressed by the Al_2O_3 surroundings. Higher temperature and larger critical grain size are needed for the phase transformation of composite powders with higher Al_2O_3 contents. For pure zirconia powder calcined at 800°C, the monoclinic content is higher than 95%(its critical grain size of transformation is equal to about 30nm at 600 C); but the monoclinic content of 80mol%alumina doped composite powder is less than 5% at 1300°C and 35% at 1550°C-3h.

INTRODUCTION

The phase transformation characteristics of ZrO_2 materials are frequently stressed as the source of the socalled phase transformation toughening of ceramic materials[1,2]. The phase transformation behavior of zirconia is influenced by its particle size, dopant level and the external environmental condition like the restrains from a second phase[3,6]. The

doping of a solid solute like Y2O3 can stabilize the
tetragonal and cubic phases at room temperature3. For undoped
zirconia powder, if a critical grain size is reached,
tetragonal ZrO2 will be stable at room temperature3,5, but if
the critical size is exceeded, the phase transformation from
tetragonal to monoclinic will take place because of the
decrease of the specific surface energy difference with the
increase of particle size. The constraint from the second
phase like Al2O3, etc. on ZrO2 particles can alter the phase
transformation behavior6,7.

The effect of the second phase on the phase formation is
greatly affected by the state of mixing of zirconia particles
and the second phase, i.e. the powder processing6. In this
paper, a new method of Al2O3-ZrO2 composite powder preparation
was developed resulting in the highly homogeneous mixing of
the two components, and the phase transformation
characteristics of zirconia particles in the composite powders
under the constraint of alumina phase were studied.

MATERIALS AND METHODS

The oxalate solution of aluminum(27.4mg/ml) and
zirconium(20.68mg/ml) were prepared in advance by the solution
of their hydroxide in excess oxalic acid(A.R., Beijing
Chemical Factory). The two kinds of complex solutions were
mixed according to the desired mol content of ZrO2 and Al2O3.
The mixed complex solution was spray-dried in a mini-spray-
dryer(Brinkmann/Buchi 190, Westbury, NY) and followed by the
calcination at various temperatures for 25min to obtain the
composite oxide powders. The decomposed powders are subjected
to the XRD(Rigaku Denki Co.Ltd, RAX-10),TEM(JEOL Co., JEM-
200CX) analysis.

RESULTS

1, Phase Transformation Behavior of the Composite Powder by
XRD
The XRD spectra of the powder containing only ZrO2 by the
decomposition of the spray-dried oxalate precursor at
different temperatures are shown in Fig.1. It can be seen that
at relative low temperature($\leqslant 600°C$), the main phase is
metastable tetragonal, at 600°C monoclinic phase starts to
occur and its content increase with the increase of
temperature. At 800°C, the tetragonal phase no longer exists
in the powder. The phase transformation behavior of the pure
ZrO2 powder is similar to that prepared by the decomposition
of other salts8.

The presence of Al2O3 in the powders results in a
considerable change of the zirconia phase transformation
behavior, which means the increasing stability of the
metastable tetragonal phase. In a previous report by Murase et
al.,the effect of Al2O3 phase on the phase transformation
behavior has not been fully brought to play becasuse of the
inhomogeneous mixing of the two components. As the composite

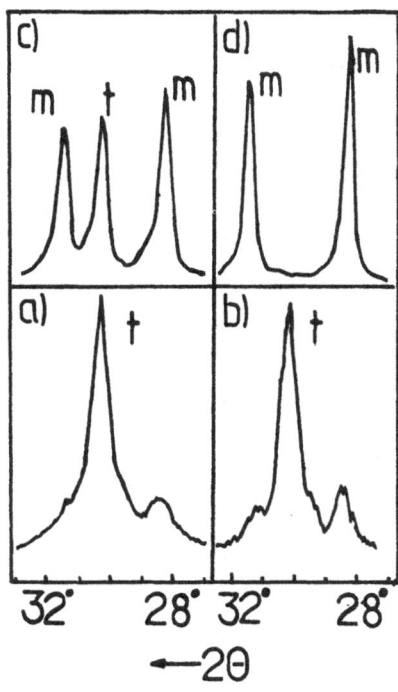 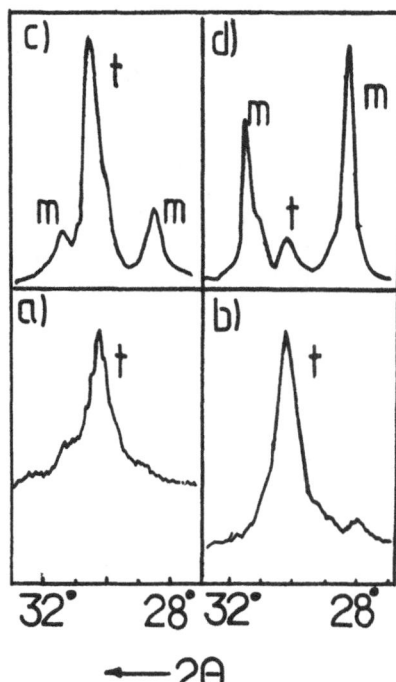

Figure 1. XRD Spectra of Pure ZrO2 Powder at (a)400°C, (b)500°C,
(c)600°C and (d)800°C(Left)

Figure 2. XRD Spectra of 20mol%Alumina-Zirconia Powder at
(a)420°C, (b)650°C, (c)800°C, (d)1000°C(Right).

powders in our experiments were obtained from the thorough
mixing the complex solutions, the presence of Al2O3 affected
ont only the stability of the metastable tetragonal phase,
also the crystallization of the ZrO2 metastable tetragonal
phase. Fig.2 shows the phase composition of a composite powder
containing 20mol%Al2O3. Tetragonal ZrO2 began to crystallize
out at about 420°C, and monoclinic phase occurred at 650°C. At
800°C, the content of monoclinic phase was only 26% while that
of pure ZrO2 powder had reached 100% at 800°C as shown in
Fig.1.

The phase transformation behavior of ZrO2 was further
altered as the Al2O3 content in the composite powder
increases. The metastable tetragonal phase crystallize out at
about 800°C and the tetragonal phase could be maintained up to
1200 ° C, as can be seen in Fig.3, for the composite powder
containing 50mol%Al2O3. So the adding of alumina into the
composite powders postpones the crystallization of the
tetragonal phase in one hand and restrains the phase
transformation from tetragonal to monoclinic.

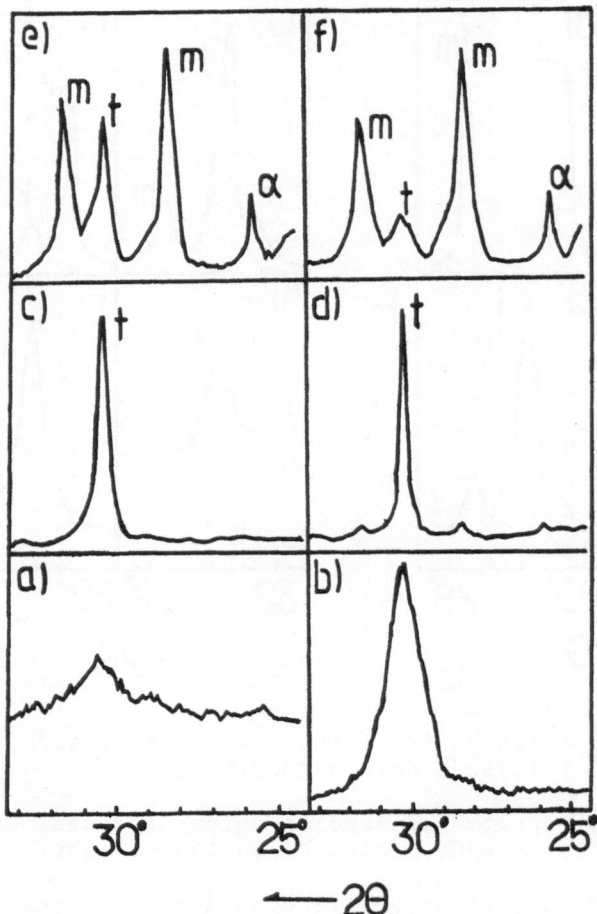

Figure 3. XRD Spectra of 50mol%Alumina-Zirconia Powder at (a)700°C, (b)800°C, (c)1000°C, (d)1200°C, (e)1420°C and (f)1500°C.

Fig.4 shows the temperature dependence of the phase composition of zirconia in the composite powders with different alumina contents. The effect of alumina content on the phase transformation behavior(expressed as the tetragonal content in the figure) of zirconia are clearly illustrated from the figure. For example, in the composite powder with 80mol%alumina, the tetragonal content still remains at 65% when calcined at 1550°C for 3h, and it reaches 76% for the composite powder containing 90mol%alumina at the same condition. Fig.5 is the XRD spectra of 80mol%alumina containing composite powder, the phase transformation behavior of zirconia in the composite powder is evidently shown. In addition, the diffraction peaks of a alumina can be seen in Fig.5 at 1000°C and beyond. The diffraction strength of alumina

is relatively low in relation to its content, implying that
the X-ray diffraction ability of the alumina crystal line

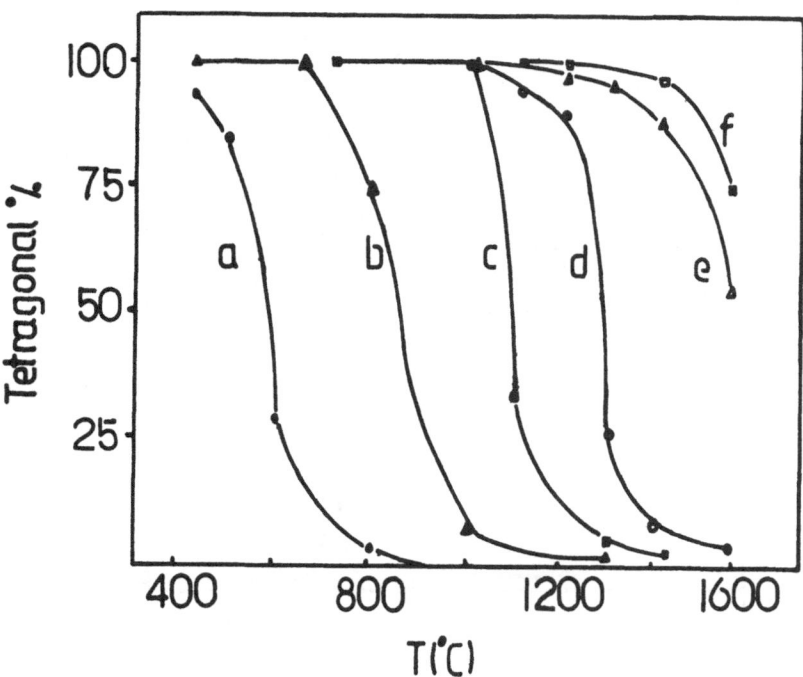

Figure 4.   Content of Tetragonal Zirconia vs Alumina content,
            (a)Pure Zirconia,   (b)20mol%Alumina,   (c)35mol%,
            (d)50mol%, (e)80mol%, (f)90mol%.

faces(α-Al2O3) is much lower than that of zirconia(this
inference is confirmed by the standard Al2O3-ZrO2 samples
deliberately compounded).

2, Powder Morphology and the Phase Transformation
Particle morphologies of the composite powder containing
50mol%alumina at various temperatures are shown in Fig.6-9. At
500 °C, the spherical particles are very homogeneous and no
contrast can be seen(figure 6(a)). The electron
diffraction(Fig.6(b)) analysis shows the amorphous diffraction
pattern, the same with the XRD analysis. Besides, the EDS
analysis shows the mixed patterns of the characteristic X-ray
strength of Zr and Al(Fig.6(c)), showing that the co-existence
of alumina and zirconia in the diffraction range.

        The particle morphology of 800°C calcined powder shown
in Fig.7 illustrates the occurence of contrast within the
spherical particles suggesting component separation,
corresponding to the crystallization of the tetragonal
phase as identified by XRD(Fig.3) and confirmed with the ED
analysis in Fig.7(b). Calcination at higher temperature of

1000 °C only leads to a larger component separation range, as shown in Fig.8, but does not change the phase structure of zirconia and the spherical shape of the particles

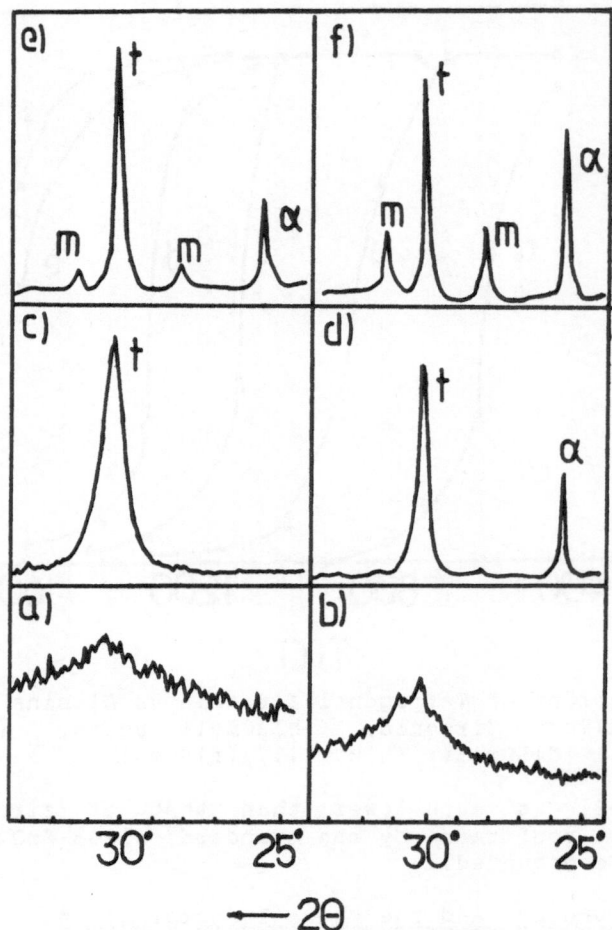

Figure 5. XRD Spectra of 80mol%Alumina-Zirconia Powder at(a)700C, (b)800 C, (c)1000C, (d)1200C, (e)1420C and (f)1550C

resulted from solution spray-drying.

Only about 10% tetragonal phase is left when the composite powder is calcined at 1350°C as shown in Fig.4. The phase transformation is accompanied by a distinct change of the particle morphology as illustrated by Fig.9---the previous existed spherical particles are no longer present in the composite powders. The ED analysis shows that the diffraction patterns of monoclinic ZrO2 and Al2O3 phase taken from regions A and B(Fig.9(b)) respectively. EDS analysis(Fig.9(c)) again shows that the region A and B are ZrO and Al O as
                                      2   2 3

illustrated. Since region A is darker in contrast than region B, so it is referred that ziconia component appears as isolated particle-like phase with darker contrast and alumina tends to be the continuous phase with less darker

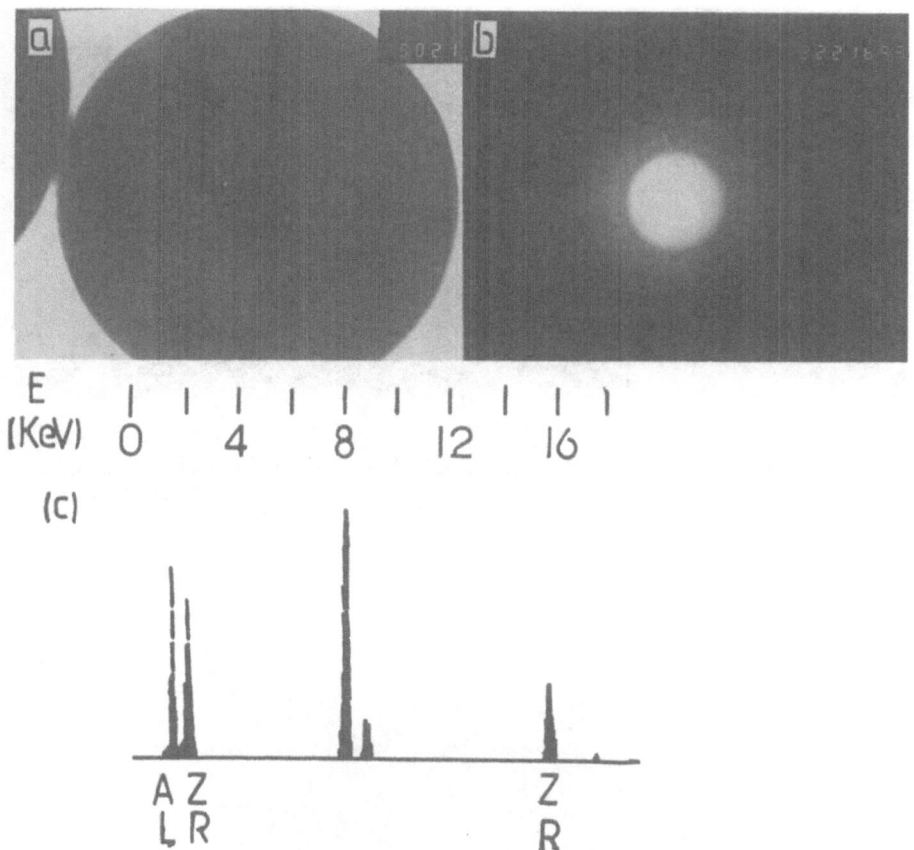

Figure 6.   Powder Characteristics of 50mol%Alumina-Zirconia at 500°C,   (a)Particle Morphology,   (b)ED Pattern and (c)EDSspectrum

contrast(ref.Fig.7 and 8). The difference in contrast between alumina and zirconia is thought to be due to their different densities.

The phase transformation behavior of zirconia is largely dependent on the alumina content in the composite powders. Fig.10 shows the particle morphology of the composite powder containing 20mol%alumina which was calcined at 800° C. It can be seen that the spherical particles are broken at 800° C and zirconia particles appeared very obviously(compared to Fig.7) and ED pattern shows that the particles arepolycrystalline(Fig.10(b)). In contrast, 80mol% alumina added composite powder shows rather different particle morphology even at 1350 C, as can be seen in Fig.11. The

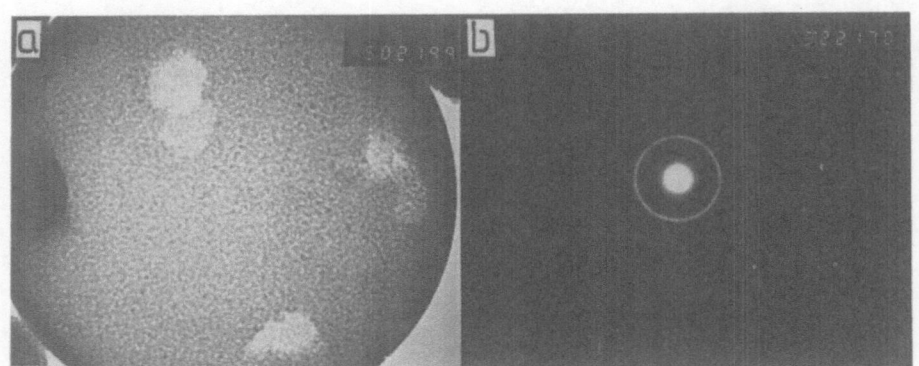

Figure 7.   Powder Characteristics of 50mol%Alumina-Zirconia at 800°C, (a)Particle Morphology and (b)ED Pattern

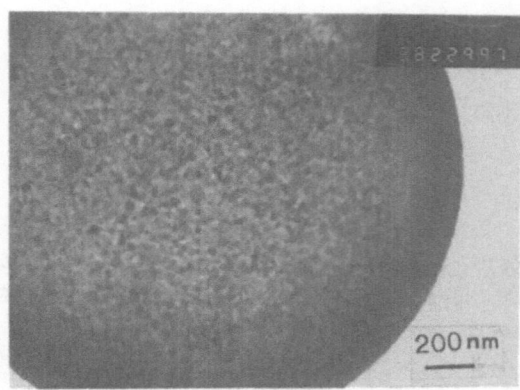

Figure 8.   Particle Morphology of 50mol%Alumina-Zirconia at 1000°C

spherically shaped particles are still maintained even at 1350° C and isolated zirconia particles(darker areas in contrast) distribute homogeneously in the alumina surroundings. Such zirconia particles are tetragonal according to Fig.4.

315

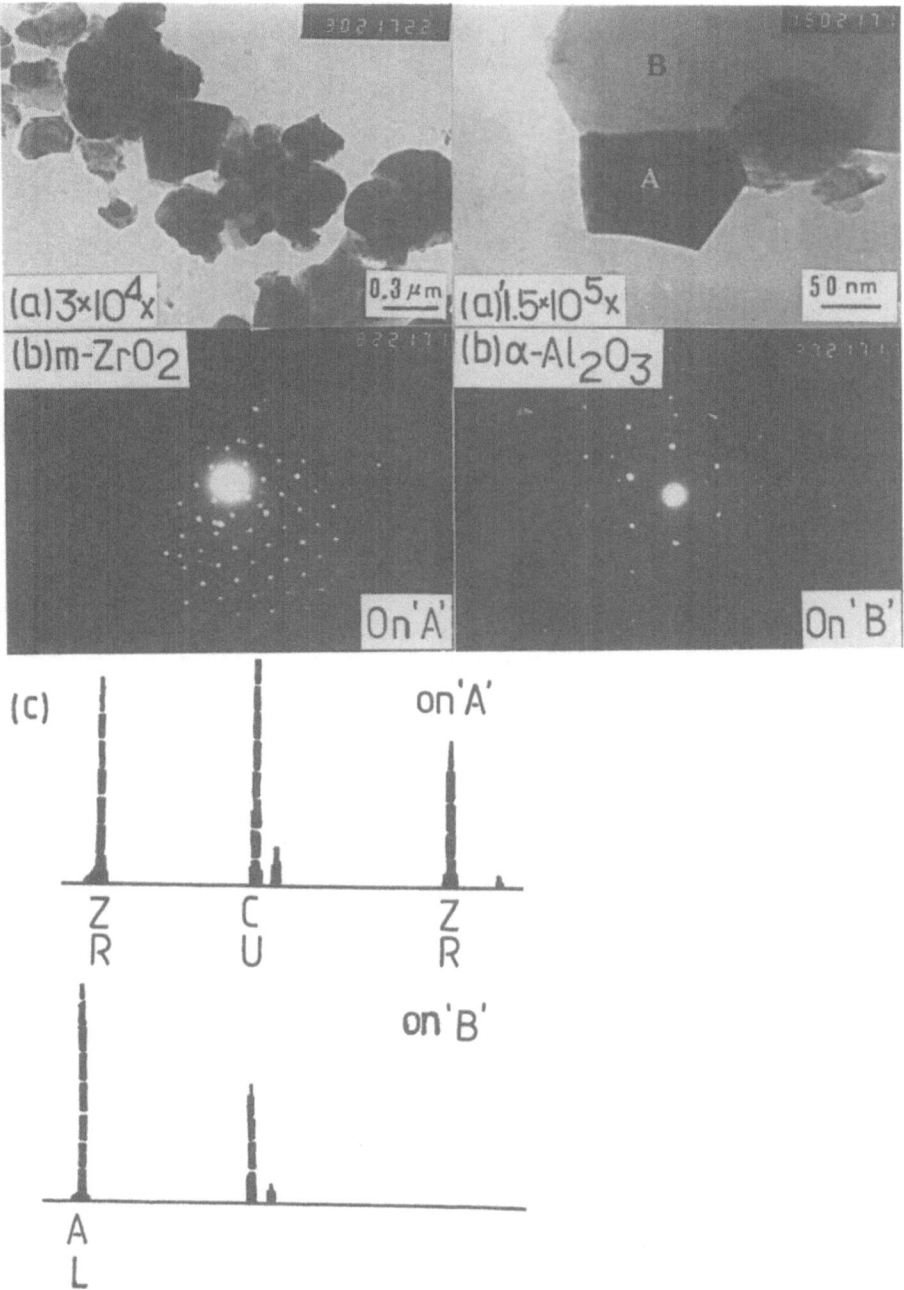

Figure 9. 50mol%Alumina-Zirconia Powder at 1350°C,(a)Particle
Morphologies, (b)ED Patterns and(c)EDSspectra on
Region A and B in (a)'

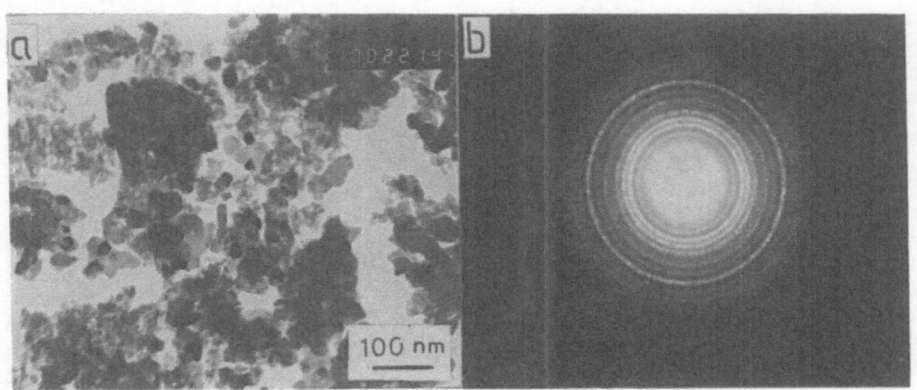

Figure 10. Powder Morphology(a) and ED Pattern(b) of 20mol%Alumina-Zirconia at 800°C

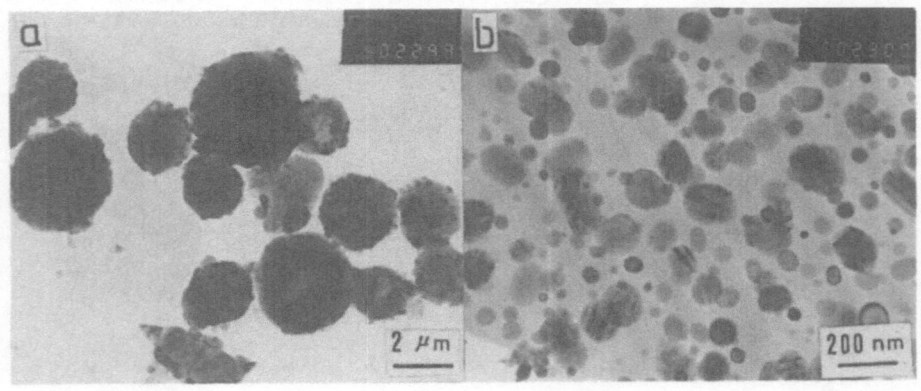

Figure 11. Powder Morphology of 80mol%Alumina-Zirconia at 1350°C, (a)5000× and (b)50000×

DISCUSSION

1, The Initial Mixing Homogeneity and the Crystallization of Tetragonal Zirconia

Tetragonal zirconia crystallized at less than 400°C(Fig.1) in pure zirconia powder. The XRD Analysis shows that the composite powders tend to crystallize at higher temperatures when alumina was added into the powders, that is, the composite powders are amorphous at temperatures higher than

the crystallization temperature of pure zirconia. Fig.12 shows
the crystallization temperature of tetragonal zirconia in the
composite powders vs alumina content(mol). The presence of
alumina in the powders apparently raised the crystallization
temperatures of tetragonal zirconia, The amorphous phase
dominating in the composite powders containing alumina at
higher temperature reffects the homogeneous mixing between
alumina and zirconia. As the diffraction patterns of the
amorphous powders always show one or more very scattered

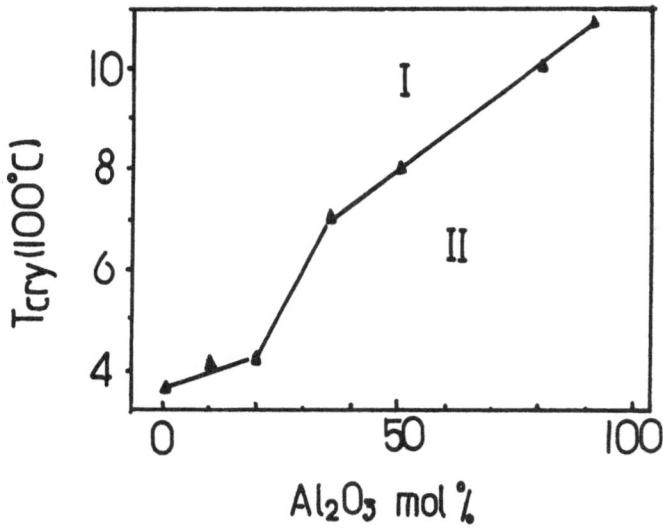

Figure 12.   Crystallization Temperature of Tetragonal Zirconia
vs Alumina Content in Composite Powders

"peak", a "grain size" of about 20nm or so is estimated with
the peak breadth at half height from Fig.3. Calcination at
lower temperature would lead to wider "peaks" and smaller
estimated size.

     A resolution of 10Å for the electron microscope(JEM-
200CX) is easy to obtain. The TEM picture of some amorphous
composite powders(e.g., Fig.6) shows no visible details by
naked eyes(the possible size of the details in the figure is
less than the resolution range of naked eyes--about 0.1mm). So
the inhomogeneity range(the separation range of components)
is smaller than 0.1mm/magnification. As the magnification in
Fig.6(a) is 100000x, thus the inhomogeneity range in the
amorphous composite powders is less than 10Å and at lower
temperature the component separation range will be smaller.
For other composite powders at the amorphous state, their
inhomogeneity ranges are equally smaller than 10Å.

     The homogeneous mixing between alumina and zirconia in
the composite powders explains the postponed crystallization
of tetragonal zirconia, which is so finely dispersed. Powders
with higher alumina content need higher calcination

318

temperature at which the ionic species of zirconia molecules
are active enough to diffuse beyond the impedance of alumina
and form a zirconia particle with a critical size for its
crystallization.

2, <u>Relation betweern Restraint, Particle Size and Phase
Transformation from Tetragonal to Monoclinic</u>
The homogeneous mixing between alumina and zirconia inhibits
the formation of crystallized zirconia particles in the

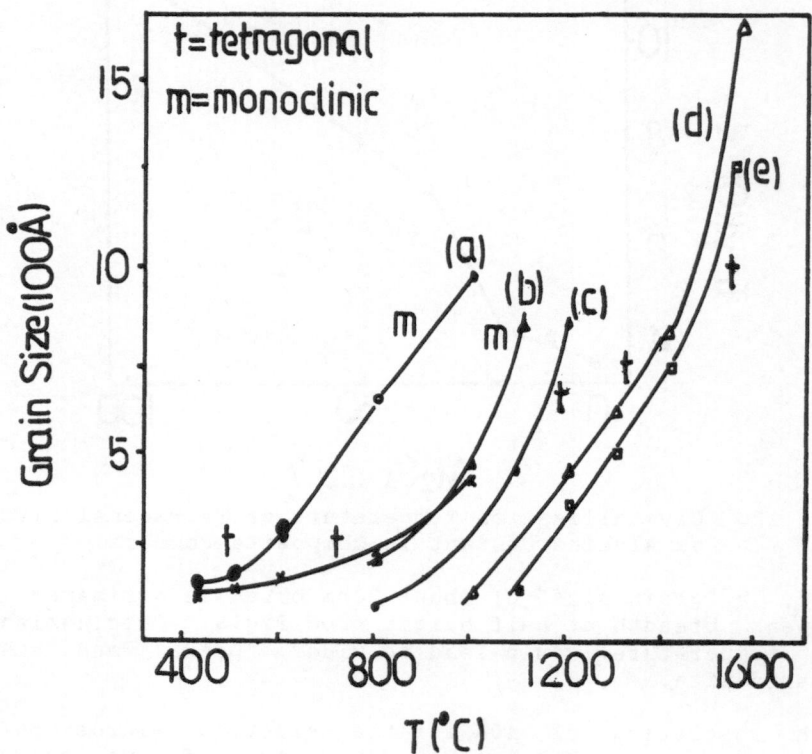

Figure 13. The Grain Growth of Tetragonal Zirconia in the
Composite Powder, (a)10mol%Alumina, (b)20mol%,
(c)50mol%, (d)80mol% and (e)90mol%

surroundings of the continuous alumina phase. The surroundings
no doubt will exert restraints(compresive force) on the
zirconia particles as the particles grow. Such an effect is
clearly shown in Fig.13.

It can be found from Fig.4 and 13, the restraint of
alumina on zirconia particles inhibits the grain growth and
increases the grain size at which the phase transformation
from tetragonal to monoclinic may take place. With more
alumina in the composite powder, the zirconia particles will

be more highly dispersed,i.e. having longer separation
distance. As discussed in the above point, higher temperature
would be needed for the grain growth of zirconia particles
when a stronger driving force for diffusion could be
prevailing. At the same time, the alumina matrix material will
be more densely sintered. It is known that the the phase
transformation of t-ZrO2 to m-ZrO2 is accompanied by 3-5%
linear expansion. Therefore, for composite powders with higher
alumina content, tetragonal zirconia will be retained stable
under such conditions of restraint. Furthermore, a larger
grain(particle) size will be needed to allow such a phase
transformation to occur. So the alumina component in the
composite powders inhibits the grain growth of tetragonal
zirconia and raise the critical size for the phase
transfromation.

CONCLUSIONS

In the initial homogeneously mixed composite Al2O3-ZrO2
powders, the crystallization temperature of ZrO2 is increased
by the amount of alumina present and its restraining effect,
The grain growth of zirconia in the composite powders is also
inhibited to different degrees by the alumina restraint
obviously. The phase transformation from tetragonal to
monoclinic in the composite powder is retarded by the
increasing of the critical size for the phase transformation
to ouuur and the inhibiting effect on the grain growth of
zirconia particles.

REFERENCES

1, R.C.Garvie, R.H.Hannink and R.T.Pascoe, "Ceramic
Steel?", Nature(London), 258(2), 703(1975).

2, A.H.Heuer and L.W.Hobbs, Edited, "Science and Technology of
Zirconia", The American Ceramic Socity, Inc., (1981).

3, C.Garvie, "Stability of the Tetragonal Structure in
Zirconia Microcrystals", J.Phys.Chem., 82(2),
218(1978).

4, V.S,Stubican and J.R.Hellmann, "Phase Equilibria in some
Zirconia Systems", in "Science and Technology of
Zirconia", ed, A.H.Heuer and L,W,Hobbs, The
American Ceramic Society, Inc., 25(1981).

5, M.Kagawa, M.Kikucki, Y.Syono and J.Nagae, "Stability of
Ultrafine Tetragonal ZrO2 Coprecipitated with Al2O3 by
Spray-ICP Techniques", J.Am.Ceram.Soc., 66(11), 751(1983).

6, Y.Murase, E.Kato and K.Dlaumom, "Stability of ZrO2 Phase in
Ultrafine ZrO -Al O Mixtures", ibid., 69(2), (1986)83.
           2   2 3
7, H.Yoshimatsu, H.Kawasaki and A.Osaka, "Stability of
Tetragonal ZrO_2 Phase in Al_2O_3 Prepared from Zr-Al

Organometallic   Compounds",   J.Mater.Sci.,   23,
332(1988).

8, M.I.Osendi, J.S.Moya, C.L.Serna and J.Soria, "Metastability
of   Tetragonal Zirconia Powders", J.Am.Ceram.Soc., 68(3),
135(1985).

Strength Evaluation of Y_2O_3-stabilized Tetragonal ZrO_2 Polycrystals with Scanning Laser Acoustic Microscopy

Manabu Oishi, and Kenichi Noguchi

Toray Research Center, Inc., Otsu, Shiga 520, Japan

Takaki Masaki, and Sadao Nakayama

Toray Industries, Inc., Otsu, Shiga 520, Japan

ABSTRACT

A scanning laser acoustic microscope, SLAM, was applied to measure an ultrasonic attenuation coefficient and detect the internal defects of Y_2O_3-stabilized tetragonal zirconia polycrystals(Y-TZP). Bend and tensile strength of the Y-TZP materials decreased with the increase of the attenuation coefficient and the number of macro-defects which were detected by SLAM, respectively, although a claer relation between bulk density and strength of Y-TZP materials was not given. SLAM can be the useful technique for nondestructive evaluation of the strength of Y-TZP .

INTRODUCTION

Nondestructive inspection using the ultrasonic wave is drawing much attention[1-5]. Quantitative ultrasonic measurements give the information on microstructure of materials[6,7]. A scanning laser acoustic microscope, SLAM, using the high-frequency ultrasonic wave through the material is the useful instrument to detect the internal defects and characterize the internal microstructure of ceramic materials. In this study, the ultrasonic attenuation coefficient of Y-TZP materials is measured and also macro-defects are detected by SLAM. Relation between the bulk density and the bend or tensile strength, the ultrasonic attenuation coefficient and the bend strength, number of the macro-defects detected by SLAM and the tensile strength of the materials is investigated, respectively.

EXPERIMENTAL PROCEDURE

Y-TZP materials examined are shown in Table I. Powders of materials A, B, and C, and materials D, E, and F were prepared by thermal decomposition[8] and hydrolysis[9], respectively. Materials A, C, D, and F were pressurelessly sintered at 1450° C for 2h. Materials B and E were presintered at 1400° C for 2h and then hot-isostatically sintered at 1400° C for 1.5h under 200MPa argon gas pressure. Sintered materials were cut into specimens by a diamond saw and were ground with a diamond wheel. The rectangular specimens for tensile test were inspected by 100 MHz SLAM to characterize the distribution of macro-defects in the specimens, and then fractured with 0.5 mm/min cross-head speed[10]. Ultrasonic attenuation coefficient of the materials was measured by 100 MHz SLAM[11]. The detail of principle of SLAM is discribed in Ref. 12. Three-point bend test for the specimens with dimensions of 3 by 4 by 36 mm was conducted under the condition of 30 mm span length with 0.5 mm/min cross-head speed. The bulk density of the material was measured by Archimedes' method.

RESULTS AND DISCUSSION

The average values of the bulk density, ultrasonic

Table I. Y-TZP materials examined in this study

Material	Powder Process*	Composition of Y_2O_3/mol%	Sintering Method**	Sintering Temp. and Time
A	TD(T-1)	2.75	PS	1400° C, 2h
B	TD(T-1)	2.75	HIP	1400° C, 2h/1400° C, 2h
C	TD(T-2)	2.75	PS	1400° C, 2h
D	H (H-1)	3.00	PS	1400° C, 2h
E	H (H-1)	3.00	HIP	1400° C, 2h/1400° C, 2h
F	H (H-2)	3.00	PS	1400° C, 2h

*)TD:Thermal Decomposition   H:Hydrolysis
 T-1, T-2, H-1, H-2:Different Lot Number
**)PS:Presureless Sintering   HIP:Hot Isostatic Pressing

Table II. Average values of the bulk density, ultrasonic attenuation coefficient, number of macro-defects detected by SLAM, tensile strength, and bend strength of the Y-TZP materials

Material	Bulk Density g/cm^3	Attenuation Coefficient dB/mm	Average number of macro-defects detected by SLAM	Strength Tensile MPa	Bend MPa
A	5.99	1.9	---*	432	707
B	6.07	0.7	3.9 [504MPa]**	572	1400
C	5.99	1.0	4.5 [533MPa]**	513	960
D	5.92	1.1	6.1 [482MPa]**	503	825
E	6.06	0.8	0.8 [768MPa]**	743	1440
F	6.00	6.1	---*	324	550

*)There were too many defects to count in the test specimens.
**)The value in brackets shows the average tensile strength of the specimens inspected by SLAM.

attenuation coefficient, number of macro-defects detected by SLAM, and strength of the materials are summarized in Table Ⅱ. Fig. 1 shows the plot of the bulk density vs. average strength for each material. A clear relation beween bulk density and strength is not given. On the other hand, the obvious relation between the attenuation coefficient and the bend strength is shown in Fig. 2 which indicates that the bend strength decreases with the increase of the attenuation coefficient. In addition, a good correlation between the average number of macro-defects detected by SLAM and the tensile strength is presented in Fig. 3. SEM observation shows that the pressurelessly sintered materials A, C, D, and F contain many pores, cracks and inclusions with the size of 1 to 30 μm distributed widely in the materials. Tensile fracture will start from the relatively large pore or inclusion. Hot-isostatically sintered materials B and E contain some inclusions with a very small amount of pores. Tensile fracture origin can be the relatively large inclusion such as Al_2O_3 and SiO_2. The bulk density will depend on the amount of volumes of internal pores and inclusions of the material. It will not reflect the distribution of the size and number of pores and inclusions. On the other hand, the ultrasonic attenuation would be attributed to pores, cracks, and inclusions which are the scatterer and scattering intensity would depend on the distribution of the size and number of the scatterers. Therefore, the attenuation coefficients of the materials would reflect the distributions of the scatterers and be the effective parameter to evaluate the bend strength of the materials. The number of macro-defecs detected by SLAM indicates the feature of the distribution

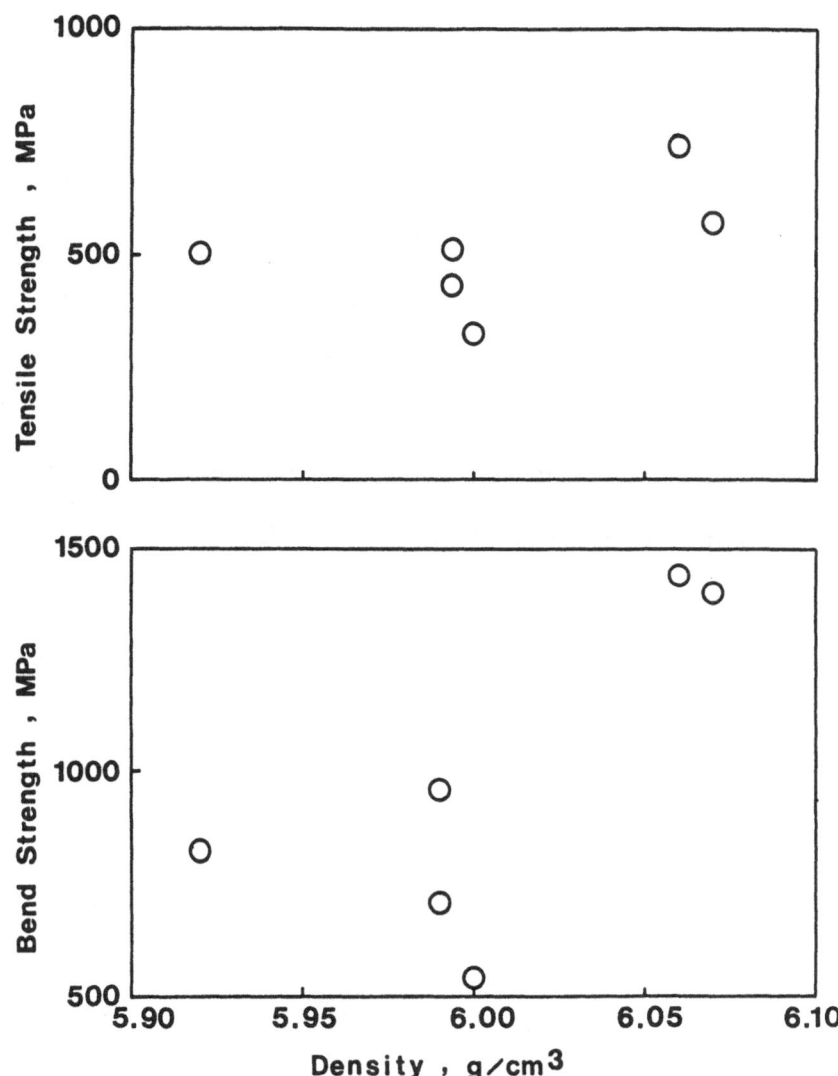

Figure 1. Plot of bulk density vs. strength of Y-TZP

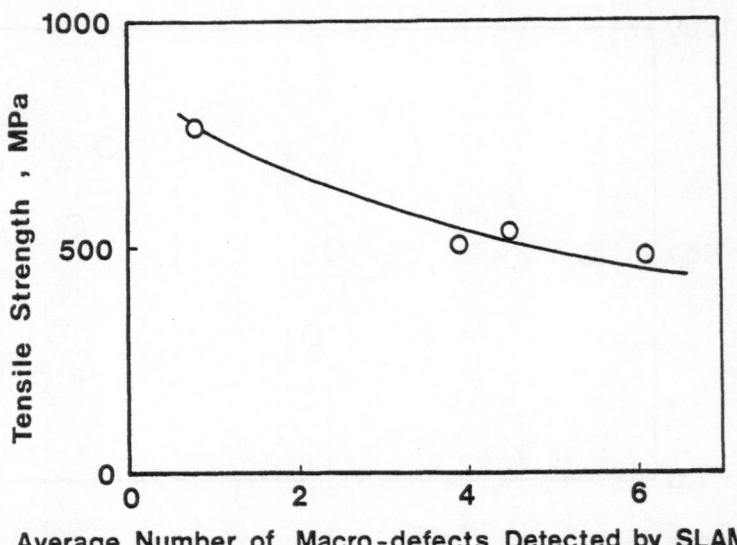

Figure 2. Plot of ultrasonic attenuation coefficient vs. bend strength of Y-TZP

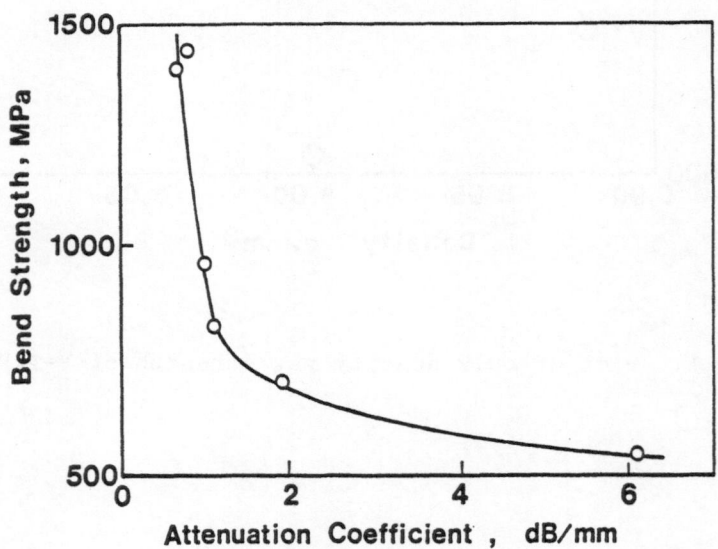

Figure 3. Plot of average number of macro-defects detected by SLAM vs. tensile strength of Y-TZP

of the macro-defects which effect on the tensile strength. Therefore, a good correlation between the average number of macro-defects detected by SLAM and the tensile strength is presented.

ACKNOWLWDGEMENTS

The authors would like to express their thanks to M. Noda of Toray Industries, Inc., and Y. Matsuda and M. Fujita of Toray Research Center, Inc., for material preparation and mechanical tests.

REFERENCES

1. D. J. Roth, E. R. Generazio, and G. Y. Baaklini, Mater. Eval., 45 [8] 958(1987).

2. D. J. Roth and G. Y. Baaklini, Advanced Ceram. Mater., 1 [3] 252(1986).

3. M. Oishi, K. Noguchi, T. Masaki, and M. Mizushina, Proc. '88 MRS Int. Meeting Advanced Mater.,(in press).

4. M. Oishi and K. Noguchi, Bull. Ceram. Soc. Jpn., 24 [3] 203(1989).

5. T. Nonaka, Bull. Ceram. Soc. Jpn., 24 [3] 208(1989).

6. A. Vary, Research Techniques in Nondestructive Testing, 4, Academic Press, London (1980) 159.

7. R. L. Smith, NDT International, 20 [1] 43(1987).

8. T. Masaki, J. Am. Ceram. Soc., 69 [8] 638(1986).

9. K. Tsukuma, Y. Kubota, and T. Tsukidate, Advances in Ceramics, 12, American Ceramic Society (1984) 382.

10. K. Noguchi, M. Fujita, T. Masaki, and M. Mizushina, J. Am.

Ceram. Soc., (in press).

11. J. D. Pohlhammer, C. A. Edwards, and W. D. O'Brien, Jr., Med. Phys., 8 [5] 692(1981).

12. L. W. Kessler and D. E. Yuhas, Proc. IEEE, 67 [4] 526(1979).

SYNTHESIS OF ZIRCON

Sridhar Komarneni and Rustum Roy
Materials Research Laboratory
The Pennsylvania State University
University Park, PA 16802

ABSTRACT

This review paper deals with the synthesis of both zircon powders and single crystals. Zircon powders can be prepared by solid state dry firing and hydrothermal reaction; solution-sol-gel derived powders can be combined with both methods. Of the various combinations, the hydrothermal method is the best for obtaining fine (~75 nm) powders at a temperature as low as 200°C under autogeneous pressure. Single crystals of zircon can be made by flux growth as well as hydrothermal methods, the latter using seeds. Although both methods yield large single crystals, the hydrothermal method is superior for obtaining pure crystals with little or no impurities.

INTRODUCTION

Zircon is a naturally occurring mineral and is an important component of igneous and metamorphic rocks. Its occurrence is common in detrital sediments or deposits derived from the above primary provenances. Because of its high stability, zircon has been extensively studied to obtain information on the geochemical and weathering conditions of sediments and soils. Single crystals of zircon, both clear and colored, have long been used as gemstones: the name zircon is very old and is believed to be derived from the Arabic zarqun, in turn derived from zar, gold, and gun color (Deer et al., 1962).[1]

Zircon is extensively used in ceramics and related applications and is obtained from major commercial zircon deposits in Australia, U.S. (Florida), South Africa, India, Brazil and the USSR.[2] The main use of zircon in ceramics is as a refractory material because of its refractoriness, thermal stability and non-reactivity with many glasses and slags. Zircon also has a low thermal expansion and good hardness which are also useful for some ceramic applications. Dense zircon shapes formed by casting or isostatic pressing are used for glass contact areas in glass tanks.[2,3] Zircon-containing bricks have been used in the most severe wear areas of steel ladles.[4,5] Alumina-zirconia-silica (AZS) composite refractories made with zircon are used in a variety of applications such as feeder parts of glass, kiln furniture, crucibles and furnace linings. Zircon is the raw material for the manufacture of zirconium oxide and zirconia-alumina abrasives. Another major use of zircon is in the whitewares industry as a glaze pacifier for tile and sanitary ware because of zircon's high index of refraction, light color, low solubility in glass and compatibility with ceramic colors. Zircon is also used in miscellaneous applications such as special glasses, porcelain enamels because of its high index of refraction, alkali resistance and radiation stability.[2] Zircon is also used as a minor component of fiber glass to improve the alkali resistance of fiber used for reinforcing concrete. Because of the above technological applications, there is a growing interest in the synthesis of zircon by different methods which can lead to superior products such as glazes. The present paper is a review of zircon synthesis studies and begins with the stability of zircon and the ZrO_2-SiO_2 phase diagram.

Stability of Zircon and the ZrO_2-SiO_2 Phase Diagram

Because of the geological and technological importance of the zircon phase, it is desirable to know the information on the thermal stability of zircon. A review of the literature shows that there is even now no consensus on the temperature or the manner of breakdown of zircon. The stability data of various investigators was summarized by Butterman and Foster[6] as given in Table I. The data presented in Table I show that there are conflicting opinions about the thermal stability of zircon. The earliest phase diagrams of Washburn and Libman[7] and Zhirnowa[8] for this system show that zircon is a congruently melting compound (Table I). However, subsequent work by Geller and Lang[9] and Cocco and Schromek[10] shows zircon as melting incongruently (Table I). Zircon was shown to decompose in the solid-state at 1540°C by Curtis and Sowman[11] and this decomposition temperature is about 135°C lower than the ZrO_2-SiO_2 eutectic. There is also controversy about the role of liquid immiscibility.[6,12] These discrepancies could be attributed to the extent of impurities in the zircon samples used.[13] Two phase diagrams, one by Levin et al.[14] and one by Butterman and Foster[6] are shown in Figures 1 and 2.

Zircon Synthesis Studies

As with any other ceramic oxide, zircon powders and single crystals can be prepared by various methods such as solid state, flux growth, hydrothermal, sol-gel, etc. This review of zircon synthesis begins with powders and is followed by single crystals.

Zircon Powder Synthesis

a. Solid State Method. According to Frondel and Collette,[15] zircon was first synthesized by Deville and Caron (1958)[16] by the action of SiF_4 on ZrO_2 and ZrF_4 on SiO_2 at red heat. Zircon has been synthesized from the oxides of zirconium and silicon by various investigators at different temperatures. Zircon synthesis was reported at 1460°C by Clark and Reynolds,[17] at 1500°C by Barlett[18] and Stot and Hilliard[19] and at 1700°C by Geller and Yavorsky.[20] These authors, however, did not investigate the maximum and minimum temperatures at which zircon can be synthesized. However, Curtis and Sowman[11] found a lower temperature limit of about 1315°C for zircon synthesis in zircon crucibles using milled zirconia and silica, the latter in the form of either quartz, cristobalite, amorphous silica, or tridymite. Frondel and Collette[15] used coprecipitated ZrO_2 and SiO_2 gels and found that they did not react when heated for 24 hours in air at 800° or $900^\circ C$ but reacted to form zircon in the temperature range of 1000° to $1500^\circ C$.

Mumpton and Roy[21] prepared zircon by dry heating 1:1 zirconia:silica gel starting material to about $1300^\circ C$ for 24 hours to study its dissociation at high temperatures.

b. Hydrothermal Method. Maurice[22] is one of the first to use hydrothermal method to synthesize zircon from $ZrO(OH)_2$ and silicic acid. Frondel and Collette[15] synthesized microcrystalline zircon over the range 150° to $700^\circ C$ by heating gelatinous ZrO_2 and SiO_2 with water in steel bombs under hydrothermal conditions. They found that the reaction is speeded by adding traces of ZrF_4 to the mixture. The hydrothermally synthesized zircon gave all unit cell dimensions identical with those of dry-sintered $ZrSiO_4$ at $1000^\circ C$ (Table II).

Mumpton and Roy[21] used concentrated NH_4OH to simultaneously precipitate $ZrCl_4$ and $SiCl_4$ as hydroxide gel and treated the 1:1 composition gels under hydrothermal conditions which consistently yielded well-crystallized zircon as low as $250^\circ C$ under 40,000 lb/in^2 water pressure in 30-day runs. However, at $200^\circ C$ and lower, in runs of less than 1 week these gels yielded poorly crystallized zircon-like material.

Zircon has been synthesized[23] from SiO_2 and ZrO_2, a large number of zirconium mineral salts such as $ZrOCl_2\cdot8H_2O$, $Zr(SO_4)_2\cdot4H_2O$, $ZrO(NO_3)_2\cdot2H_2O$, $Zr(NO_3)_4\cdot5H_2O$, and $ZrF_4\cdot H_2O$ and an organic compound of zirconium, $Zr(C_5H_7O_2)_4$ serving as the ZrO_2 source under hydrothermal conditions. Well-crystallized zircon was obtained at $200^\circ C$ or above with 350 or 700 bars of confining pressure.

Vilmin, Komarneni, and Roy[24] prepared a zircon sol hydrothermally by treating at 175°-$200^\circ C$ for 48 hours in a mixture of the appropriate volumes of a $ZrOCl_2\cdot8H_2O$ solution and tetraethylorthosilicate (TEOS) in ethanol. The $ZrSiO_4$ crystallites prepared at $175^\circ C$ are non-uniform in size (Fig. 3A) while those prepared at $200^\circ C$ are lense shaped, 60 nm thick and 300 nm diameter (Figs. 3B, 3C). The above hydrothermal experiments conducted with autogeneous pressure led to ultrafine powders of zircon. These powders are potentially useful in glazes. It is obvious from the above results that zircon fine powder can be prepared at a much lower temperature under hydrothermal conditions than possible under solid state dry firing conditions.

c. Solution-Sol-Gel Method. A sol-gel method was reported by Yokihiro and Yamato[25] for the preparation of zircon at low temperatures of about 1180° to $1200^\circ C$. This method involves the mixing of an aqueous solution of $Si(OC_2H_5)_4$ and $ZrOCl_2\cdot8H_2O$ with 0.06 mole of HCl per mole of $Si(OC_2H_5)_4$ as catalyst, drying the solution over a period of several days to form the gel from the solution and calcining the gel at different temperatures to crystallize zircon.

Recently, Vilmin, Komarneni, and Roy[24] prepared zircon by using different sol-gel precursors and to see their effect on zircon crystallization. A monophasic gel was prepared by mixing appropriate volumes of tetraethoxysilane and a zirconium oxychloride solution in ethanol and heating the mixture at 40°C. The addition of hydrothermally prepared zircon seeds to the same solution prior to gelation yielded a structurally diphasic gel. A compositionally diphasic gel was made using a monoclinic zirconia sol and a commercial silica sol and gently heating at 70°C to gel. The zircon gels which are both compositionally and structurally diphasic were made by simply mixing the crystalline zircon seed sol with the mixture of silica and zirconia sols and gelling at 70°C. All these zircon precursor gels were calcined to determine the crystallization temperature of zircon (Table III) which showed that both the compositionally and structurally diphasic gels yielded zircon at the lowest temperature.

d. Combined Sol-Gel and Hydrothermal Methods. In this method, Komarneni et al.,[26] first made single and diphasic gels of zircon composition and then treated them hydrothermally. The single phase zircon gels were made by mixing stoichiometric amounts of $ZrOCl_2 \cdot 8H_2O$ and tetraethoxysilane and aging at 60°C while the diphasic gels were made by mixing stoichiometric amounts of zirconia and silica sols and aging at 60°C. These gels were then treated hydrothermally in gold capsules in cold-seal vessels at 450° to 600°C under a confirming pressure of 100 MPa. The crystallization behavior of single and diphasic gels was found to be different under these hydrothermal conditions. The single phase gels yielded zircon under all the treatment conditions (Table IV) while the diphasic gels resulted in only zircon at the highest temperature of treatment, i.e., 600°C but yielded zircon and baddeleyite at 400° and 500°C. The silica and zirconia components in the diphasic gel appear to react independently at low temperatures and thus forming the zirconia phase, baddeleyite. On the other hand, the single phase gels resulted in zircon at all temperatures due to atomic scale mixing of the two components. Zircons formed from the two types of gels showed different morphologies, i.e., the single phase gels formed aggregates of ~ 75 nm zircon particles as vermicular morphological entities (Fig. 4A) and the diphasic gels formed thin plates of ~ 75 nm in size (Fig. 4B). Thus, it appears, one can control the morphology of zircon by combining both sol-gel and hydrothermal methods.

Synthesis of Zircon Single Crystals

Single crystals of zircon are possible candidate materials for polarizer use[27] in high-power laser systems and hence there is a need for their synthesis. Single crystals of zircon cannot be grown at ordinary pressures using the Czochralski method because it melts incongruently. There are two main methods for the growth of single crystals, i.e., flux and hydrothermal methods.

a. Flux Method. This method uses the cooling of molten salt fluxes for the growth of single crystals of zircon. Ballman and Laudise[28] grew single crystals of zircon as large as 0.7 cm from molten alkali metal vanadates and molybdates. They reported the growth of the best zircon crystals in platinum crucibles by slow cooling $Na_2O \cdot 3MoO_3$ or $Li_2O \cdot 3MoO_3$ solutions saturated with $SiO_2 + ZrO_2$ from $1400°$ to $900°C$ at a rate of $2°$ per hour. Chase and Osmer[29] also used a similar procedure to grow well-formed, large single crystals. Zircon single crystals were also grown in platinum capsules by Dharmarajan et al.,[30] from a lithium tungstate melt. Although these flux methods are useful in obtaining large single crystals, one of the disadvantages of this method is the incorporation of impurities during growth.[30]

b. Hydrothermal Method. Using natural zircon seed crystals and sintered ZrO_2 and SiO_2 as nutrient Dharmarajan et al.[30] grew single crystals of zircon in a 3-molar aqueous solution of KF under hydrothermal conditions. They found the optimum hydrothermal conditions to be $500°-600°C$ and 30000 psi.

Uhrin et al.,[31] grew single crystals of zircon (from 0.5 to 1.0 cms) hydrothermally at $700°C$ and 25000 psi using seeds of zircon (100) and 2m KF or 1 m LiF as mineralizers and a thermal gradient of $20°C$. Caruba et al.,[23] recently reported a new method of hydrothermal growth of zircon on natural and synthetic seeds using $Zr(SO_4)_2 \cdot 4H_2O$ and SiO_2 gel at $750°C$ under a confining pressure of 0.5 to 1×10^8 Pa. All the above hydrothermal methods use seeds in order to obtain large crystals. The crystals synthesized hydrothermally without any seeds can only attain a size of several tens of micrometers.[32]

CONCLUSION

The hydrothermal method is the preferred method for the synthesis of either zircon powders or single crystals. It is superior to other methods because of the low temperatures involved and the purity of the resulting zircon phase.

ACKNOWLEDGEMENTS

The authors gratefully acknowledge financial support for this work by Johnson Matthey Technology Center.

REFERENCES

1. W.A. Deer, R.A. Howie, and J. Zussman, "Rock-Forming Minerals," Vol. 1, Longmans, London, p. 59 (1962).

2. A.J. Hathaway and J.B. Munro, "Zircon and Zirconia," Am. Ceram. Soc. Bull. 63[5], 690-691 (1984).

3. P. Robyn, J. Moreau, and G. Soumoy, "Selection of Zircon Bricks for the Superstructure of Container-Glass Furnaces by Laboratory Simulative Testing," Interceram., Special Issue (1986).

4. R.J. O'Brien and E. Tauber, "The Use of Zircon in Steel Ladle Lining," 2nd Ind. Miner. Int'l. Cong., Munich, West Germany, 245-249 (1977).

5. T. Kawakami, "Properties of Zircon-Roseki Bricks for Steel Ladles," Taikabutsu Overseas 1[2], 29-36 (1981).

6. W.C. Butterman and W.R. Foster, "Zircon Stability and the ZrO_2-SiO_2 Phase Diagram," Am. Mineral. 52, 880-885 (1967).

7. E.W. Washburn and E.E. Libman, "An Approximate Determination of the Melting Point Diagram of the System Zirconia-Silica," J. Am. Ceram. Soc. 3[8], 634-640 (1920).

8. N. Zhirnowa, "Melting Diagram of System ZrO_2-SiO_2," Z. Anorg. Allgem. Chem., 218:192-300 (1934).

9. R.F. Geller and S.M. Lang, "System SiO_2-ZrO_2," J. Am. Ceram. Soc. 32[12], 157 (1949).

10. A. Cocco and N. Schromek, "Stability of $ZrSiO_4$ at High Temperatures," Ceramica (Milan) 12:45-48 (1958).

11. C.E. Curtis and H.G. Sowman, "Investigation of the Thermal Dissociation, Reassociation, and Synthesis of Zircon," J. Am. Cer. Soc. 36[6], 190-193 (1953).

12. F.R. Glasser, I. Warshaw, and R. Roy, "Liquid Immisibility in Silicate Systems," Phys. Chem. Glasses 1:39-45 (1960).

13. B. Brezny and R. Engel, "Evaluation of Zircon Brick for Steel Ladle Slag Lines," Am. Ceram. Soc. Bull. 63[7], 880-883 (1984).

14. E.M. Levin, C.R. Robbins, and H.F. McMurdie, "Phase Diagrams for Ceramists," Am. Ceram. Soc., Columbis, OH, pp. 601 (1964).

15. C. Frondel and R.L. Collette, "Hydrothermal Synthesis of Zircon, Thorite, and Huttonite," Am. Mineral. 42[6], 759-765 (1957).

16. H.S.C. Deville and H. Caron, "New Method of Synthesis of Crystals of Various Substances," C.R. 46, 764; Ann. Chim. Phys. 5:109 (1865).

17. G.L. Clark and D.H. Reynolds, "Chemistry of Zirconium Dioxide; X-ray Diffraction Studies," Ind. Eng. Chem. 29[6], 711-715 (1937).

18. H.B. Barlett, "X-ray and Microscopic Studies of Silicate Melts Containing ZrO_2," J. Am. Ceram. Soc. 14[11], 837-843 (1931).

19. V.H. Stott and A. Hilliard, "Variation in the Structure of Zircon," Min. Mag. 27, 198-203 (1945).

20. R.F. Geller and P.J. Yavorsky, "Effects of Some Oxide Additions on Thermal-Length Changes of Zirconia," J. Res. Natl. Bur. Standards 35[1], 87-110 (1945).

21. F.A. Mumpton and R. Roy, "Hydrothermal Stability Studies of the Zircon-Thorite Group," Geochimica et Cosmochimica Acta 21, 217-238 (1961).

22. O.D. Maurice, "Transport and Deposition of the Non-Sulphide Vein Minerals: V. Zirconium Minerals," Econ. Geol. 44:721-731 (1949).

23. R. Caruba, A. Baumer, and G. Turco, "Nouvelles Syntheses Hydrothermales du Zircon: Substitutions Isomorphiques; Relation Morphologie-Milieu de Croissance," Geochimica et Cosmochemica Acta 39[1], 11-26 (1975).

24. G. Vilmin, S. Komarneni, and R. Roy, "Lowering Crystallization Temperature of Zircon by Nanoheterogeneous Sol-Gel Processing," J. Mat. Sci. 22:3556-3560 (1987).

25. K. Yukihiro and Y. Yamate, "Synthesis of Zircon by the Sol-Gel Method," Yogyo-Kyokai-shi, 93[6], 74-76 (1985).

26. S. Komarneni, R. Roy, E. Breval, M. Ollinen, and Y. Suwa, "Hydrothermal Route to Ultrafine Powders Utilizing Single and Diphasic Gels," Advanced Ceramic Materials 1[1], 87-92 (1986).

27. V.O. Nicolai, "Light Polarizing Prism of Zircon," U.S. Patent 3,700,308, October 24, 1972.

28. A.A. Ballman and R.A. Laudise, "Crystallization and Solubility of Zircon and Phenacite in Certain Molten Salts," J. Am. Ceram. Soc., 48[3], 130-133 (1965).

29. A.B. Chase and J.A. Osmer, "Growth and Preferential Doping of Zircon and Thorite," J. Electrochem. Soc. 113:198-199 (1966).

30. R. Dharmarajan, R.F. Belt, and R.C. Puttbach, "Hydrothermal and Flux Growth of Zircon Crystals," Journal of Crystal Growth 13/14, 535-539 (1972).

31. R. Uhrin, R.F. Belt, and R.C. Puttbach, "The Hydrothermal Growth of Zircon," Journal of Crystal Growth 21, 65-68 (1974).

32. R. Caruba, A. Baumer, and P. Hartman, "Crystal Growth of Synthetic Zircon Round Natural Seeds," J. Crystal Growth, 88:297-302 (1988).

TABLE I.
Data on the Thermal Stability of $ZrSiO_4$*

Investigators	Year	Manner of Breakdown	Temperature
Washburn and Libman	1920	congruent melting	2550°C
Zhirnowa	1934	congruent melting	2430°C
Geller and Lang	1949	incongruent melting	1775°C
Curtis and Sowman	1953	solid state decomposition	1540°C
Cocco and Schromek	1957	incongruent melting	1720°C

*After Butterman and Foster[6]

TABLE II.
Hydrothermal Synthesis of Zircon*

Temperature	Pressure in Bars	Duration in Hours	Reagents	Results
700°C	1000	48	ZrO_2, SiO_2, mixed gels	Zircon
400°C	1000	168	ZrO_2, SiO_2, gels	Zircon
350°C	165	48	ZrO_2, SiO_2, gels	Zircon
325°C	121	240	ZrO_2, SiO_2, gels	Zircon
300°C	86	120	ZrO_2, SiO_2, gels, NaCl	Zircon
240°C	34	72	ZrO_2, SiO_2, gels, ZrF_4	Zircon
150°C	4.8	500	ZrO_2, SiO_2, gels, ZrF_4	Zircon

*After Frondel and Collette[15]

TABLE III.
Lowest Temperature at Which Zircon Formed in Different
Mono- and Diphasic Precursors*

Structural Diphasicity	Compositional Diphasicity	
	No	Yes
No	1325°C	1175°C
Yes	1100°C	1075°C

*After Vilmin et al., (1987)[24]

TABLE IV.
XRD and TEM Analyses of Single-Phase and Diphasic Zircon
Gels After Hydrothermal Treatment with Deionized Water.*

Sample	Hydrothermal treatment	Reaction products by XRD	Morphology by TEM
Single-phase gel	450°C/100 MPa/4 h	Zircon	
Single-phase gel	500°C/100 MPa/12 h	Zircon	Vermicular aggregates (≈75-nm crystals)
Single-phase gel	600°C/100 MPa/12 h	Zircon	
Diphasic gel	450°C/100 MPa/4 h	Zircon + baddeleyite	
Diphasic gel	500°C/100 MPa/12 h	Zircon + baddeleyite	Platy crystals (≈75 nm)
Diphasic gel	600°C/100 MPa/12 h	Zircon	

*After Komarneni et al.[26]

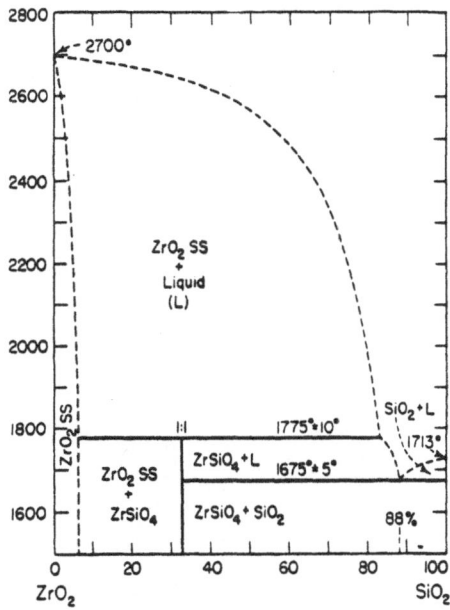

Figure 1.   System SiO_2-ZrO_2 Revised and corrected phase diagram after Levin at al., (1964).[14]

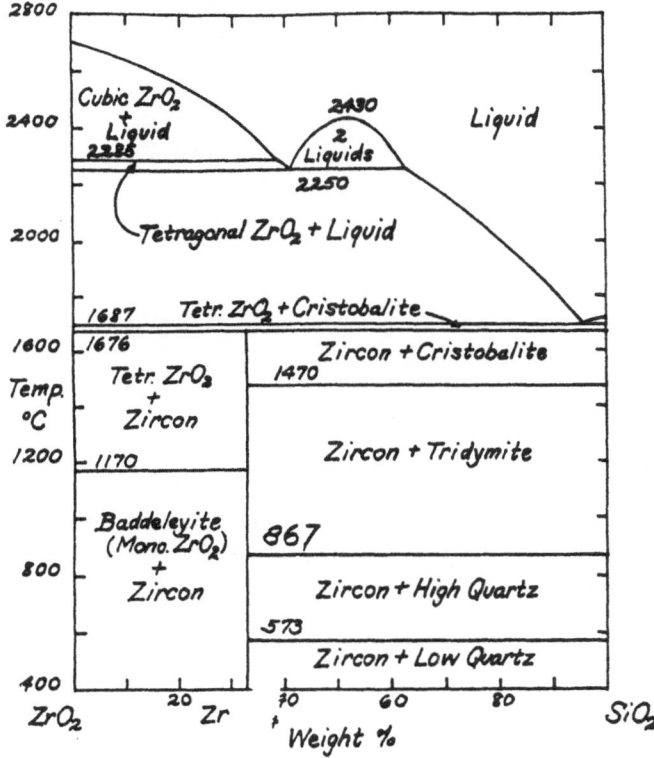

Figure 2.   Phase diagram for ZrO_2-SiO_2 system after Butterman and Foster.[6]

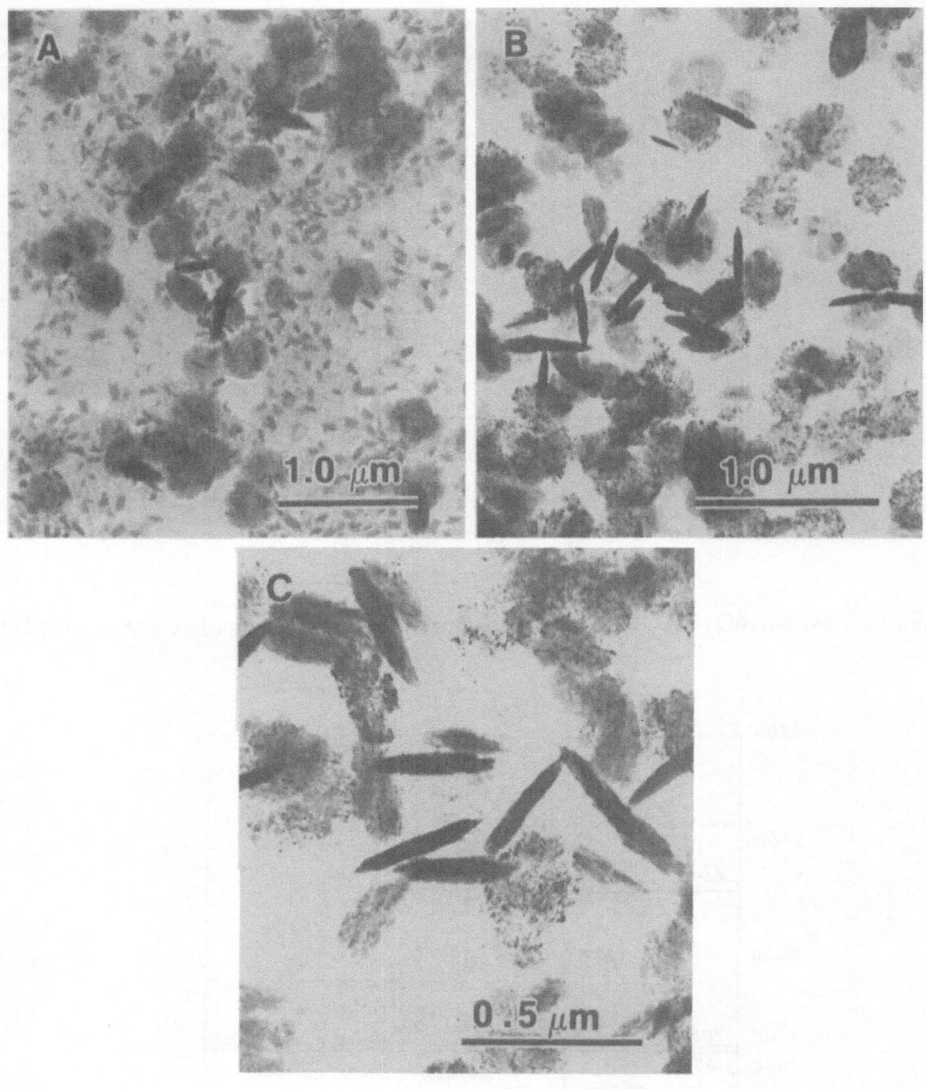

Figure 3.   Transmission electron micrographs of hydrothermally prepared $ZrSiO_4$ powders: (a) non-uniform crystallites at 175°C; (b) uniform crystallites at 200°C; (c) same as in (b) but at a higher magnification (after Vilmin et al., 1987[24]).

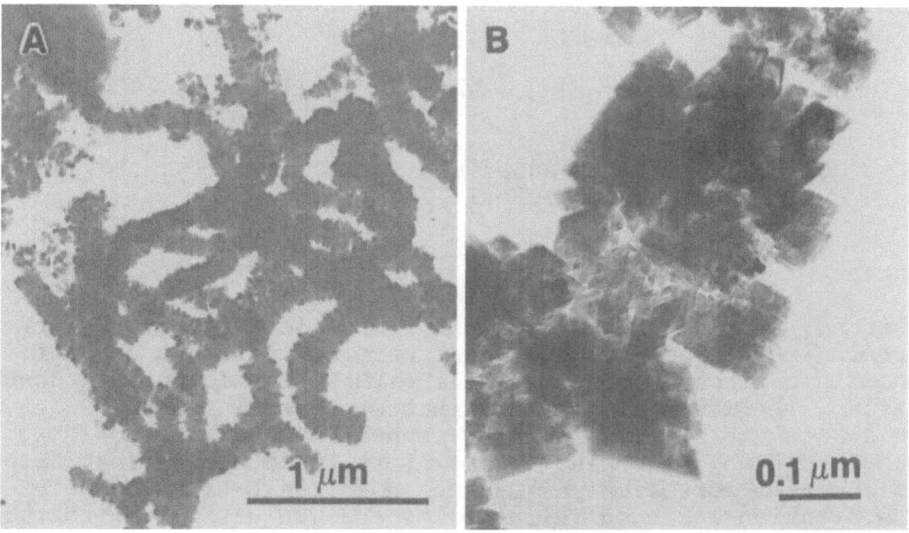

Figure 4. Transmission electron micrographs of ZrSiO$_4$ powders prepared hydrothermally at 500°C and 100 MPa: (a) vermicular form of ZrSiO$_4$ made from single phase gel; (b) platy ZrSiO$_4$ crystals made from diphasic gel (after Komarneni et al., 1986[26]).

Design and Application of a Zircon Advanced Refractory Ceramic

R.C Garvie, J.Drennan, M.F. Goss, S. Marshall, and C.Urbani

CSIRO, Division of Materials Science and Technology
Clayton, Victoria 3168, Australia

Abstract

A strong, thermal shock resistant zircon advanced refractory ceramic (ZS10) was designed by dispersing 10 weight percent monoclinic zirconia polycrystalline (MZP) particles in a dense matrix. The dispersion of MZP induced thermal shock resistance via a new toughening mechanism unrelated to transformation toughening. Laboratory tests showed that ZS10 had superior corrosion resistance in molten E-glass compared to conventional zircon refractories and was equivalent to them in thermal shock resistance. Accordingly, a field test was conducted on a ZS10 bushing block. The fabrication and performance of the block are described.

Introduction

Traditional refractories cannot be simultaneously dense and thermal shock resistant.[1] The reason is that the pores in these materials induce stable crack propagation by virtue of the fact that they are stress concentrators.[2] As such they generate a network of microcracks in regions of material subject to thermal shock; this is a recognised toughening mechanism. Unfortunately, using a dispersion of pores as a toughening mechanism greatly weakens the material and reduces its resistance to erosion and corrosion; in this sense a refractory is a badly degraded ceramic.

The traditional trade off in the properties of refractories can be overcome by dispersing solid particles in the matrix which can perform the same function as pores. Such particles should display highly localised self-stresses which can be transmitted to the surrounding matrix. One material of choice for this purpose is monoclinic zirconia whose transformational strains have long been used to toughen a variety of refractories and ceramics. However, the chief difficulty with a toughening mechanism based on the tetragonal/zirconia phase transformation is that it is strongly and inversely dependent on temperature.[3]

A better choice is monoclinic polycrystalline zirconia particles. The reason is that these particles are self-stressed due to their severe anisotropy of thermal expansion; moreover, the intrinsic thermal stresses are amplified by virtue of the polycrystalline morphology of the particles.[4-5] A considerable advantage of a system designed in this way is

that the toughening mechanism functions at least up until the
monoclinic to tetragonal transformation temperature at about
1200°C. However, tetragonal zirconia itself has an
appreciable thermal expansion anisotropy so that some
toughening can be expected from this source until some
temperature such that intrinsic stress relieving mechanisms
are activated in the matrix. In this condition, the matrix
itself would display intrinsic thermal shock resistance.

The purpose of this work is to describe the preparation,
characterisation and application in the glass industry of an
advanced zircon refractory ceramic toughened with a dispersion
of MZP.

Experimental

The MZP was comprised of aggregates with an overall size of
13μm.` An aggregate, itself, consisted of microcrystals about
1-2μm in diameter which were strongly bonded together. This
fine structure is stable; no change was observed after it was
heated at 1600°C/1hr. A screening test revealed that the
optimum concentration of MZP was 10 wt %; samples of this
composition are designated ZS10.

Laboratory samples were prepared by making a slurry of MZP and
zircon powder`` in isopropyl alcohol together with a fugitive
wax binder; the latter amounted to 4 wt % of the dry solids.
The powder had a mean particle size of 1.5μm. After mixing in
a Glen Creston mill with plastic balls for 15 minutes, the
slurry was evaporated to dryness. Then the powder batch was
granulated successively through 20- and 50- mesh screens.
After being formed by die pressing, test billets were
isostatically pressed at 210MPa. The billets then were fired
at 1600°C/1hr to attain a density of about 92% of the
theoretical value. The billets were machined by diamond
grinding to a standard size of about 3x3x40mm.

Similarly, billets of pure zircon (ZS) and also zircon
containing a dispersion of 10 wt. % baddelyite particles"
(ZB10) were prepared for purposes of comparison. The single
crystal baddelyite particles were milled to obtain a particle
size distribution close to that of the MZP aggregates. A
second set of larger billets of ZS and ZS10, 25x25x75mm, was
prepared in the same way except that they were not diamond
ground.
The thermal shock behaviour of ZS, ZB10 and ZS10 was
determined by quenching the small test billets, which had been
progressively heated to various temperatures, into water at

`S-grade, Magnesium Elektron Ltd., Manchester, U.K.

``Opacifine 5, Curumbin Minerals Ltd., Curumbin, Qld.,
Australia.

"400-K, Mandoval Ltd., Lightwater, Surrey, U.K.

room temperature. Also the thermal shock fatigue behaviour of
ZS and ZS10 was determined using an industry standard test.[6]
One fatigue cycle consisted of inserting a cold billet into a
furnace for 15 minutes which had been preheated to 1400°C and
then allowing it to cool outside the furnace for 15 minutes.
A sample failed when its weight loss due to spalling amounted
to 5%. The score achieved by a billet was the number of
cycles attained at failure.

ACI Ltd. performed corrosion tests on ZS10 in which a
cylindrical sample, 25x150mm, was immersed in molten E-glass
at 1330°C and rotated for 235 hours. A photograph of the test
facility is shown in Figure 4. The figure of merit used to
assess the rate of corrosion was the percent decrease in the
diameter of the sample measured at a point ½ the distance
between the glass line and the immersed end of the sample.
Other commercial zircon refractories were tested
simultaneously for purposes of comparison. The results are
summarised in Figure 6.

On the basis of combined thermal shock and corrosion data for
ZS10, the ACI Fibreglass, Dandenong, Victoria agreed to
install a prototype bushing block made of ZS10 in their
production line. A scaled up version of the laboratory
process was used to make the block as follows. An aqueous
slurry containing about 71 % solids and 2 wt % (solids basis)
of an organic binder was spray dried. The powder batch was
then isostatically pressed at 200MPa to form an oversized
rectangular block weighing about 40 kgm. The block was fired
at 15°C/hr until 500°C and then at 50°C/hr from 500°C to
1400°C; the block was held at this temperature for 2hr. Then
heating was resumed at the same rate to 1600°C where it was
held again for 2 hr. At the end of the soak period the block
was furnace cooled to room temperature. Finally it was
diamond ground to tolerance and the central channel formed by
trepanning with a diamond tool.

A 'post-mortem' examination of the ZS10 block was performed
using transmission optical microscopy (TEM), optical
microscopy and x-ray energy dispersion analysis.

Results and Discussion

1. Laboratory tests
The thermal shock behaviour of the three zircon based
materials is shown in Figure 1 which is a plot of the strength
retained in damaged billets (MOR) as a function of the
quenching temperature difference (T°C) The pure matrix
phase, ZS, showed behaviour typical of a dense brittle
material (Figure 1a); i.e., after an initial plateau a
critical quenching temperature difference (330°C) is reached
which activates preexisting surface flaws resulting in a
catastrophic decrease in strength (unstable crack propagation)
to a second plateau. At the second plateau there is a second
critical quenching temperature difference at about 660°C which
heralds the onset of stable crack propagation; i.e., with

increasing severity of thermal shock there is only a gentle
decline in retained strength. Stable crack propagation is
synonymous with thermal shock resistance.

The thermal shock resistance of the ZB10 billets (Figure 1b)
is improved by the dispersion of baddelyite particles in that
the critical quenching temperature difference for the onset of
unstable crack propagation is increased to about 360°C and the
degradation of the strength from to unstable crack propagation
is reduced. The modest increase in toughness could be due to
crack bowing/ deflection by the single crystal zirconia
particles.

The dispersion of MZP has considerably enhanced the thermal
shock resistance of ZS10 because it displays only stable crack
propagation (Figure 1c). There must be an additional
contribution to the toughening increment in addition to crack
bowing/deflection.
One plausible mechanism is based on the severe thermal
expansion anisotropy of monoclinic zirconia. During cooling
of a sample of ZS10 to room temperature after sintering,
considerable thermal stresses would be generated in the
aggregates. Moreover these stresses would be amplified by up
to a factor of 3 because of the polycrystalline morphology of
the particles. These stresses would be transmitted
efficiently to the matrix because the dilatational strain
during the tetragonal to monoclinic transformation guarantees
intimate contact between particle and host at their interface.
In any region of ZS10 subject to thermal shock, stresses so
generated would be superimposed onto the preexisting stresses
arising from the anisotropy with the possibility of nucleating
microcracks.[4-5] In this way, the dispersed MZP particles behave
like pores with the considerable advantage of preserving much
of the original strength of the material.

The improvement in mechanical properties possible with a
dispersion of MZP as a toughening agent rather than pores is
demonstrated by the data in table 1. The strength and Young's
modulus of ZS10 are improved by several hundred percent
compared to the values for a commercial zircon refractory.[7]
The reason is that the porosity of the former is 8% whilst
that of the latter is more than 20%. Yet the thermal shock
resistance parameter, R_{st} ($=\tau_{WOF}/\tau_{NBT}$) is about the
same for both materials.

Direct evidence of a toughening mechanism operating in ZS10 is
shown by Figure 2 which features optical micrographs of
thermally shocked ZS and ZS10. The former material displays
long straight through cracks characteristic of unstable crack
propagation. The latter shows strong crack/particle
interaction with many instances of crack branching and crack
arrest, indicative of stable crack propagation.

Figure 3 shows the results of the industry standard thermal
fatigue test applied to ZS and ZS10. The pure matrix
material, ZS, shattered upon its first insertion in the

furnace thereby acquiring a score of ½; ZS10 showed no weight
loss by spalling after 20 fatigue cycles and received a score
of 20.   Experience with this test indicates that little
further damage occurs after 20 cycles so that there is no
point in continuing the test beyond this point.
Similar thermal fatigue data for various commercial zircon
refractories are summarised in Figure 6 and is discussed
below.

The test rig used to obtain estimates of the relative rates of
corrosion of the zircon refractories in E-glass is shown in
Figure 4.   Figure 5 is a photograph of the ZS10 and zircon
samples after the corrosion test in E-glass. The results are
summarised in Figure 6 together with the results of the
thermal fatigue test; the corrosion rate and the thermal
shock fatigue data are plotted together as a function of
porosity. The corrosion rate increases almost exponentially
as a function of the porosity of the various refractories. A
similar dependency has been observed for slag corrosion tests.[1]
ZS10, as the densest material, had the lowest rate of
corrosion. The datum point labelled 'zircon 20' refers to the
present industry standard material for the bushing block
application. The thermal shock behaviour of the commercial
refractories follows the expected pattern which is opposite to
the trend observed for the corrosion data; the porous
materials have a high value of the figure-of-merit for thermal
shock resistance which decreases sharply with increasing
density. The thermal shock data for the dense ZS10 is
anomalous in that it is equivalent to the porous materials,
due to its toughening mechanism.

2. Field Test
The results of the laboratory thermal shock and corrosion
tests indicated that ZS10 could be a candidate material for
industrial applications. A suitable application is the
bushing block used in the fibre glass industry to meter molten
E-glass to the Pt/Rh bushing which forms the fibres. A
schematic of the fibre glass process is shown in Figure 7.

The bushing is usually replaced annually because the forming
holes have become too enlarged. The replacement is straight
forward requiring only a short down time. The bushing block
is replaced at the same time but requires a down time of
several hours and several personnel so that it represents a
serious loss of production. There would be a considerable
increase in productivity if the blocks only needed replacement
biannually instead of annually. This was the motivation for
testing the ZS10 block in the field.

Two blocks were fabricated with the assistance of local
industries (Figure 8). One block was installed in the
production line of a local fibre glass producer (Figure 9) on
4 June 1987. After fixing the block and bushing in place the
temperature of the ensemble was increased to the operating
value according to the following schedule; from an initial
value of around 430-480°C, the rate was 6.5°C/min. to 660°C

with a soaking time of 30 minutes; then heating was resumed at the same rate to 980°C for another soak of 30 minutes. Heating is done electrically by passing a current through the bushing. The unit operated until 6 May, 1988 after metering 386 tonnes of glass; it was shut down for technical reasons related to the melter. The performance of the block was marred by enhanced stoning which was observed during all stages of the test; the reason for this will be discussed below. The shut down involves circulating cooling water through the bushing. After the bushing is removed the block is hosed down with cold water until it can be removed manually using jack hammers, etc.

Figure 10 is a photograph of the used ZS10 block which shows that it has reasonable mechanical integrity. Probably the integrity could be improved further by encasing the block in a steel band around its circumference.

Unfortunately the chemical integrity of the ZS10 block was compromised by the micronising process which used alumina grinding media. The resulting alumina contamination (table 2) allowed the formation of a glassy phase in the grain boundaries as shown by the TEM micrograph in Figure 11. Further evidence for the formation of a high alumina grain boundary phase was obtained by x-ray energy dispersion analysis of a grain boundary and also pristine bulk material (Figure 12).

It the presence of alumina was known at the beginning of the project but it was thought to be harmless because it ought to form isolated mullite grains according to the well known reaction;

$$2ZrSiO_4 + 3Al_2O_3 = 2ZrO_2 + 3Al_2O_3.2SiO_2$$

The presence of a glassy grain boundary phase allowed the easy penetration of E-glass into the ZS10 block. The E-glass dissolved the dispersed zirconia phase preferentially which probably contributed to the enhanced stoning noted in the later stages of the test. The optical micrograph in Figure 13 illustrates the penetration and preferred dissolution phenomena; glass has entered a thermal shock crack and penetrated the refractory adjacent to the crack boundaries. Unfortunately, the laboratory corrosion test did not reveal the stoning problem. With hindsight, the zircon powder should have been micronised with steel grinding media and then washed with acid.

Conclusions

It has been shown that a zircon/zirconia alloy, ZS10, a member of new category of dense, advanced refractory ceramics can be used in an industrial application involving severe thermal shock.

The performance of the block warranted further tests although chemical contamination of the zircon used in the bushing block application prevented a definitive conclusion as to whether use of the new material would enhance productivity of fibre glass production.

Acknowledgements

We thank Mr. Rodd Judd, Laboratory Services Glass Packaging Divsion, ACI Ltd. for performing the glass corrosion tests and Dr. Paul Flavel, Technical Manager, Reinforcements, ACI Fibreglass for supervising the field test. Also we are grateful to Magnesium Elektron Ltd. for partial financial support of this work.

References

1. W.S. Trettner, "Refractories Technology", Bull. Am. Ceram. Soc., 58, 715-18 (1979).

2. R.D. Smith, H.U. Anderson and R.E. Moore, "Influence of Induced Porosity on the Thermal Shock Characteristics of Al_2O_3", ibid, 55, 979-82 (1976).

3. M.V. Swain, R.H.J. Hannink and R.C. Garvie, "The Influence of Precipitate Size and Temperature on the Fracture Toughness of Calcia- and Magnesia-Partially Stabilized Zirconia", Fract. Mech. of Ceram., (Ed. R.C. Bradt, A.G. Evans, D.P.H. Hasselman and F.F. Lange), 6, 339-354 (1983).

4. R.C. Garvie and M.F. Goss, "Thermal Shock Resistant Alumina/Zirconia Alloys", published in Advanced Ceramics II, 69-87. Ed. S. Somiya, Elsevier App. Sci., London, 1988.

5. R.C. Garvie, M.F. Goss, S. Marshall and C. Urbani, "Dense, Thermal Shock Resistant Advanced Refractories", Proc. Metall. Soc. Can. Inst. Min and Metall., 4, "Adv. Refr. Metall. Ind.", Ed. M.A.J. Rigaud, Permagon, New York, pp. 53-69, (1988)

6. T.M. Wehrenberg and J.R. Stein, Glass International, December, 19-20 (1984).

7. R.N. Enderfield, B.C. Hocking and M.G. Oxlade, "Recent Advances in Refractories for the Teeming of Steel", Refract. J., Sept./Oct., 10-19, (1975).

Table 1.   Property Data for ZS10 and a Zircon Refractory.

Property	Material	
	ZS10	Zircon
MOR, MPa	149	21.9
Young's Modulous, MPa	188	55.6
K_{1c}, MPa m	2.9	1.5
τ_{WOF}, J/m^2	73.1	20.9
τ_{NBT}, J/m^2	22.2	20.8
τ_{WOF}/τ_{NBT}	3.3	1.0
Total Porosity, %	8	21

Table 2.   Chemical Analysis of Zircon 20 and ZS10.

Oxide	Amount (wt%)	
	Zircon 20	ZS10
ZrO_2	64.7	64.4
SiO_2	34.2	32.0
Al_2O_3	0.25	1.7
TiO_2	0.7	0.12

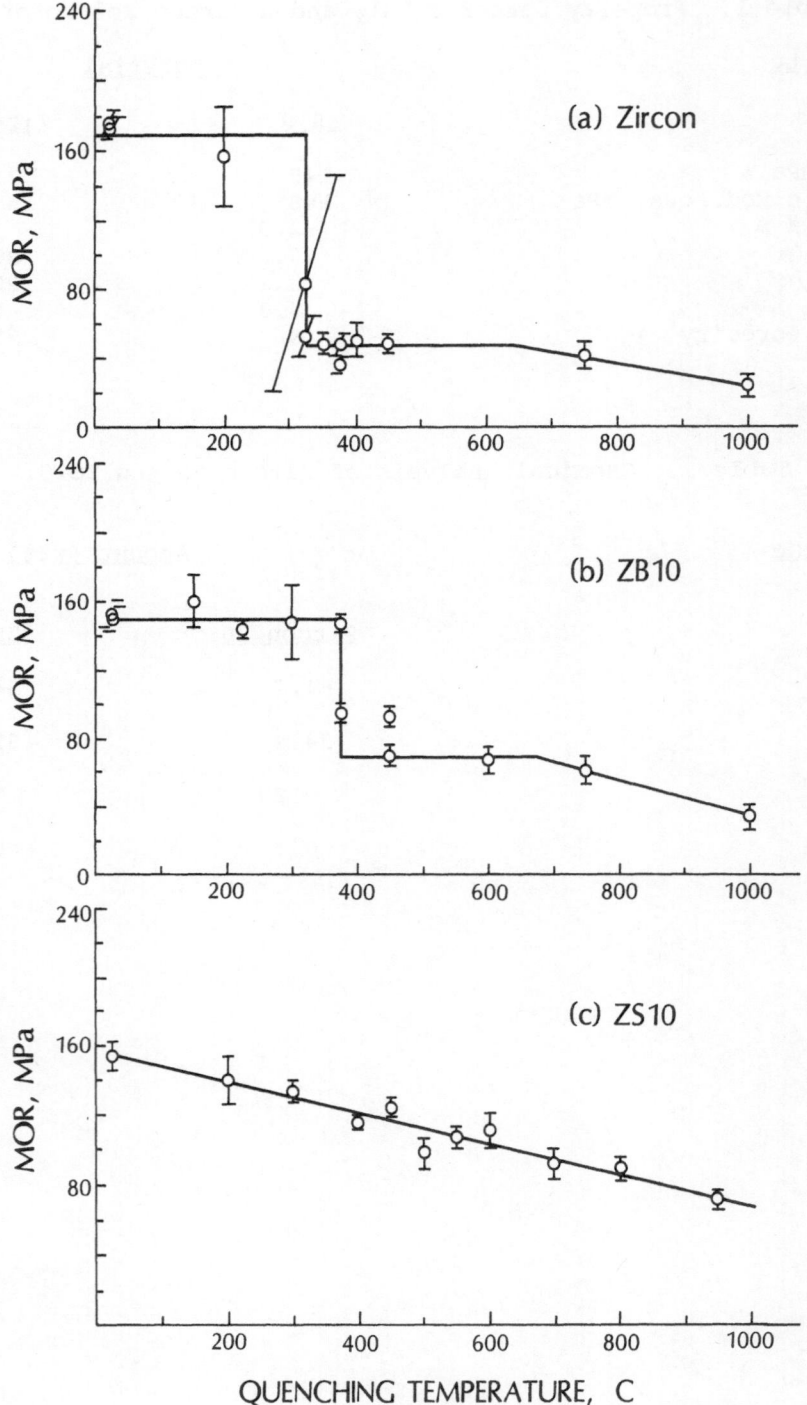

1. Thermal shock behaviour of (a) zircon, (b) the zircon/baddelyite alloy, ZB10 and (c) the zircon/MZP alloy, ZS10.

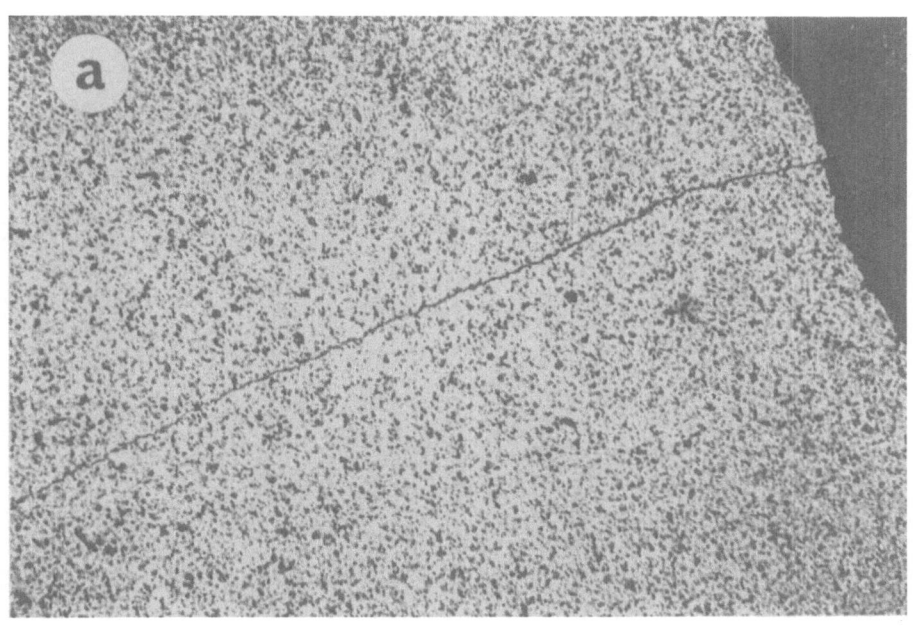

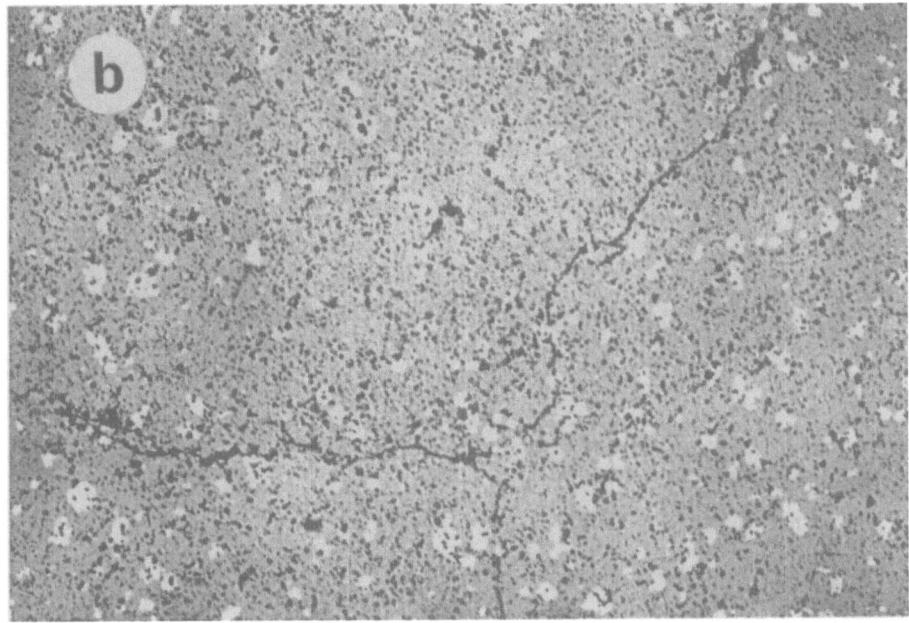

2. Optical micrographs of (a) zircon and (b) ZS10 quenched
from 1000°C into water at room temperature.

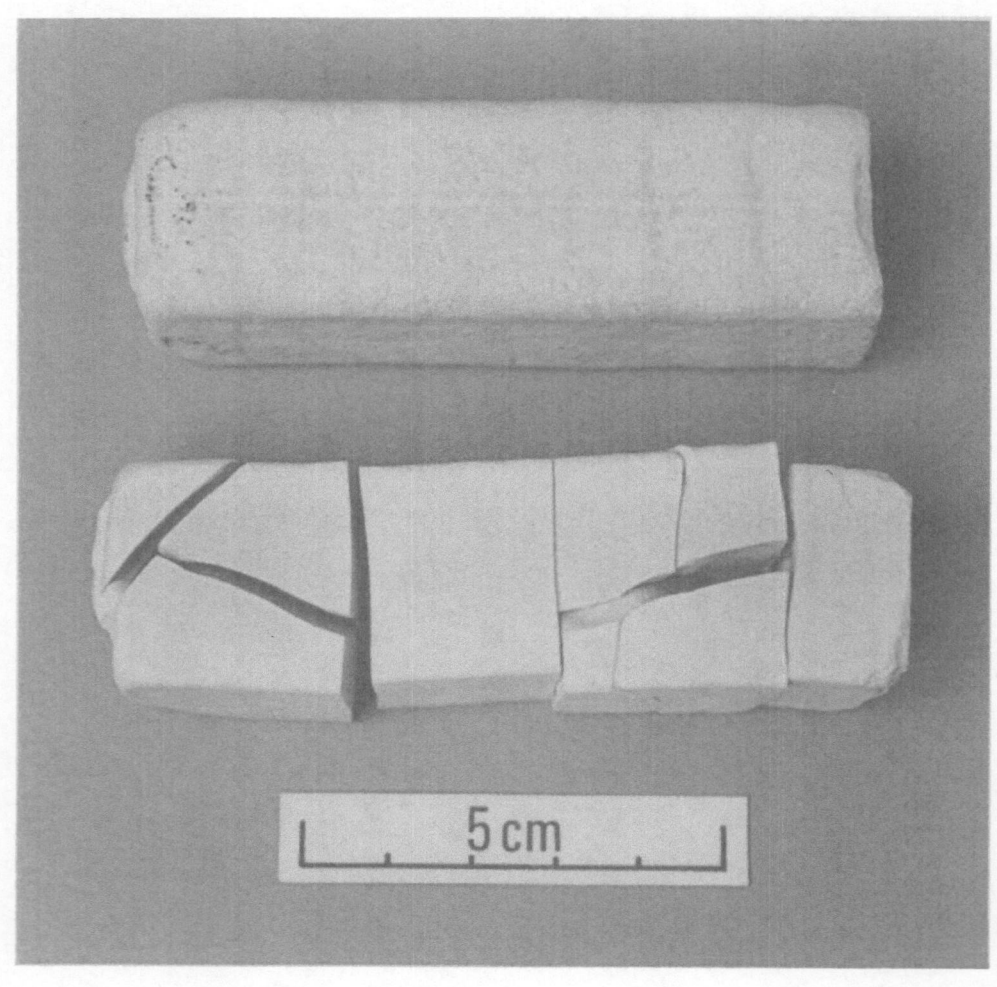

3. Photograph of ZS10 after 20 thermal fatigue cycles (top sample) and zircon after ½ cycle; a cycle is comprised of 1400°C/15min.-room temperature/15min.

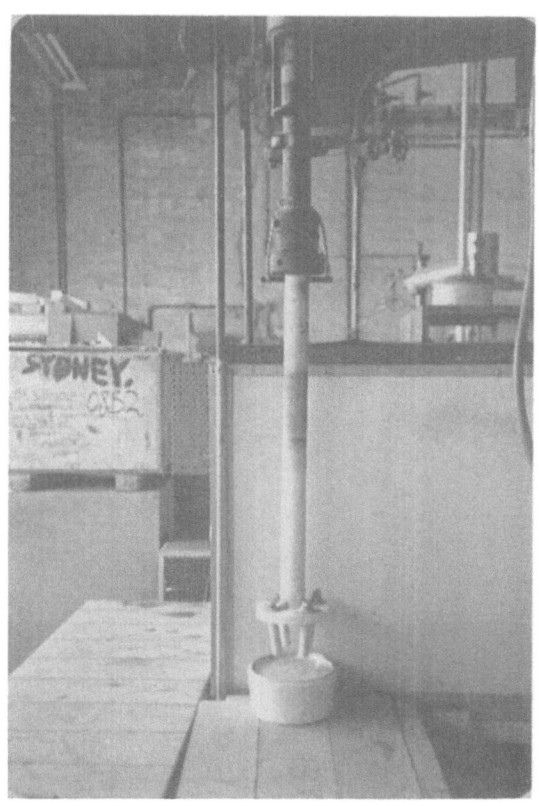

4. Photograph of test rig used to assess glass corrosion resistance of refractories at ACI Ltd.

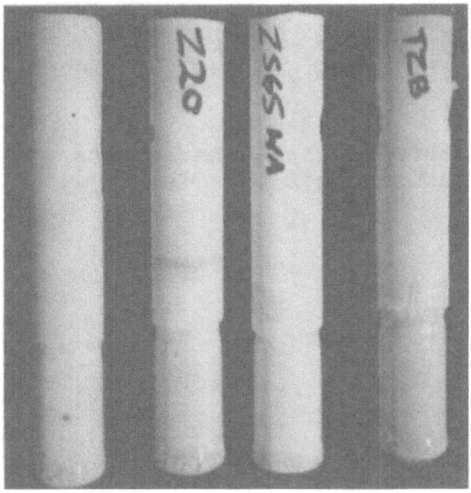

5. Photograph of ZS10 and commercial zircon refractories after testing in molten E-glass at 1330°C/235hr.

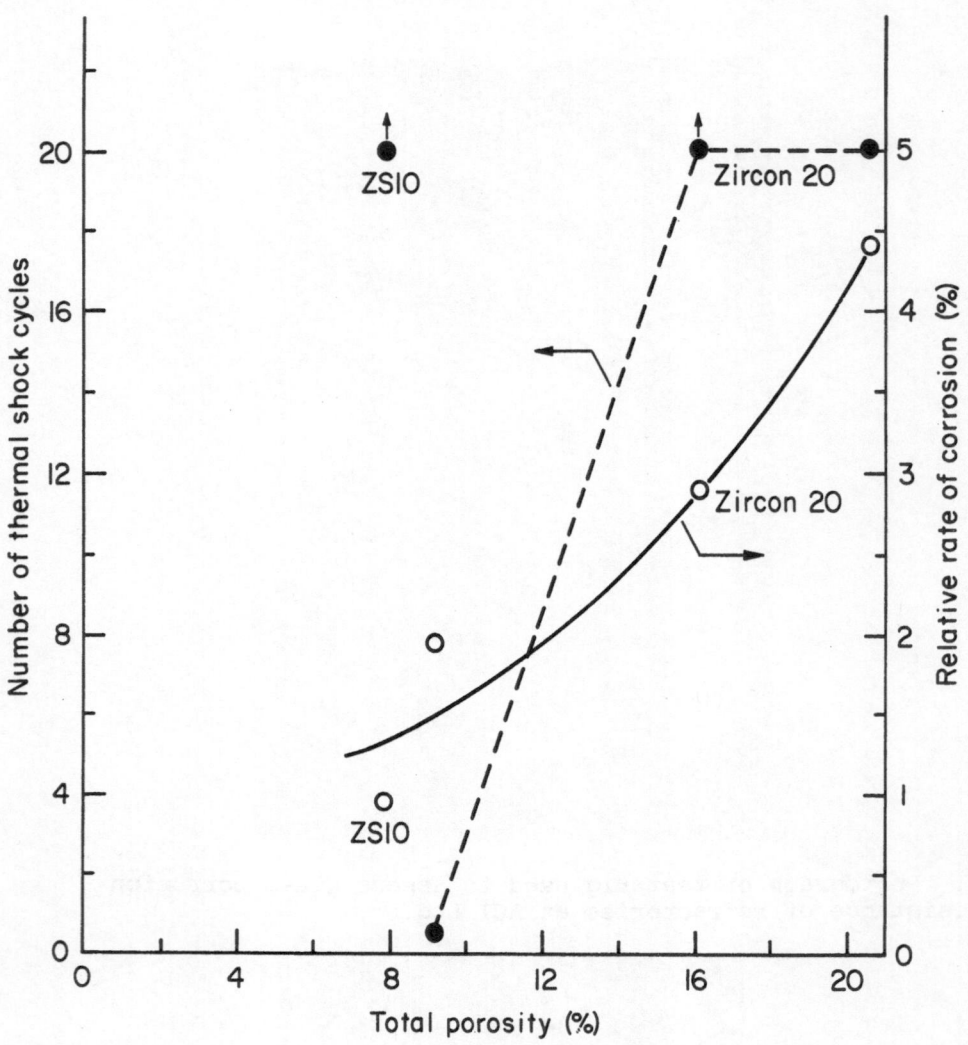

6. Porosity dependence of the thermal shock and corrosion
resistance of various zircon materials.

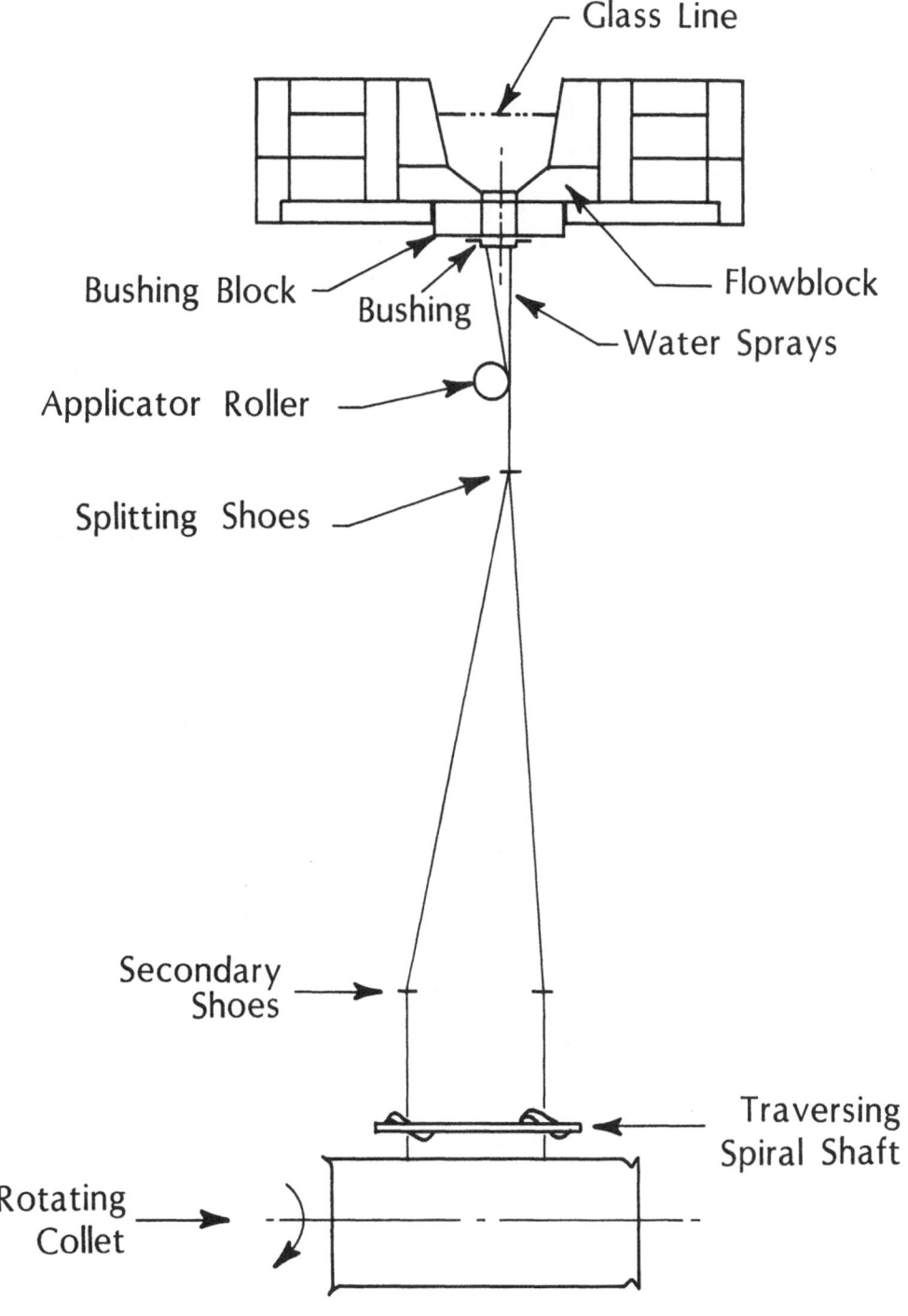

7. Schematic outline of the production process for fibreglass.

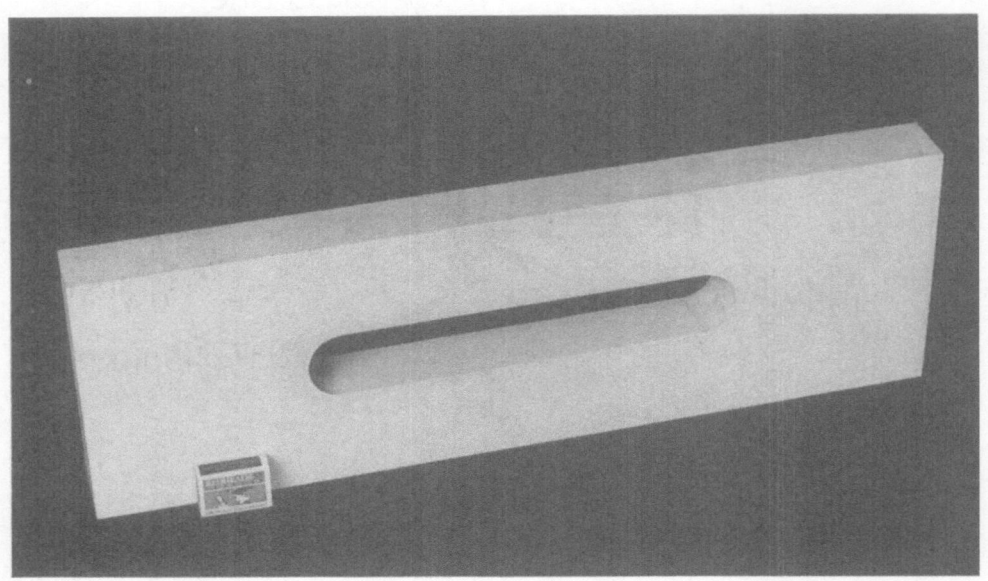

8. Photograph of a prototype bushing block made of ZS10.

9. Photograph of the ZS10 bushing block installed in position 16, number 1 tank at ACI Fibreglass.

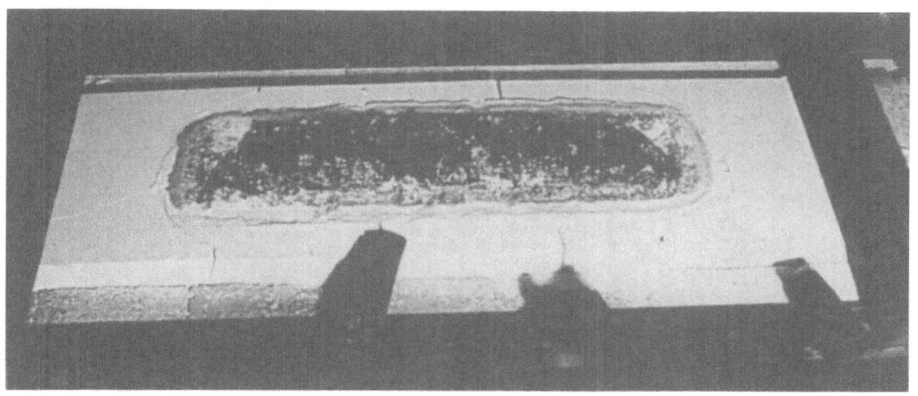

10. Photograph of the ZS10 bushing block after being in
operation for 11 months and processing 386 tonnes of E-glass.

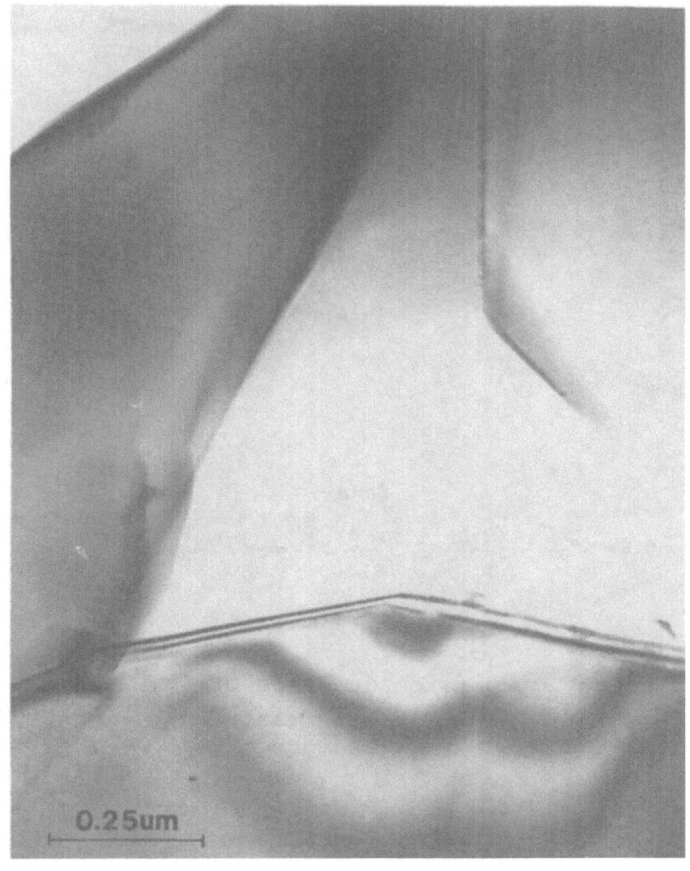

0.25um

11. Bright field TEM micrograph of a triple point in a
pristine sample of ZS10 showing a grain boundary glassy phase.

356

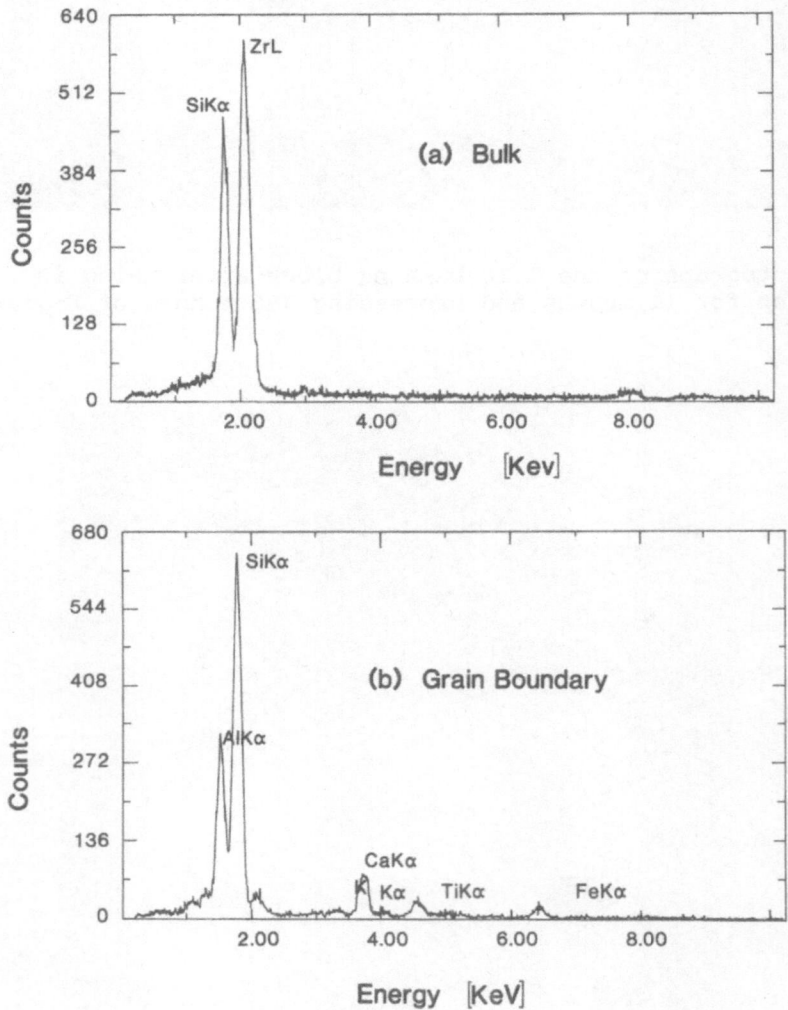

12. X-ray energy dispersion analyses of (a) bulk zircon and
(b) a grain boundary in ZS10; the latter confirms the
presence of alumina in the grain boundary.

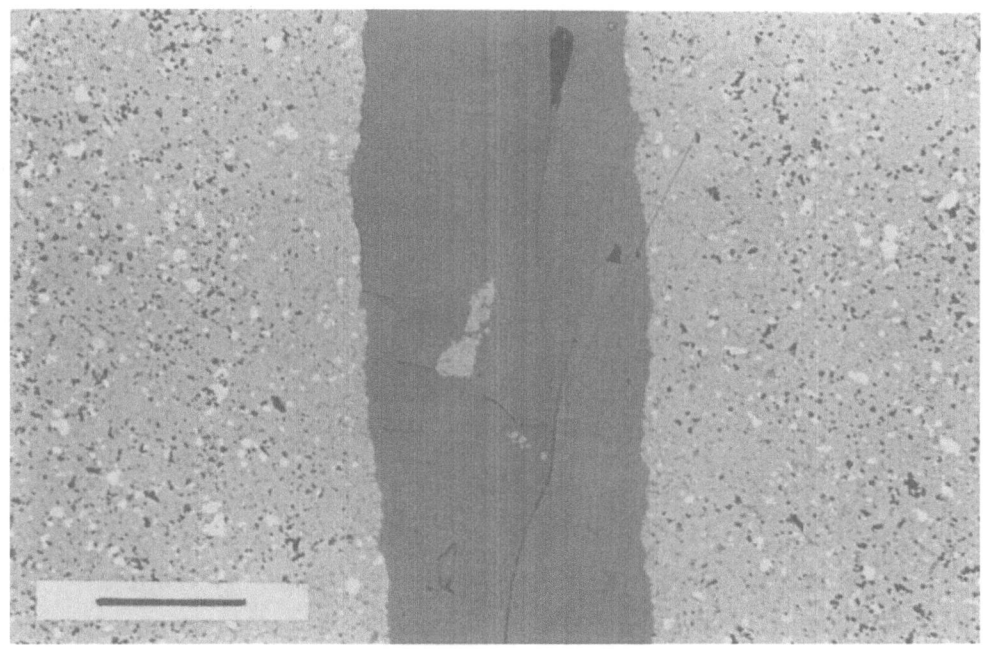

13. Optical micrograph of the used ZS10 block showing the
presence of E-glass in a crack; the glass has penetrated the
bulk refractory along its grain boundaries. The bar = 50μm.

HYDROTHERMAL PROCESSING OF ZIRCON

H. Kido and S. Komarneni*

Materials Research Laboratory
The Pennsylvania State University
University Park, PA 16802

Abstract

Zircon was prepared hydrothermally from zirconium oxychloride, tetraethoxysilane or tetramethoxysilane, water and/or alcohol by heating in teflon-lined hydrothermal vessels at 150°C for 6 hr or at 200°C for 4 hr. The effects of starting materials, temperature, heating time and catalysts on the crystallization of zircon were studied.

The simultaneous hydrolysis of Si alkoxide and $ZrOCl_2$ leading to the formation of a single phase gel of intimately mixed Si and Zr atoms was found to be the most important process which controlled the formation of zircon. Insufficient amount of water led to a segregation of the components during the hydrothermal treatment as evidenced by the formation of monoclinic zirconia. Increased heating duration at a constant temperature of treatment led to softer powder compared to a smaller heating duration which resulted in hard agglomerates. Transmission electron microscopy revealed doughnut-like or red blood cell-like morphology for the particles which are dense at the edge and thin at the core. The individual particles may be assemblies of small crystals with the same crystallographic orientation rather than single crystals.

*Also with the Department of Agronomy

Introduction

Zircon has a very high thermal shock resistance because of its low thermal expansion coefficient. Zircon single crystals are used in jewelry and zircon powder is used in glazes (1). However, the preparation of zircon needs a heat treatment of 1450°-1500°C in air by the solid state method and usually transition metal oxides are used as additives (2-4). Frondel and Collette (5) have prepared zircon hydrothermally from gelatinous ZrO_2 and SiO_2 with water in steel bombs. Single phase zircon has been obtained by heating at 325°C for 240 hrs (5). The addition of a trace amount of ZrF_2 decreased the crystallization temperature to 150°C, but heating for 500 hrs was needed.

In this study, pure zircon was prepared hydrothermally from zirconium oxychloride, Si alkoxides, water and/or alcohol.

Experimental

Zircon was prepared from tetraethoxysilane (TEOS), $Si(OC_2H_5)_4$, or tetramethoxysilane (TMOS), $Si(OCH_3)_4$, $ZrOCl_2 \cdot 8H_2O$, H_2O and/or ethanol (ETOH) or methanol (MeOH) by using the hydrothermal technique. Either 10.830 g (for TMOS system) or 7.2237 g (for TEOS system) of $ZrOCl_2 \cdot 8H_2O$ was dissolved in the desired amount of deionized water (x ml) and mixed at room temperature to get transparent (clear) solution. Then 5 ml of TMOS (or TEOS) was added to the above transparent solution and mixed for about 30 min by stirring. The mixture was sealed in the hydrothermal vessels, Parr bombs and heated at 100°-200°C and for 3-48 hr. Either NH_4OH or HCl was used as a catalyst in some cases. The hydrothermal products were characterized by X-ray powder diffraction using a Picker-Siemens diffractometer with Ni-filtered CuKα radiation and by transmission electron microscopy using a Philips 420 instrument.

Results and Discussion

    (I) Synthesis from TMOS-$ZrOCl_2 \cdot 8H_2O$

First, the effect of water content on the crystallization of zircon was studied. In Figure 1, X-ray diffraction patterns for samples with different amounts of water (x) are shown. At x=5 ml, the dominant cyrstalline phase formed at 200°C was monoclinic zirconia under these hydrothermal conditions. When gels made from TEOS-$ZrOCl_2 \cdot 8H_2O$ were heated in air from 120°-1350°C, the tetragonal form of zirconia formed at 320°C and the monoclinic phase appeared only at >800°C (4). Thus the present hydrothermal conditions yielded monoclinic zirconia at much lower temperature than the heat treatment in air. In the present study, at x<20 ml the crystalline phases were zircon and monoclinic zirconia. With increasing amounts of water, the quantity of zircon phase increased. At x=20 ml, single phase of zircon was obtained probably as a result of the complete hydrolysis of Si alkoxide along with $ZrOCl_2$ leading to the formation of a single phase gel followed by its crystallization. Thus simultaneous hydrolysis of alkoxide and $ZrOCl_2$ to form a single phase gel appears to be the most important factor which is controlling the formation of zircon. The peaks of (200) and (400) are relatively sharp, but other peaks are broad (Fig. 1). This result shows that the crystallite size is rather small and has the a-axis orientation. The zircon crystallites prepared hydrothermally from TEOS-EtOH system were reported to have lens-shaped morphology (6) and this morphology may have led to the a-axis orientation.

When methanol (5 ml) was added to the above TMOS-$ZrOCl_2$ solution of x=20 ml and treated hydrothermally at 200°C the XRD peaks became sharper when compared with the sample without methanol (Fig. 2). The addition of methanol appears to promote the crystallinity of the zircon phase slightly.

The effect of acid or base additions was also examined, and the results are shown in Figure 3. There was little or no effect of acid on the crystallization of zircon (Fig. 3). On the other hand, the addition of base led to a sharpening of the XRD peaks of zircon as can be seen from Figure 3.

(II) Synthesis from TEOS-ZrOCl$_2$·8H$_2$O-EtOH-H$_2$O

X-ray diffraction patterns of hydrothermally reacted samples resulting from different amounts of EtOH and H$_2$O are shown in Figure 4. As expected from the results for TMOS, zircon phase increased as the amount of water increased. At low water contents, crystalline phases were again zircon and monoclinic zirconia just as in the case of TEOS-ZrOCl$_2$ starting materials and these results can be explained in the same way.

The effect of time and temperature on the crystallization of zircon was studied using fixed volumes of ethanol and water (20 ml for each) along with constant amounts of TEOS and ZrOCl$_2$·8H$_2$O. The results are shown in Figure 5. Under these conditions, the boundary between crystalline and amorphous phases is rather sharp which is indicated by a broken line (Fig. 5). This boundary is located at lower temperature and time when compared to the earlier work of Frondel and Collette (5). The low temperature synthesis of zircon achieved here can be attributed to the higher reactivity of the starting materials compared to those of Frondel and Collette (5) who used mixed ZrO$_2$-SiO$_2$ diphasic gels.

Transmission electron microscopy (Fig. 6) was used to determine the particle size and morphology of the above hydrothermal products. Figure 6A reveals massive amorphous phase after treatment at 135°C for 6 hrs which is consistent with the XRD results (Fig. 5). However, treatment for a longer duration i.e. 24 hrs at 135°C led to the crystallization of zircon phase (Fig. 5). Treatment of the mixture at 150°C for a short duration resulted in non-uniform sized zircon (Fig. 6B) while treatment for a longer duration led to uniform sized zircon (Fig. 6C). Increasing the temperature of treatment to 175°C also resulted in uniform sized powders but at a shorter duration (Fig. 6D). The size of zircon powder increased upon treatment for a longer duration at 175°C (Fig. 6E). Figure 6F shows the zircon powders which resulted at 200°C upon treatment for 48 hrs. There is virtually no difference between the zircon powders made at 175° and 200°C when treated for the same length of time. Thus it appears, a critical size (~650 nm) for the zircon powder was attained in the approximate temperature range of 175°-200°C. Although the zircon powders appear spherical they are not spherical as revealed by their morphology on the edge (Fig. 6). Furthermore, the core of the spherical-shaped particles is less dense than the edge as can be clearly seen in Figs. 6E and 6F. Thus the morphology of the particles is either red blood cell-like or somewhat doughnut-like although there is no hole in the middle. It is not known for certain whether the individual particles are single crystals or an assembly of many small crystals with the same crystallographic orientation. The latter case is more likely than the former based on the morphology of the zircon in Fig. 6B which shows very small zircon crystals along with large non-uniform sized particles.

References

1. R. Carter, "Zircon-Ceramic Pigments," Ceram. Eng. Sci. Proc., **8**, 1156-61 (1987).

2. V.I. Matkovich and P.M. Corbett, "Formation of Zircon from Zirconium Dioxide and Silicon Dioxide in the Presence of Vanadium Pentoxide," J. Am. Ceram. Soc., **44**, 128-130 (1961).

3. A.A. Ballman and R.A. Laudise, "Crystallization and Solubility of Zircon and Phenacite in Certain Molten Salts," J. Am. Ceram. Soc., **48**, 130-33 (1965).

4. Y. Kadogawa and T. Yamate, "Synthesis of Zircon by the Sol-Gel Method," Yogyo-Kyokai-Shi, **93**, 338-40 (1985).

5. C. Frondel and R.L. Collette, "Hydrothermal Synthesis of Zircon, Thorite, and Huttonite," Am. Mineralogist, **42**, 759-65 (1957).

6. G. Vilmin, S. Komarneni, and R. Roy, "Lowering Crystallization Temperaure of Zircon by Nanoheterogeneous Sol-gel Processing," J. Mat. Sci., **22**, 3556-60 (1987).

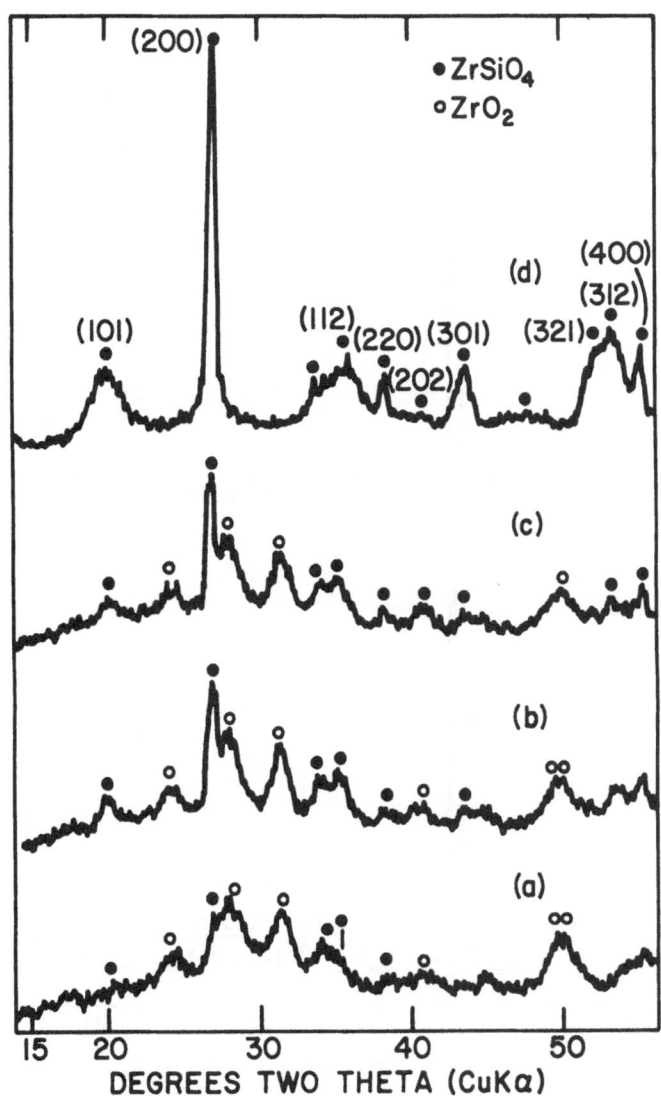

Figure 1. X-ray diffraction patterns showing the effect of water content on the crystallization of TMOS-ZrOCl$_2$·8H$_2$O-H$_2$O mixtures which were hydrothermally treated at 200°C for 2 days: (a) 5 ml H$_2$O; (b) 7 ml H$_2$O; (c) 10 ml H$_2$O; and (d) 20 ml H$_2$O.

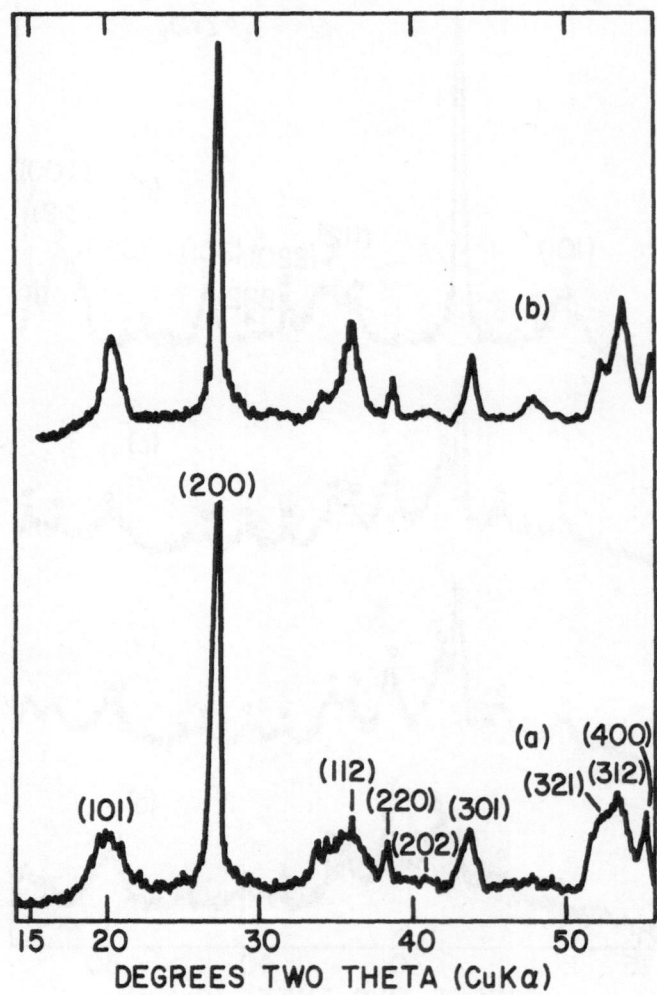

Figure 2. X-ray diffraction patterns of zircon from TMOS (5 ml)-
ZrOCl$_2$·8H$_2$O (10.83 g)-H$_2$O (20 ml) without (a) and with (b)
methanol (5 ml) treated hydrothermally at 200°C for 2 days.

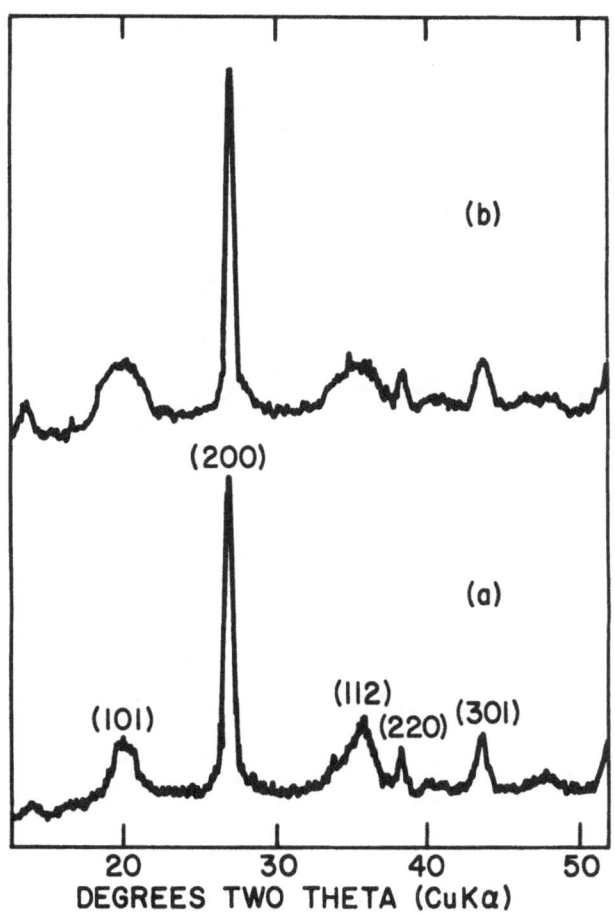

Figure 3. X-ray diffraction patterns showing the effect of (a) acid and (b) base on the crystallization of zircon. Zircons were prepared from TMOS-$ZrOCl_2 \cdot 8H_2O$ and (a) NH_4OH (1.0×10^{-2} mol/l) or (b) HCl (1.0×10^{-2} mol/l).

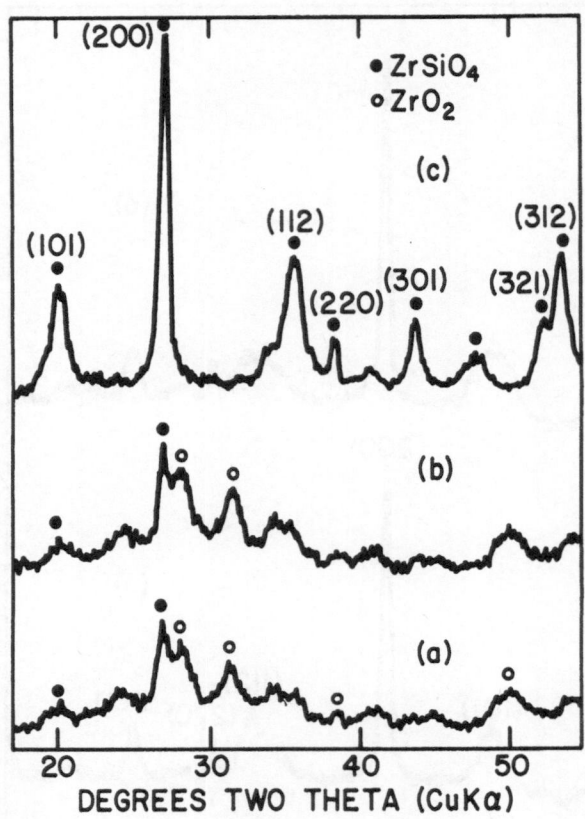

Figure 4. X-ray diffraction patterns of different samples hydrothermally treated at 200°C for 2 days. Starting compositions were TEOS (5 ml)-$ZrOCl_2 \cdot 8H_2O$ (7.2237 g), and (a) EtOH (40 ml); (b) EtOH, 40 ml-H_2O (5 ml); (c) EtOH, 20 ml-H_2O (20 ml).

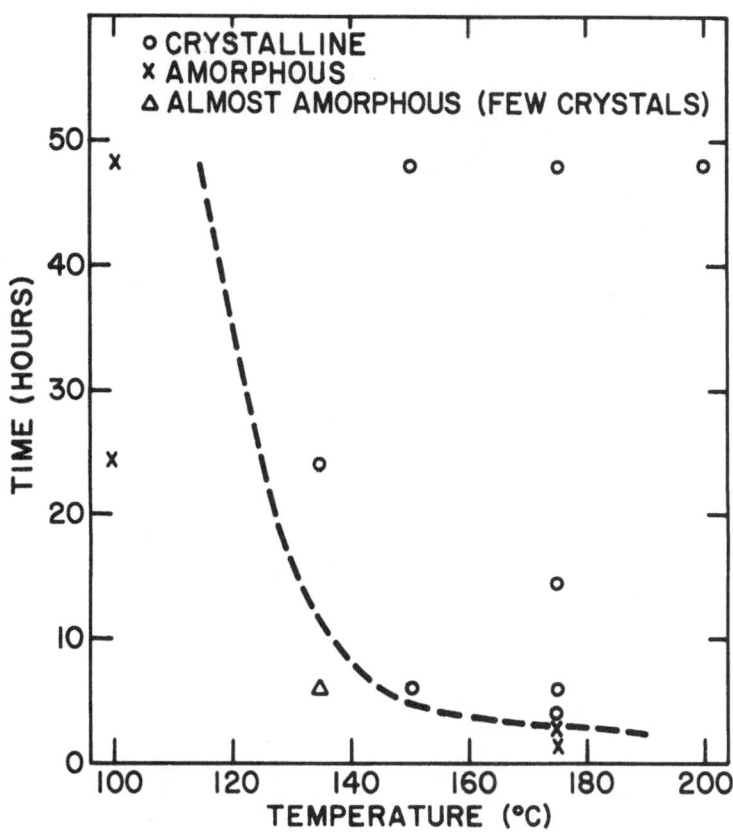

Figure 5. Effects of temperature and treating time on the crystallization of zircon from TEOS (5 ml)-$ZrOCl_2 \cdot 8H_2O$ (7.2237 g)-EtOH (20 ml)-H_2O (20 ml).

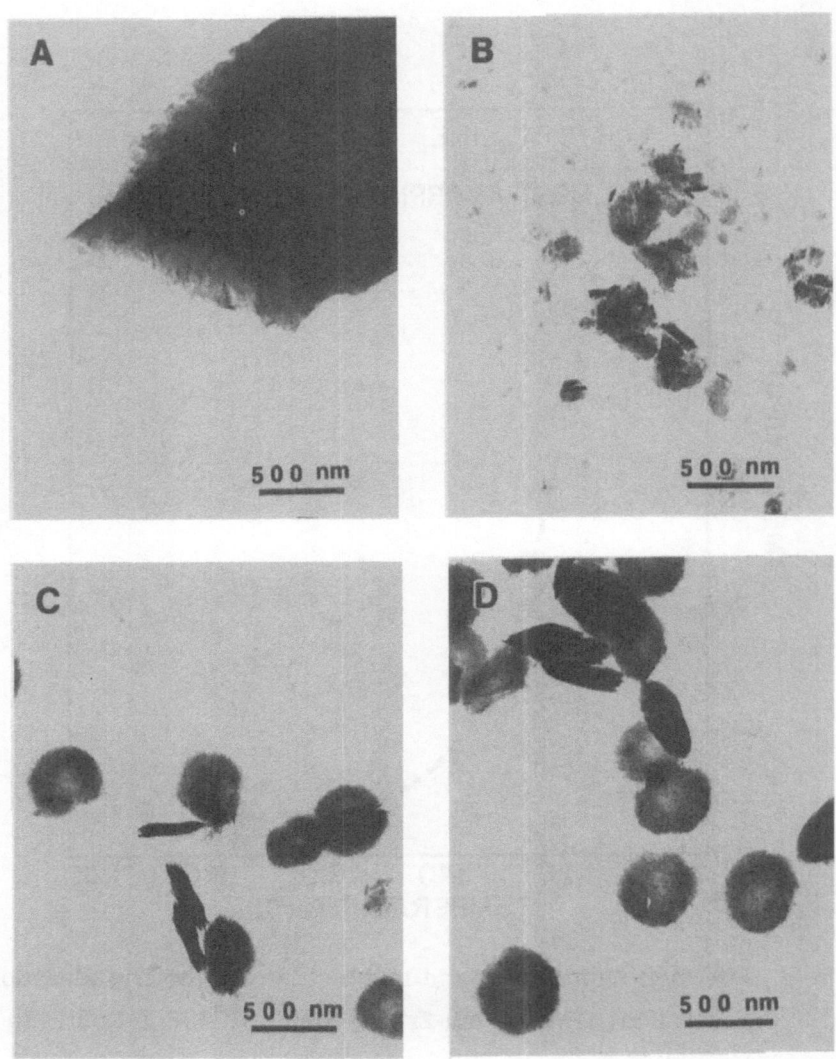

369

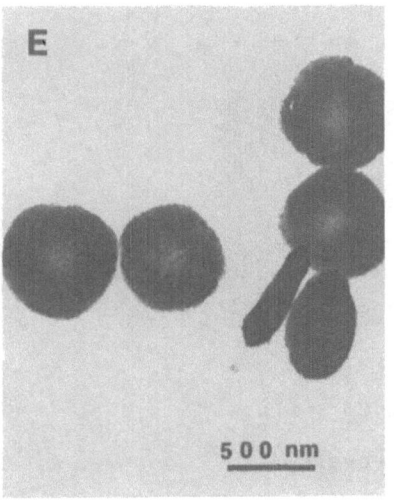 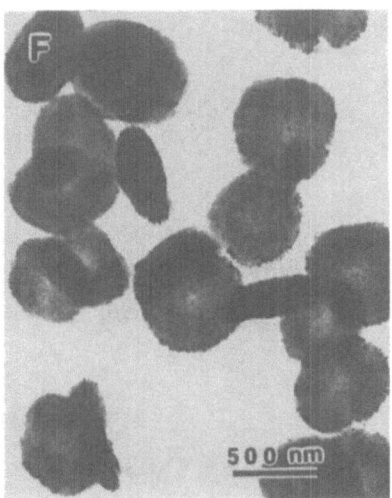

Figure 6. Transmission electron micrographs of zircon crystallization from TEOS (5 ml)-$ZrOCl_2 \cdot 8H_2O$ (7.2237 g)-EtOH (20 ml)-H_2O (20 ml) under hydrothermal conditions: (A) Amorphous mass with few crystals after treatment at 135°C for 6 hrs; (B) non-uniform sized zircon after treatment at 150°C for 6 hrs; (C) uniformsized zircon after treatment at 150°C for 48 hrs; (D) uniform sized zircon after treatment at 175°C for 6 hrs; (E) uniform sized zircon after treatment at 175°C for 48 hrs; (F) uniform sized zircon after treatment at 200°C for 48 hrs.

Preparation of Yttrium Carbonate by Hydrothermal Precipitation Method

Kazumitsu Hishinuma*, Takao Kumaki*, Zenjiro Nakai*,
Tokuji Akiba*, Yukio Suwa* and Sigeyuki Sōmiya**

*Chichibu Cement Co., Ltd.
**Nishi-Tokyo University

Abstract

Fine yttrium carbonate (hydrate and hydroxide) powders were
prepared by hydrothermal precipitation from the mixed solution of
yttrium chloride and urea. The mixed solutions were treated under
various hydrothermal conditions in a zirconium-lined autoclave.
Under the hydrothermal condition, urea acts as a precipitation
agent dy decomposing into CO_2 and NH_3.

The products were washed with distilled water and ethanol by
centrifugation to remove Cl^- and NH_4^+ ions. The products obtained
were crystalline yttrium carbonate powders involved water. The
water content were depended on the hydrothermal temperatures.
Some unkown phases were obtained by hydrothermal precipitation at
220°C.

Observation under SEM indicated that the plate-like or
foliated powders. The particle morphology was depended on the
hydrothermal conditions. THe crystalline yttrium carbonate was
transformed to amorphous by calcination at 500°C and yttrium oxide
above 700°C.

1. Introduction

Yttrium has played an important role as a fluorescent substance and a ceramic raw material so far. $YBa_2Cu_3O_{7-\delta}$ has recently gained attention as a high-temperature superconducter[1] and accodingly yttrium compounds have increasingly become the object of public attention. Considering the background of the yttrium compounds, the authours have tried to synthesized yttrium carbonate by the hydrothermal precipitation method.

Although yttrium is generally used as oxide, the demands of chloride, nitrate and carbonate have been increasing as the rang of use of yttrium has increased. Yttrium carbonate is produced by adding ammonium hydrogencarbonate to a water-soluble yttrium salt solution. Long time aging or hydrothermal treatment is, however, required to prepare cryatalline yttrium carbonate powder.

Crystalline yttrium carbonate is known as hydrate (tengerite) and hydroxide (ancylite). Perttunen[2] and Tareen et al.[3] presented X-ray diffraction data.

Nagashima et al.[4] prepared yttrium carbonate hydrate in such a way that trichloroacetic acid was added to an yttrium chloride aqueous solution, Which was neutralized with ammonia water, and the heated solution was aged for a long times. Beall[5] and Tareen et al.[3] prepared yttrium hydroxide carbonate by the hydrothermal process. Beall et al. synthesized it by the hydrothermal treatment of commercially available yttrium carbonate at 360°C for 168 hours under 76 MPa in an ammonium chloride solution saturation with carbon dioxide. Tareen et al. prepared it by the hydrothermal treatment of yttrium hydroxide gel using formic acid as the mineralizer at 220 to 260°C for 80 to 90 hours under 35.3 MPa.

The authers previously prepared fine crystalline zirconia powder by the hydrothermal treatment of a zirconium oxychloride solution containing urea at approximately 200°C[6,7]. Urea was hydrolyzed in the hydrothermal process into ammonia and carcon dioxide. While most of ammonia was cosumed in the neutralization reaction, carbon dioxide remained unchanged ia an autoclave, keeping the partial pressure considerably high. Just as it was estimated that yttrium carbonate would be also prepared in the same way as that, yttrium carbonate was successfully synthesized by this hydrothermal process.

2. Experimental method

Fig.1 illustrates the flow sheet of the experiment. An yttrium chloride aqueous solution was obtained by dissolving 99.9 %-yttrium oxide[*1] in 4N-hydrochloric acid at 70°C. The aqueous solution was diluted to 0.25, 0.5 and 1.0 mol/L with pure water and these diluted solutions were used as the stating solutions Urea 1.2 times as much as the amount required for neutralization was added to each solution and the solution was hydrothermally treated at a temperature of 140 to 220°C for one to five hours. The solution was heated up to the specified temperature at the rate of 100°C/h and stirred at 500 rpm during the hydrothermal treatment process. An 1L-zirconium-lined autoclave with a stirrer was used for the hydrothermal treatment and 600mL of the solution was chaged into it.

The reaction products were washed in a centrifugal separater untill neither chlorine nor ammonium ions were detected and dried at 120°C for 24 hours in an air bath.

The reaction products were analyzed by the powder X-ray diffractiometry (XRD; RU-200, Rigaku Electric Co., Ltd., Tokyo,

*1) Made by Nippon Yttrium Co., Ltd.

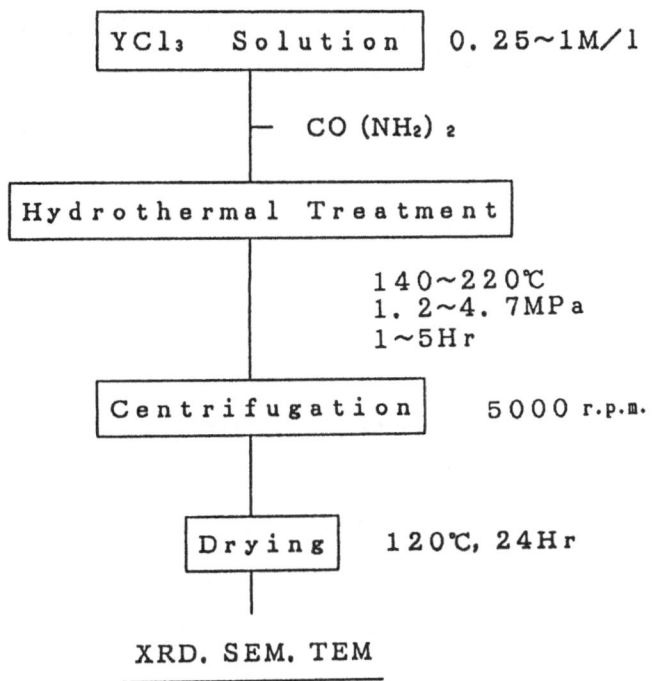

Fig. 1 Processing flow sheet of hydrothermal treatment.

Japan). As occasion demands, the reaction products were analyzed
by the thermogravimetry and differential thermal analysis (TG and
DTA; TG/DTA 300, Seiko Instruments & Electronics Ltd., Tokyo,
Japan), and the particle shape was determined by the scanning
electron microscopy (SEM; JSM-255, JEOL Ltd., Tokyo, Japan) and
transmission electron microscopy (TEM; JEM-2000EX, JEOL Ltd.,
Tokyo, Japan).

3. Results and Discussion

The pressure during the hydrothermal treatment process
depended upon the temperature and concentration of solution and
it was increased up to the pressure higher than the satureted
vapor pressure by the production of carbon dioxide and ammonia.
For instance, the pressure of an 1 mol/L-solution was 1.2 MPa at
140°C and 4.7 MPa at 220°C. Precipitable white powders were
produced under all the hydrothermal treatment conditions in the
experiments and the dried powders were bulky.

Fig.s 2 and 3 illustrate the powder X-ray diffraction patterns.
Fig. 2 shows the X-ray diffraction patterns of the products obtained
by the hydrothermal treatments of 1mol/L-solutions at various
temperatures for 5 hours. Considerably crystalline yttrium
carbonate hydrate $[Y_2(CO_3)_3 \cdot nH_2O$ (n = 2 or 3)], YI, was produced
at 140°C. Yttrium hydroxide carbonate $[Y(OH)CO_3]$, YII, was mixed
in the hydrate with the increase of temperature and the mixture
of YII and an unknown phase, XI, was produced at 220°C.

Another unknown phase, XII, was produced from a 0.25 mol/L-
solution hydrothermally treated at 220°C for 5 hours as shown in
Fig.3. The diffraction pattern of XII is sharp and the highest peak
is observed at a position of 2 of 12.04 (d = 7.3) on a lower angle
side in Fig.3. The peak cannot be, therfore, identified by the

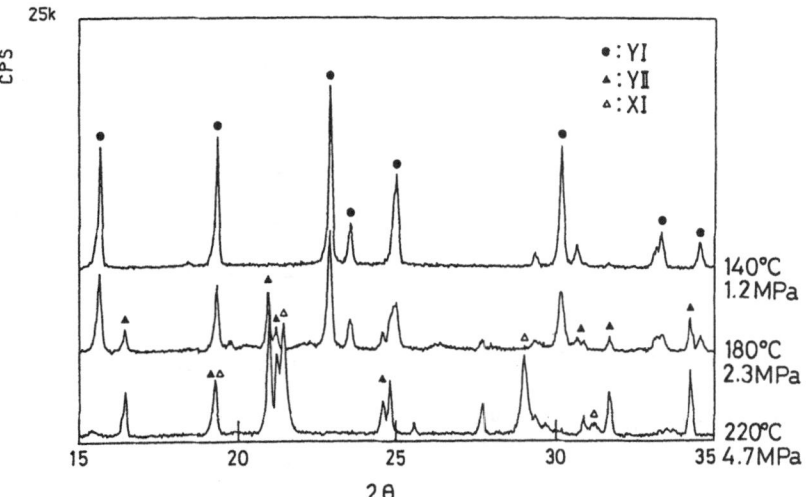

Fig.2   X-ray diffraction patterns of products by hydrothermal
precipitation at various temperatures (1M/1,5Hr).

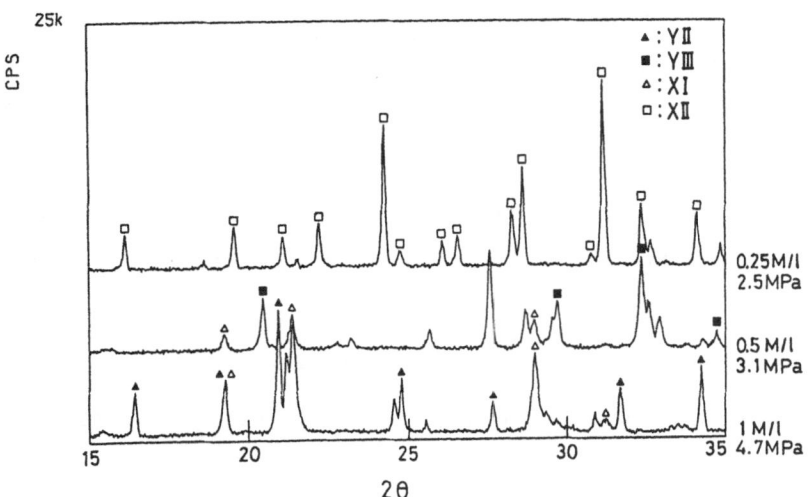

Fig.3   X-ray diffraction patterns of products by hydrothermal
precipitation at various concentrations (220°C,5Hr).

JCPDS card. The chemical composition of the unknown phase, XII, was Y : CO_2 = 1 : 0.24. The reaction product from a 0.5 mol/L-solution was composed of YII, XI and yttrium hydroxide [$Y(OH)_3$], YIII. It is considered that such a veriety of phases are produced because the amount of urea added and consequently the partial pressure of carbon dioxide vary acorrding to the concentration of solution.

The results of powder X-ray diffractiometry of the reaction products obtained under various hydrothermal treatment conditions are shown in Table 1. Only YI was produced from the 0.5 and 1.0 mol/L-solutions at 140 and 180°C, and YII and XI were produced at 220°C. YII was produced even at 180°C for a long reaction time.

Fig.4 shows SEM photgraphs of the reaction products obtained by hydrothermal treatment of 1 mol/L-solutions at various temperatures for 5 hours. Photo A showing the products at 140°C reveals anisotropic plate paticles 5 μm long and 1 μm wide each. Photo B showing the product at 180°C reveals hexagonal plate particles and Photo C showing the product at 220°C reveals flaky particles approximately 1 μm in size. It is known that hexiagonal plate particles are produced by hydrothermally treating the orthorhombic hydrate and hydroxide[3.8.9]. Chai et al. [9] produced yttrium carbonate [$Y_2O_2CO_3$ and $Y_2O(OH)_2CO_3$] by the hydrothermal process though X-ray diffraction date were not disclosed. Although they mentioned that the crystal of $Y_2O_2CO_3$ is hexagonal and it fomes also hexagonal plate particles, it is considered difficult to estimate the product from the shape of particle.

The thermal decomposition process of yttrium carbonate has been investigated by several researchers. A study by Nagashima et al. [4] reveals that large adsorption peaks are observed at 360°C

Table. 1   Products prepared by hydrothermal
           precipitation under various
           conditions

1Hr

Conc. (M/1) Temp.(℃)	140	180	220
0. 25	Y I	Y Ⅱ	Y Ⅱ
0. 5	Y I	Y I	Y I + Y Ⅱ + X I
1. 0	Y I	Y I	Y I + Y Ⅱ + X I

5Hr

Conc. (M/1) Temp.(℃)	140	180	220
0. 25	Y I	Y Ⅱ + X Ⅱ	X Ⅱ
0. 5	Y I	Y Ⅱ	Y Ⅲ + X I
1. 0	Y I	Y I + Y Ⅱ + X I	Y Ⅱ + X I

Y I : Y$_2$ (CO$_3$) $_3$ · 3H$_2$O   X I : ? Unknown
Y Ⅱ : Y (OH) CO$_3$ X Ⅱ : ? Unknown
Y Ⅲ : Y (OH) $_3$

Fig.4 SEM photographs of the products by hydrothermal
precipitation at various temperaturea (1M/L, 5Hr).
(A) 140°C, 1.2MPa   (B) 180°C, 2.3MPa
(C) 220°C, 4.7MPa   (bar=10um)

*2) and 560 and 610°C on the DTA curve of synthesized yttrium
carbonate hydrate, corresponding to the dehydration and decarbonation reactions, respectively. D'assunção[10] reported that water
of crystallization in yttrium hydroxide carbonate is released at
220°C and the dehydroylation and decarbonation reactions of it
take place at 520 to 640°C.

Fig.5 illustrated the DTA curves of reaction products obtained
in the experiment. The figure reveals that an adsorption peak
corresponding to the dehydration and those corresponding to the
decarbonation are observed at 361°C and 466, 546 and 574°C,
respectively, on the DTA curve of yttrium carbonate hydrate, and
that no absorption peak corresponding to the dehydration is
observed and absorption peaks corresponding to the dehydroxylation
and decarbonation are observed at 443, 496, 550 and 631°C on the
DTA curve of yttrium carbonate hydroxide. These curves are almost
equivalent to the past experimental data. As mentiond above, two
or more absorption peaks corresponding to the decarbonation are
observed in the DTA curves. Maybe this is because òxo-carbonate
was produced on the way of the reaction. Oniy an absorption peak
is observed at 488°C on the DTA curve of unknown phase, XII. This
is different from the above-mentioned DTA curves. It was found,
therefore, that XII is decarbonated in an one-step process.

Fig.6 shows powder X-ray diffraction patterns of the products
obtained by calcining yttrium carbonate hydrate. Amorphous product
was formed at 500°C and yttria was produced at 700°C. All the
products including unknown phases obtained by the hydrothermal
treatment were transfomed into yttrium oxide by calcining at 700°C

*2) Value from DTA curve

The paper describes that the valu is 260°C

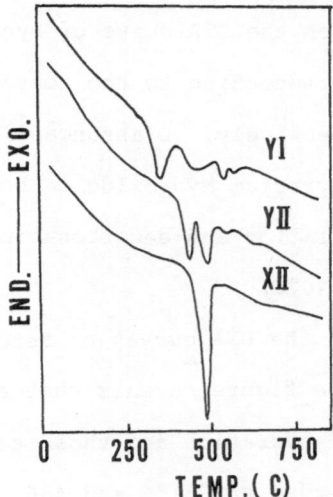

Y I : $Y_2(CO_3)_3 \cdot 3H_2O$ X II : ? Unknown
Y II : $Y(OH)CO_3$

Fig.5 DTA curves of the products prepared by hydrothermal

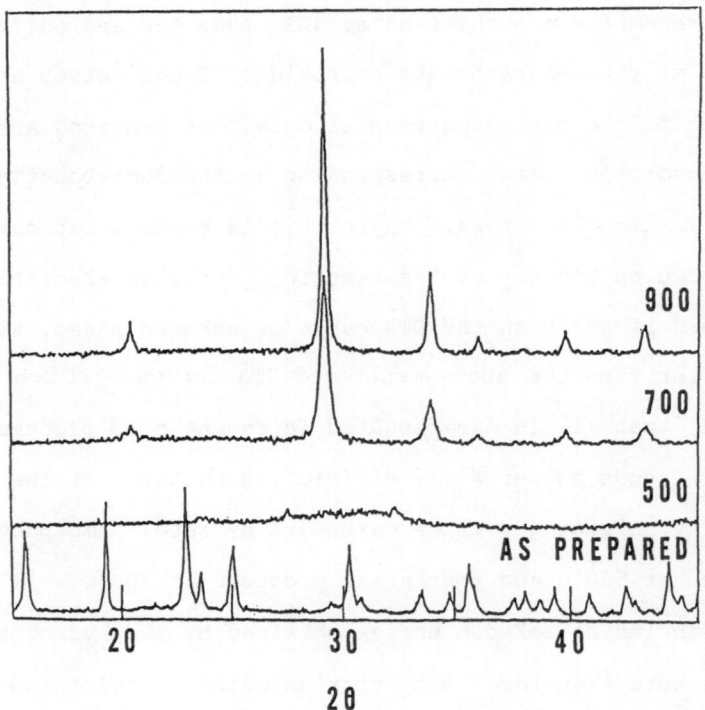

Fig.6 X-ray diffraction patterns of the products

calcined at various temperatures.

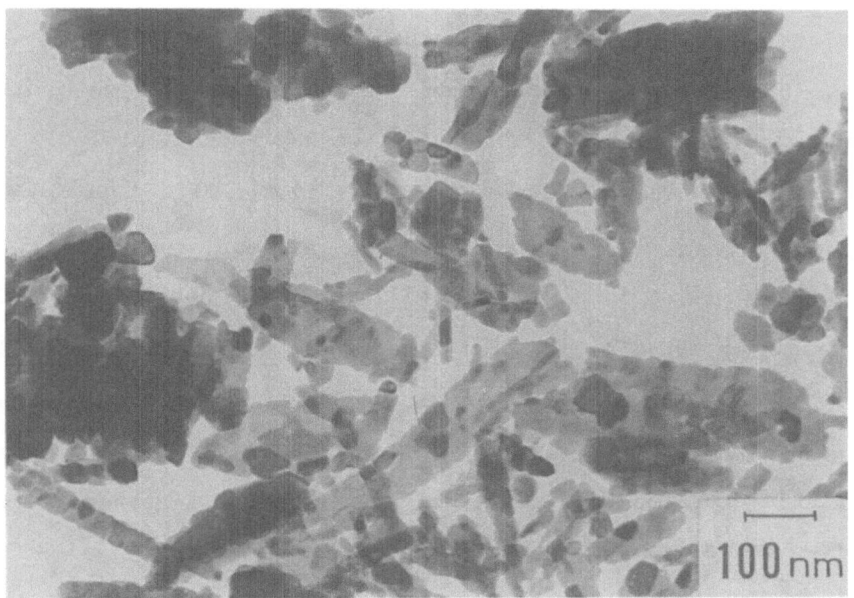

Fig.7　TEM photograh of the polycrystalline yttrium oxide
by calcined at 900°C.

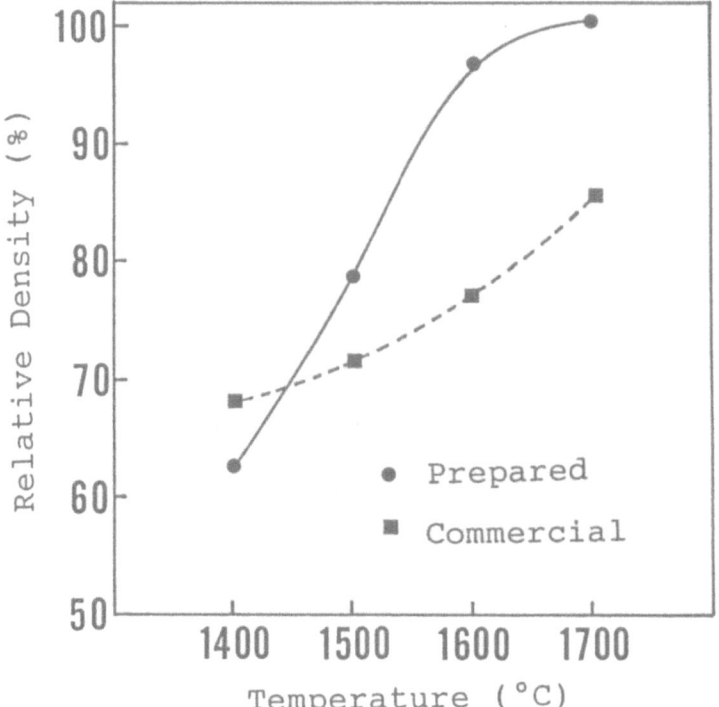

Fig.8　Effct of temperature on sintered density.

for one hour. Fig.7 shows a TEM photograph of yttrium oxide obtained by calcining yttrium carbonate hydrate at 900°C for one hour. Skeletal powder was prepared by calcining yttrium carbonate hydrate and the diameter of crystallite was approximately 30 nm. The skeletal powder can be easily pulverized. Affter calcining yttrium carbonate hydrate at 900°C for one hour, the skeletal powder was milled in a wet ball mill for three hours and then molded under 198 MPa in a pelletized. The sintering curves of the pellet are shown in Fig.8. Yttrium oxide produced in the experiment has superir sintering characteristics to those of commercially available one, and the density of yttrium oxide sintered at 1700°C for two hours reached neary the theoretical density[*3].

The indentication of the unknown phases and investigation of sintering characteristics will be further made.

4. Conclusions

It became possible to produce crystalline yttrium carbonate by the hydrothermal precipitation method in ashort time. The water content varied according to the hydrothermal treatment conditions. Hydrate was produced at low temperatures, while hydroxide was produced at high temperature. Moreover, it was founed that unknowm phases were produced under certain hydrothermal precipitation conditions. Those reaction products were transfomed into yttrium oxide by the calcination at 700°C or higher. The yttrium oxide produced by calcining yttrium carbonate hydrate has excellent sintering characteristics and the density of yttrium oxide produced by sintering at 1700°C reached nearly the theretical value.

*3) 5.031 g/cm^3

Reference

1) M. K. Wu, J. R. Ashburn, C. J. Torng, P. H. Hor, R. L. Meng, L. Cao, Z. J. Huang, Y. Q. Wang and C. W. Chu: "Super-conductivity at 93 K in a New Mixed-Phase Y-Ba-Cu-O Compound System at Ambient Pressure", Physical Review Letters, 58, 9, 903-910 (1987).

2) V. Perttunen: "Lokkaite, A New Hydrous Re-Carbonate from Pyoronmaa Pegmatite in Kangasala, SW-Finland", Bull. Geol. Soc. Finland, 43, 67-72 (1971).

3) J. A. K. Tareen, M. N. Viswanathiah and K. V. Krishnamurthy: "Hydrothermal synthsis and Growth of $Y(OH)CO_3$-Ancylite Like Phase", Revue de Chimie Minérale, t. 17, 50-57 (1980).

4) K. Nagashima and H. Wakita: "Composition of Tengerite", Bull Chem. Soc. Japan, 856-859 (1968).

5) G. W. Beall, W. O. Milligan and Stanley Mroczkowski: "Yttrium Carbonate Hydroxide", Acta Cryst., B32, 3143-3152 (1976).

6) S. Sōmiya, M. Yoshimura, Z. Nakai, K. Hishinuma and T. Kumaki: "Hydrothermal Processing of Ultrafine Single-Crystal Zirconia and Hafnia Powders with Homogeneous Dopants", pp.43-55 in Advances in Ceramics, Vol. 21, Edited by G. L. Messing, K. S. Mazdiyasni, J. W. McCauley and R. A. Haber, The American Ceramic Society, inc., 1987.

7) K. Hishinuma, T. Kumaki, Z. Nakai, M. Yoshimura and S. Sōmiya: "Characterization of Y_2O_3-ZrO_2 Powders Synthesized under Hydrothermal Conditions", pp.201-209 in Advances in Ceramics, Vol. 24, Edited by S. Sōmiya, N. Yamamoto and H. Yanagida, The American Ceramic Soc. Inc., 1988.

8) J. A. K. Tareen, T. R. Narayanan Kutty and K. V.
   Krishnamurty: "Hydrothermal Growth of $Y_2(CO_3)_3 \cdot nH_2O$
   (Tengerite) Single Crystals", J. Cryst. Growth, 49,
   761-765 (1980).

9) B. H. T. Chai and S. Morczkwski: "Synthesis of Rere-Earth
   Carbonates under Hydrothermal Conditions", J. Cryst.
   Growth, 44, 84-89 (1978).

10) L. M. D'assuncao, I. Giolito and M. Ionashiro: "Thermal
    Decomposition of The Hydrated Basic Carbonate of Lanthanides
    and Yttrium", Thermochimica Acta, 137, 319-330 (1989).

HIGH PRESSURE HYDROTHERMAL GROWTH OF BERYL SINGLE CRYSTALS

KOHEI KODAIRA and TSUYOSHI FURUSAKI
Department of Applied Chemistry, Faculty of Engineering,
Hokkaido University, Sapporo, 060 JAPAN

ABSTRACT

Beryl crystals were grown under high pressure hydrothermal condition of 200 MPa and 1 GPa. Transparent and smooth crystals were grown on the seed at 550° – 700°C from 0 – 0.3N NaOH solutions. The optimum crystal growth was observed at 600° C from 0.1N NaOH for 1GPa and from 0.025N NaOH solution for 200 MPa. Growth rates along [$11\bar{2}0$], which were considerably influenced by NaOH concentration, were 40 μ m/day for 0.025N and 25 μ m/day for 0.05N NaOH at 650°C under 200 MPa. The beryl crystals incorporated water molecules and alkali cations in the channels of the structure. The crystals showed same refractive indices and density as those of natural emerald crystals.

INTRODUCTION

Beryl, $Be_3Al_2Si_6O_{18}$, is found with a variety of impurity cations leading to its well-known beautiful colors and its favorable use as a gem stone. The structure consists of six-membered rings of silica tetrahedra cross-linked by Be-containing tetrahedra and Al-containing octahedra to form tetrahedral frameworks with open channels that parallel the c-axis. Natural beryls commonly contain varying amounts of alkali cations, as well as water molecules in the channels. Natural beryls and synthetic ones are clearly characterized by alkali cations in the channels; cations are present in natural crystals and absent in synthetic ones(1).

The most spectacular growth of beryl was conducted by hydrothermal method(2), but there are no reports on synthesis of beryl single crystals containing alkali cations in the channels. We have already tried to establish a high pressure hydrothermal method under 1 GPa, by which we have prepared beryl single crystals which is similar to natural ones(3). In this report, properties of beryl single crystals grown under 200 MPa and 1 GPa are described in detail with the preparation method.

EXPERIMENTAL

Reagent grade Be(OH)$_2$(Kishida chemical Co., Ltd.) and Al(OH)$_3$(Wako pure chemical industries Ltd.) powders, and synthetic quartz blocks(made by Toyo communication equipment Co., Ltd.) were used as starting materials. These powders and blocks with the stoichiometric composition of beryl (3BeO·Al$_2$O$_3$·6SiO$_2$) were placed with 0 – 0.3N NaOH solutions in a growth chamber. An internal heating autoclave was used for the experiments under 200 MPa. A piston cylinder type vessel was also used for the experiments under 1 GPa. As a seed crystal, synthetic emerald was supported with Pt wire at an upper part of the chamber, in the middle of which a baffle plate with 20% opening space was attached to moderate thermal convection. The details of these equipments have been described by the authors(4,5).

The experiments were conducted at 500° to 700°C under the pressure of 200 MPa and 1 GPa. After heating at the desired temperature and duration, the growth chamber was rapidly cooled to room temperature. The grown beryl crystal was cleaned with hot water. Infrared spectra were obtained in the wave number of the 4000 to 1200 cm^{-1} by using a double beam spectrometer(Japan spectroscopic Co., Ltd., A-202). Specific gravities were determined by a water displacement method. Refractive indices were also measured by an immersion method.

RESULTS AND DISCUSSION

Beryl crystal was grown on a seed crystal by dissolution-precipitation of each starting material. Table 1 shows the representative results obtained under high pressure hydrothermal conditions of 200 MPa and 1 GPa. Crystal growth of beryl was observed at 500° – 650°C and from 0 – 0.3N NaOH solutions. At higher concentration of 0.5N NaOH solution, only albite(NaAlSi$_3$O$_8$) was produced instead of beryl. A clear and smooth beryl crystal was grown under the optimum condition at 600°C and from 0.1N NaOH for 1 GPa and from 0.025N solution for 200 MPa. In the experiment under 1GPa, Pt wire was included by transparent grown layer in 1 mm thick as shown in figure 1. Representative p(10$\bar{1}$1) plane for hydrothermal growth was developed. The p(10$\bar{1}$1) plane seems to have relatively high growth rate, compared with those of m(10$\bar{1}$0) and c(0001) planes. In the experiments under 200 MPa, growth rates along [11$\bar{2}$0], which were considerably influenced by NaOH concentration, were 40 μ m/day for 0.025N and 25 μ m/day for 0.05N NaOH solution at 650°C (bottom). Many pyramidal growth patterns were observed on the surfaces of the crystals grown on the seed crystals with a(11$\bar{2}$0) and s(11$\bar{2}$1) planes as shown in figure 2.

In natural emeralds(4), water molecules and alkali cations are incorporated in the open channels as shown in figure 3. According to Wood(5), in the absence of alkali cations, water molecules are oriented in type I configuration. If alkali cations are also present in the

TABLE 1

Representative results by high pressure hydrothermal growth under 200 MPa and 1 GPa

NaOH (N)	Pressure (MPa)	Temperature (°C)	Durations (days)	Products
Water	1000	500	3	Be, Qu
Water	1000	650	3	Be
0.1	1000	600	6	Be
0.1	1000	700	5	Be, Ph
0.2	1000	600	5	Be, Ab
0.2	1000	650	4	Ab, Be
0.3	1000	600	4	Ab, Be
0.025	200	630	4	Cb, Ph*
0.025	200	700	3	Cb, Be
0.05	200	540	2	Cb, Be
0.05	200	650	4	Cb, Be
0.1	200	700	2	—

Be:Beryl(Be$_3$Al$_2$Si$_6$O$_{18}$), Ab:Albite(NaAlSi$_3$O$_8$), Qu:Quartz
Cb:Chrysoberyl(BeAl$_2$O$_4$), Ph:Phenacite(Be$_2$SiO$_4$), *:trace

388

channels, its electrostatic attraction rotates the adjacent
water molecules in type II configuration. These two types of
water molecules can be distinguished the origin of beryl by
IR spectra. There are large differences between IR spectra of
synthetic hydrothermal emerald and that by the authors.
Synthetic hydrothermal emeralds show only one absorption at
3690 cm^{-1}(7). In the beryl crystals obtained from 0.1 - 0.3N

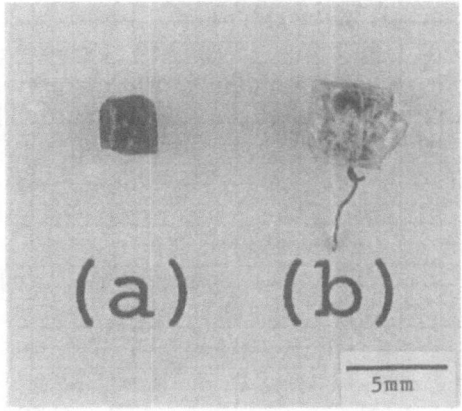

FIGURE 1. Flux grown seed(a) and beryl crystal(b) grown at
600°C under 1 GPa from 0.1N NaOH solution

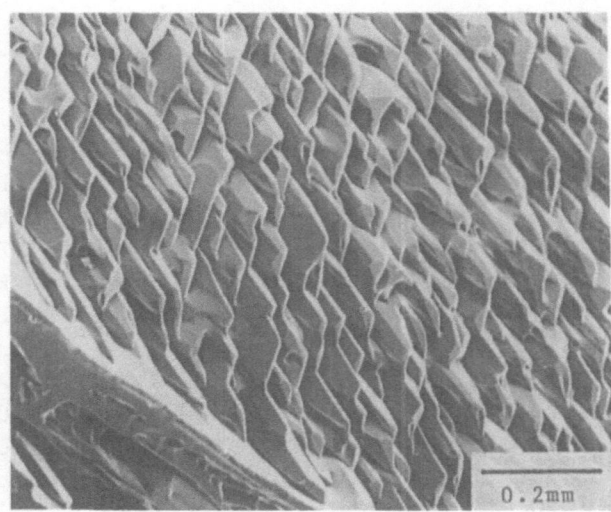

FIGURE 2. Scanning electron micrograph of the crystal grown
on s(11$\bar{2}$1) under 200 MPa from 0.025N NaOH solution for 4 days

NaOH solutions under high pressure hydrothermal conditions, two sharp absorption spectra corresponding to O-H vibration were observed at 3690 cm^{-1} (type I) and 3590 cm^{-1} (type II). The absorption belong to 3590 cm^{-1} was deeper with increasing the concentration of NaOH solution as shown in figure 4. Accordingly, the present beryl crystal (0.3N NaOH solution) was identical with natural Colombian emeralds in IR spectra(6).

The presence of alkali cations in the channels was also recognized from line analysis of seed and growth region in the crystal by EPMA(JEOL, JSCM-733) as shown in figure 5. The channels were also somewhat large for the direction of a-axis in the present beryl, compared with emerald grown by flux method due to the presence of alkali cations and water molecules.

As indicated in table 2, there are so many different type emerald distinguishing physical properties. The refractive indices were determined as ω =1.5880 and ε =1.5790. These values were intermediate between Linde hydrothermal beryls (ω =1.570-1.580 ε =1.562-1.572) and natural emeralds (ω =1.565-1.602, ε =1.558-1.596)(5). The density, 2.68 gcm^{-3}, was also identical with that of natural emeralds (2.67-2.70 gcm^{-3}).

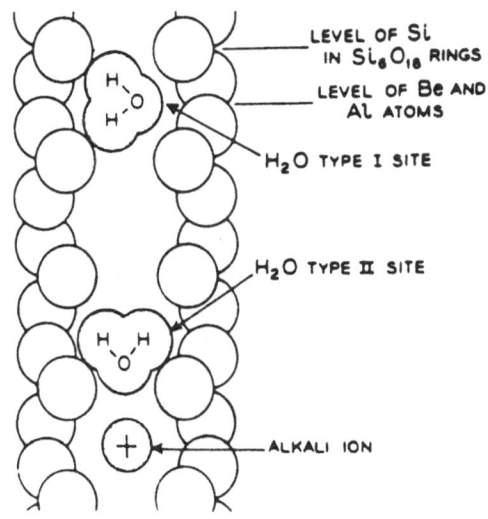

FIGURE 3. The location of water molecules and alkali cations in the channels of beryl structure

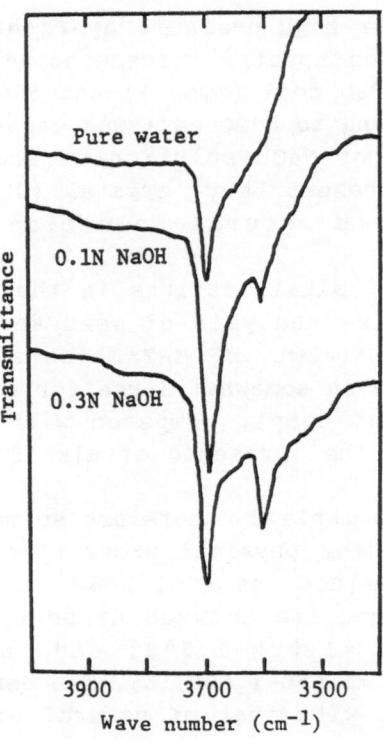

FIGURE 4. Infrared spectra of high pressure hydrothermal
beryl from pure water, 0.1N NaOH and 0.3N NaOH

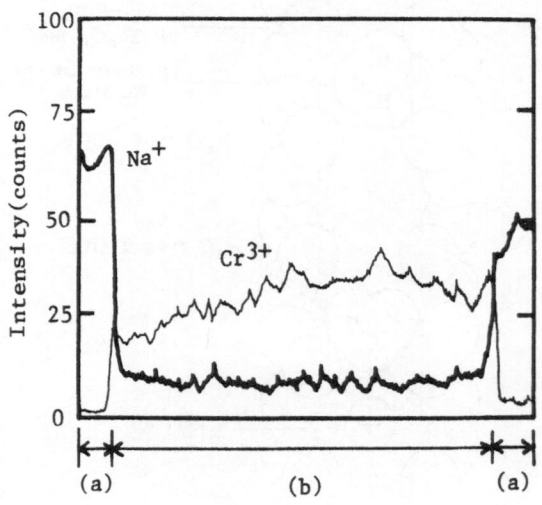

FIGURE 5. EPMA spectra for Na$^+$ and Cr^{3+} in the seed and
growth region

TABLE 2
Properties of different types of emeralds and beryl

	High pressure hydrothrmal beryl	Flux emerald	Hydrothermal emerald	Natural emerald
Specific gravity	2.70	2.56-2.67	2.67-2.69	2.69-2.74
Refractive index ω	1.588	1.560-1.563	1.566-1.576	1.565-1.586
Refractive index ε	1.579	1.563-1.566	1.571-1.587	1.570-1.593
Types of water	I, II	-	I	I, II
Alkali present	Na	Li	-	Na, Li, Cs
Inclusions	Very few	Phenacite Pt	Phenacite	Tremolite Mica

CONCLUSION

Beryl single crystals were grown at 500° -650℃ and from 0-0.3N NaOH solutions under high pressure of 200 MPa and 1 GPa. The optimum growth was observed at 600℃ and from 0.1N NaOH solution. Water molecules and alkali cations were incorporated in the open channels of the beryl structure, as in natural emeralds. The refractive indices of the crystals with the density of 2.68 gcm^{-3} were ω =1.5880 and ε =1.5790.

In a survey of high pressure hydrothermal growth, though sizable crystals are not yet obtained, the beryl crystals are identical with natural ones inasmuch as they incorporated water molecules and alkali cations in the channels.

REFERENCES

1. K. Nassau, Synthetic emerald :The confusing history and the current technologies. J. Cryst. Growth, 35, 211-222 (1976).
2. K. Nassau, Emerald and the beryl. Gems Made by Man, Chilton, Pennsylvania, 1980, pp 125-158.
3. K. Kodaira, Y. Iwase, A. Tsunashima and T. Matsushita, High pressure hydrothermal synthesis of beryl crystals. J. Cryst. Growth, 60, 172-174(1982).
4. T. Furusaki, Y. Bando, K. Kodaira and T. Matsushita, Properties of beryl single crystals grown by a high pressure hydrothermal method. Mat. Res. Bull., 24, 967-972(1989).

5. T. Furusaki, Y. Bando and K. Kodaira, Growth of beryl
   single crystals by internal heating hydrothermal apparatus
   J. Ceram. Soc. Jpn., 97, 1462-1465(1989).
6. D. L. Wood and K. Nassau, The characterization of beryl
   and emerald by visible and infrared absorption
   spectroscopy. Am. Mineralogist, 53, 777-806(1968).
7. E. M. Flanigen, D. W. Breck, N. R. Mumbach and A. M.
   Taylor, Characteristics of synthetic emeralds. ibid, 1967,
   52, 744-772.

Microstructure-designed Hydroxyapatite Ceramics

Prepared by Hydrothermal Hot-Pressing

Koji IOKU, Tokio KAI, Mamoru NISHIOKA,

Kazumichi YANAGISAWA, Nakamichi YAMASAKI.

Res. Lab. Hydrothermal Chemistry,

Faculty of Science, Kochi University,

2-5-1 Akebono-cho, Kochi 780, Japan.

Abstract

   Microstructure-designed hydroxyapatite ceramics, i.e. dense ceramics and
porous ceramics, were prepared by hot-pressing under hydrothermal conditions.
Fully dense hydroxyapatite ceramics could be obtained from ultra-fine
hydroxyapatite single crystals by sintering at 1050 ℃ for 3h in air, after
hydrothermal hot-pressing at 300℃ under 30MPa of mechanical pressing for 2h. This
ceramics had homogeneous grain size of about 0.5μm with few pores. Porous
hydroxyapatite ceramics with 42% porosity, about 300μm pores in diameter, was
prepared from mixture of $Ca(OH)_2$ and $(NH_4)_2HPO_4$ by sintering at 1050℃ for 3h in
air, after hydrothermal hot-pressing at 300 ℃ under 30MPa of mechanical
pressing for 2h. This ceramics had comparable compressive strength of 150 ± 20MPa
to that of cortical bone.

1. Introduction

Hydroxyapatite, $Ca_{10}(PO_4)_6(OH)_2$:(HAp), is expected as implant materials because of HAp's bio-compatibility due to similarity to natural bones and teeth mineral[2]. The implant materials for human bones and teeth require not only bio-compatibility but also mechanical strength at a certain level and porosity to promote the connection with tissues. Thus HAp ceramics as the implant materials requires various microstructure ; dense one and/or porous one[3-6]. The dense HAp ceramics is suitable for the bones substitute in high stress-bearing situations. While the porous HAp ceramics is applied to filling up defects of the bones. The pore size of this ceramics is required to be larger than the 100 μm above we may expect ingrowth of bone tissue[7].

While several ceramics can be prepared by pressing powders mechanically under hydrothermal conditions by hydrothermal hot-pressing technique[8-11], therefore this technique should be effective for preparation of the ceramics with hydroxyl group. This paper describes preparation of dense HAp ceramics and porous HAp ceramics with designed microstructure by hydrothermal hot-pressing.

2. Experimental methods

The autoclave for hydrothermal hot-pressing used in this work (Fig.1) was a cylinder made of steel. A starting sample and cast rods were surrounded by a case consisted of three pieces, thus the sample could be taken out easily. The starting sample with distilled water in the chamber was compressed uniaxially by cast rods from above and below. The cast rods had a space for water retreat, into which water included in the starting sample was released. Gland packing made of Teflon between cast rod and piston prevented leakage.

Preparation of the HAp ceramics by hydrothermal hot-pressing is shown following process. Dense HAp ceramics was obtained from ultra-fine HAp single crystals[12]. The crystals were prepared hydrothermally in an autoclave at 200℃ for 10h from the precipitate of the solutions of $Ca(NO_3)_2$ (reagent gade, Wako Pure Chemical Industries, LTD.) and $(NH_4)_2HPO_4$ (reagent grade, Wako) in the HAp

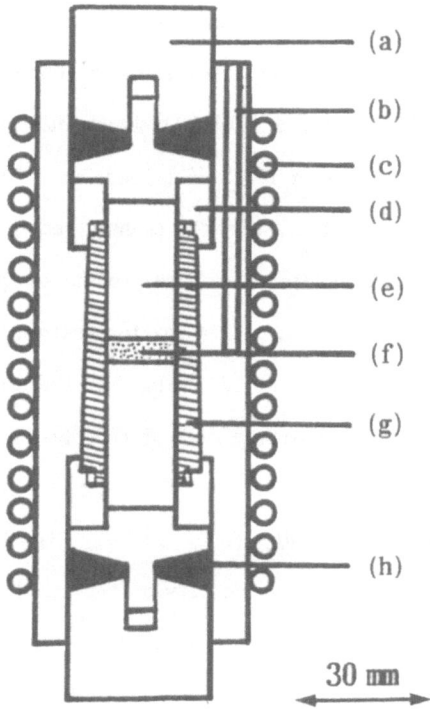

Fig.1 Autoclave for hydrothermal hot-pressing; (a) piston for packing, (b) well for thermocouple, (c) heater, (d) space for water retreat, (e) cast rod, (f) sample, (g) sample case, and (h) gland packing.

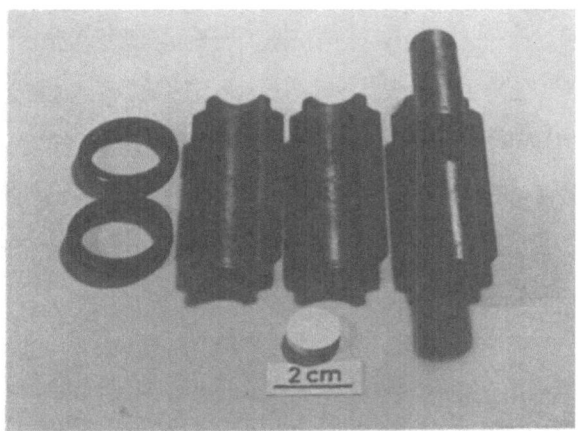

Fig.2 Sample case for hydrothermal hot-pressing and the solidified disk of HAp hot-pressed hydrothemally at 300 ℃ under 30MPa for 2h from ultra-fine HAp single crystals (15mm φ × 3mm).

stoichiometry (Ca/P=1.67). The crystals were kneaded with 10wt% distilled water for 30min by mortar. Porous HAp ceramics were obtained from the mixture of Ca(OH)$_2$ (reagent grade, Wako) and (NH$_4$)$_2$HPO$_4$ in the HAp stoichiometry (Ca/P=1.67). The mixture were kneaded with 10wt% distilled water for 30min by mortar. These samples kneaded with distilled water were transferred into the sample case of the autoclave of hydrothermal hot-pressing, and pressed mechanically under 30MPa. The samples were heated to 300℃, and kept the temperature for 2h. The heating rates was kept constant for all runs at 15℃/min. Then the samples were cooled to room temperature in air. The solidified disk of HAp obtained was sintered at 900 ～ 1100℃ for 3h in air to bring about densification.

The solidified disk of HAp (15mm φ×3mm, Fig.2) was used for following measurement and observation. Bulk density of the samples was evaluated from the weight and the size of them. Relative density of the ceramics was calculated as ratio against theoretical density (3.16g/cm^3) of HAp[13]. Porosity of the porous ceramics was measured by the Archimedean method with water[14]. Crystalline phases of the samples were identified by powder X-ray diffractometry (XRD ; RAD-RC, Rigaku Denki Co.) with carbon monochromated Cu-Kα, operating at 40KV and 100mA. The microstructure of the sintered ceramics was observed by a scanning electron microscope (SEM ; S530, Hitachi), after the polished surface was etched thermally at the temperature below 10 ℃ of the sintering temperature. Densification of the solidified disk of HAp prepared by hydrothermal hot-pressing was investigated by dilatometry (high-temperature type TMA ; Rigaku) in air at a heating rate of 10℃/min. Compressive strength of the porous ceramics (15mm φ×7mm) was measured by mechanical machine(RH-100, Shimazu).

3. Results and discussion

3-1. Preperation of dense HAp ceramics

No phases other than HAp were revealed by XRD in starting samples. TEM observation of the HAp powders demonstrated that the particles were fine single crystals with hexagonal prismatic shape (25nm ×90nm) (Fig.3). Hydrothermal

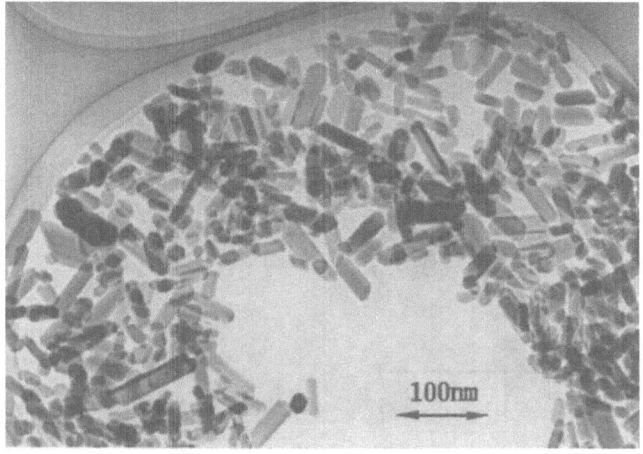

Fig. 3   TEM photograph of the HAp fine crystals synthesized hydrothermally at 200℃
under 2MPa for 10h.

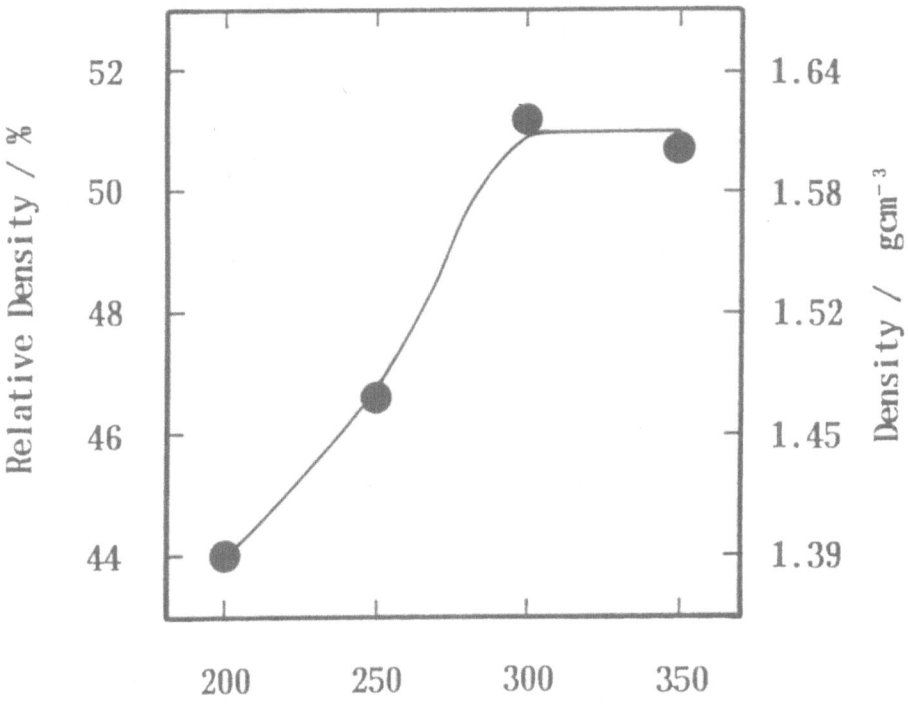

Fig. 4   Relative density of the solidified disk of HAp hot-pressed hydrothermally
at 200 °～350 ℃ under 30MPa for 2h from ultra -fine HAp single crystals.

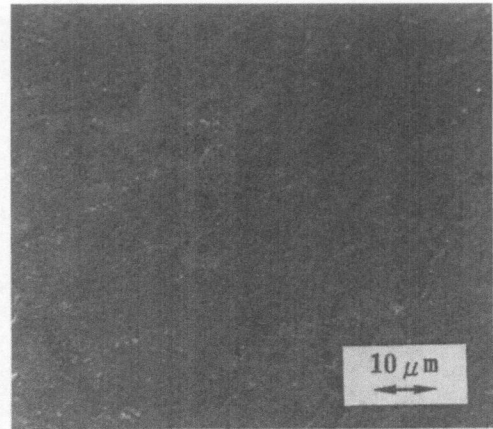

Fig.5 SEM photograph of the solidified disk of HAp hot-pressed hydrothermally at
300℃ under 30MPa for 2h from ultra-fine HAp single crystals.

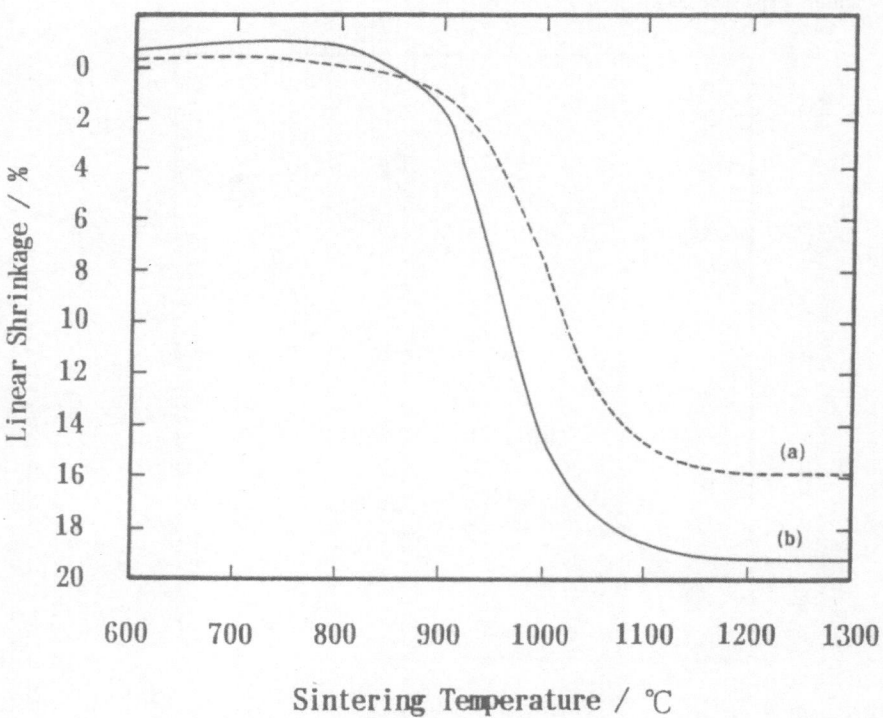

Fig.6 Densification of solidified disk of HAp prepared from ultra-fine HAp single
crystals (heating rate : 10℃/min.) :
(a) cold isostatic pressed sample (CIP'ed under 200MPa for 5min.), and
(b) hydrothermal hot-pressed sample (at 300℃ under 30MPa for 2h).

hot-pressing at the temperature from 200° to 350 ℃ under 30MPa of mechanical pressing for 2h brought about densification with increasing temperature to result in a maximum relative density (Fig. 4) of about 51% (1.61g/cm³) at 300 ℃. Therefore hydrothermal hot-pressing temperature was decided to be at 300℃. SEM observation of the solidified disk of HAp hot-pressed hydrothermally at 300 ℃ under 30MPa of mechanical pressing for 2h demonstrated that the microstructure was homogeneous. The micro pores with pore size of about $0.1\mu m$ were dispersed homogeneously (Fig. 5).

According to the dilatometry, densification of the solidified disk of HAp started at about 800℃ and proceeded gradually with increasing temperature to result in linear shlinkage of about 19% at 1150 ℃ (Fig. 6). The densification progressed at lower temperature than that of HAp compacts pressed isostatically (CIP) at room temperature. The solidified disk of HAp shows more homogeneous microstructure with micro-pores dispersion than that of the CIP'ed pellet (Fig. 7), because HAp particles are packed efficiently by water between particles. The CIP'ed compact of HAp, however, has pores in various sizes dispersed inhomogeneously because there are no lubricant media between particles. Therefore in the case of hydrothermal hot-pressing, efficient packing by water between particles caused the progress of the densification at lower temperature. HAp compacts with relative density of about 73% were preparedby hot isostatic pressing (HIP) at 550℃ under 140MPa for 3h from HAp powders added distilled water by Hirota et al. [15]. This indicated that hot-pressing under hydrothermal conditions was effective for densification of HAp. Accordingly hydrothermal hot-pressing should improve not only the packing of HAp particles but also the bonding between particles.

Sintering brought about densification of the solidified disk of HAp by hydrothermal hot-pressing at 300℃ to result in a relative density of about 99% (3.13g/cm³) at 1050 ℃ for 3h (Fig. 8). SEM observation demonstrated that HAp ceramics sintered at 1050 ℃ had the dense microstructure with few pores and the grain size of about 0.5 μm (Fig. 9). Dehydrated HAp ($Ca_{10}(PO_4)_6(OH)_{2-2x}O_x\square_x$,

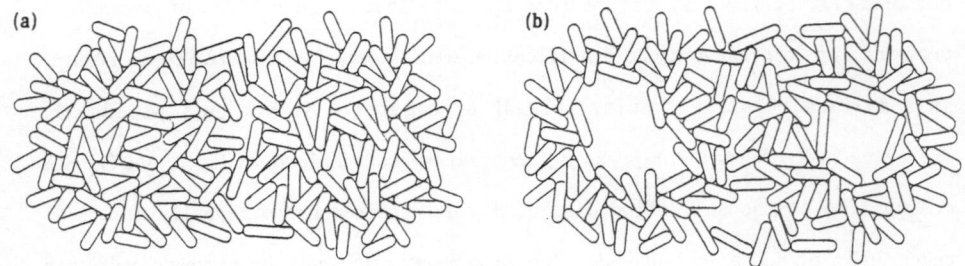

Fig. 7 Packing models of the HAp particles

   (a) packed by hydrothermal hot-pressing at 300 ℃ under 30MPa for 2h.

   (b) packed by cold isostatic pressing under 200MPa for 5min.

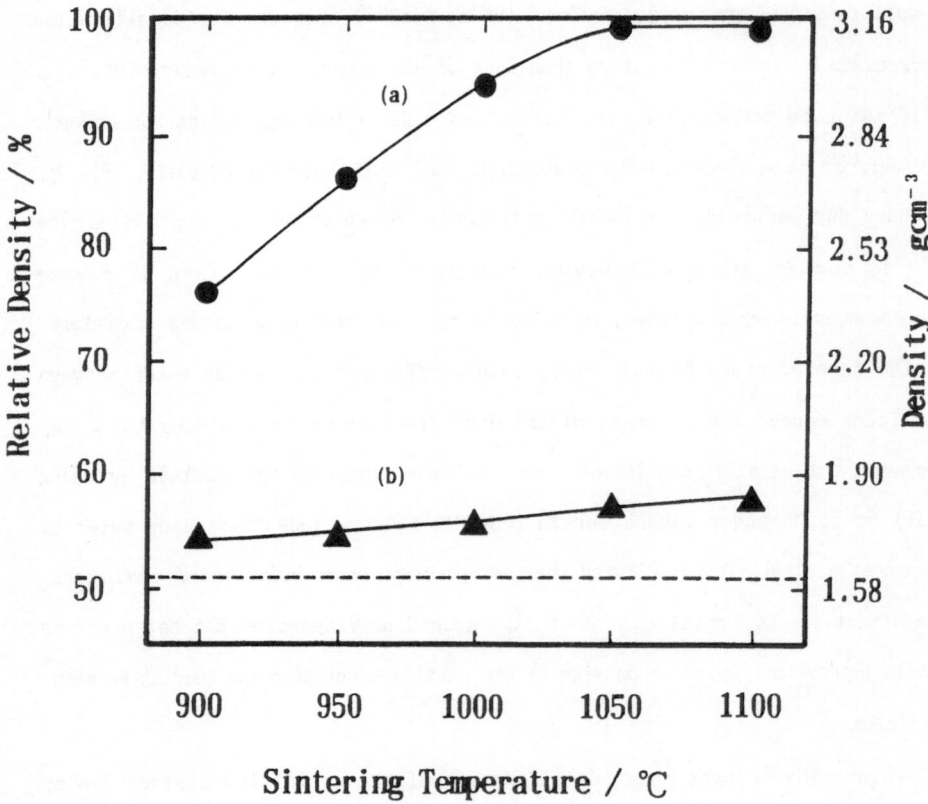

Fig. 8 Densification by sintering of the solidified disk of HAp prepared from

   (a)ultra-fine HAp single crystals and (b)mixture of Ca(OH)$_2$ and (NH$_4$)$_2$HPO$_4$

   by hydrothermal hot-pressing (heated at the indicated temp. for 3h in air).

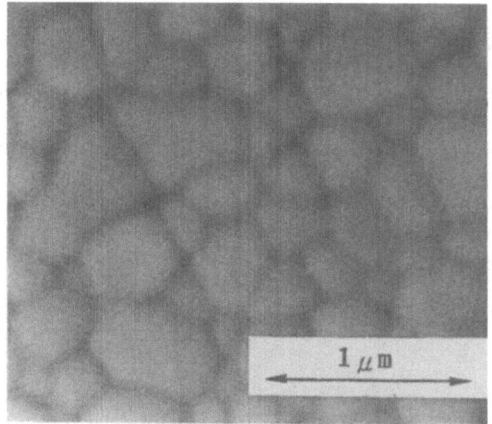

Fig. 9   SEM photograph of the dense HAp ceramics sintered at 1050℃ for 3h in air.

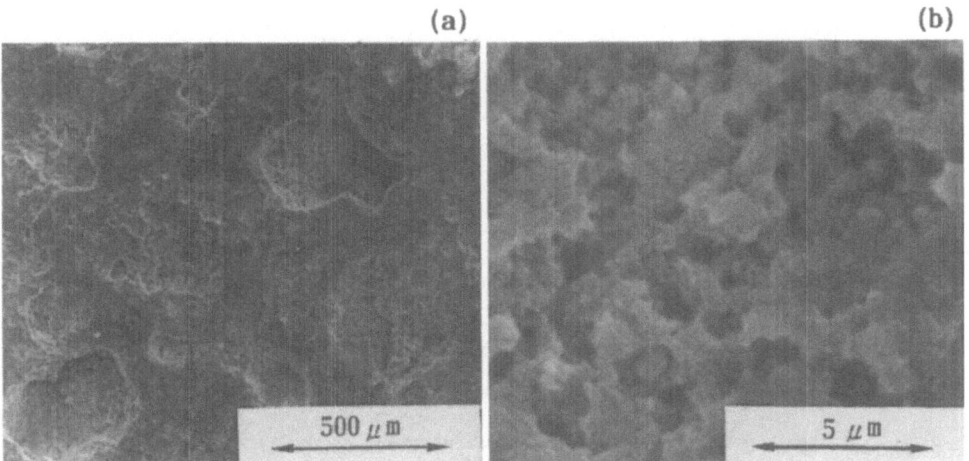

Fig. 10   SEM photograph of the solidified disk of HAp hot-pressed hydrothermally at
300 ℃ under 30MPa for 2h from mixture of $Ca(OH)_2$ and $(NH_4)_2HPO_4$.
(a) low magnification (×50) and (b) high magnification (×5000).

□ : vacancy, X < 1) ceramics can be regarded as being composed of oxyhydroxyapatite with vacancies located on OH sites and was believed to have a slightly lower bioactivity[16]. The previous paper reported that HAp synthesized under hydrothermal conditions did not dehydrate at lower temperature below 1050 ℃ [7]. Therefore sintering temperature should be lower than 1050℃ to prepare the pure HAp ceramics without dehydration. On this study, the dense HAp ceramics without dehydration could be prepared by sintering at 1050℃. The hydrothermal hot-pressing technique should be suitable for preperation of the dense HAp ceramics with stoichiometric composition.

3.2 Preparation of porous HAp ceramics

The reaction process into HAp from $Ca(OH)_2$ and $(NH_4)_2HPO_4$ expressed as follows was characterized by production of NH_3 gas to make large pores in the HAp and to keep the reaction atmosphere basic. HAp is the stable orthophosphate of calcium in neutral and alkaline media[17].

$$10Ca(OH)_2 + 6(NH_4)_2HPO_4 \rightarrow Ca_{10}(PO_4)_6(OH)_2 + 18H_2O + 12NH_3 \uparrow$$

The solidified disk of HAp prepared by hydrothermal hot-pressing at 300 ℃ under 30MPa for 2h showed about 51% relative density (1.61g/cm³) of a porous structure with 40% open porosity (Fig.10(a)). The pore size was 100-500 μm from scanning electron micrographs (SEM), which was larger than the 100 μm above we may expect ingrowth of bone tissue[3]. Figure 10(b) showed micro-pores with the sizes in the order of 1-5μm, which were seemed closed pores. The HAp ceramics sintered at 1050℃ for 3h in air had about 58% relative density (1.83g/cm³), therefore the open porosity scarcely decreased by sintering. In fact, the macro-pore size was almost same as that of the solidified disk of HAp before sintering (Fig.11). While the micro-pore in the solidified disk of HAp decreased extremely by sintering with increasing temperature.

The compressive strength (Fig.12) of the solidified disk of HAp was about

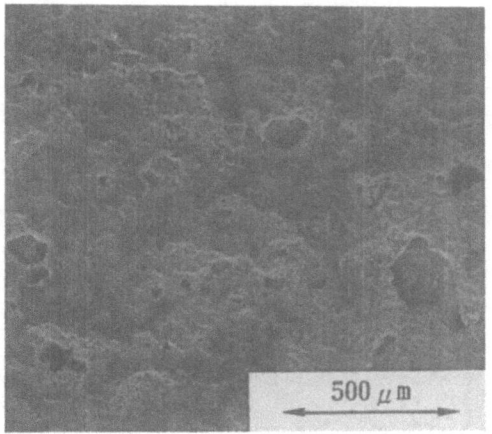

Fig. 11 SEM photograph of the porous HAp ceramics sintered at 1050℃ for 3h
in air.

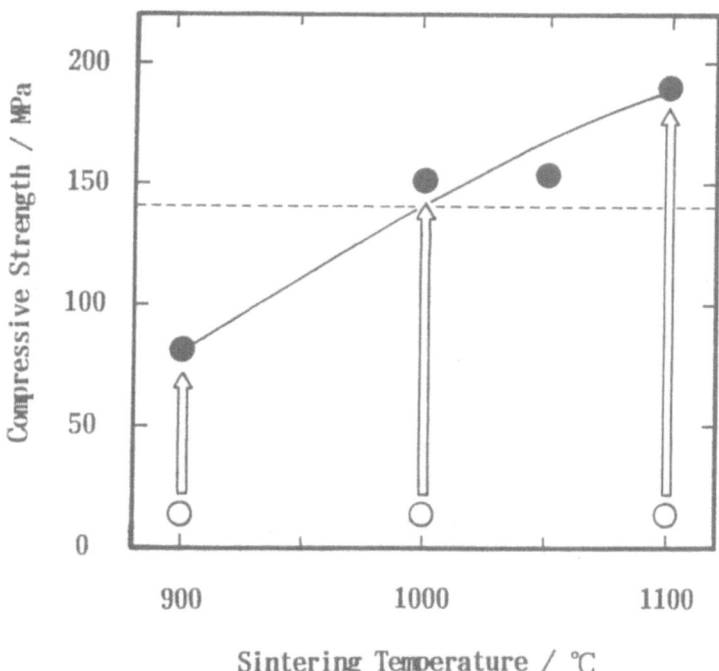

Fig. 12 Compressive strength of the porous HAp ceramics.

○: before sintering (hot-pressed hydrothermally at 300 ℃ under 30MPa
for 2h),

●: after sintering at 900〜ι 00℃ for 3h in air,

--- : compact bone.

15MPa. It was quite low compared with bone tissues (88.3~163.8MPa for cortical bone[1], 40 ~60MPa for cancellous bone[18]). The HAp ceramics sintered at 1050 ℃ for 3h showed comparable strength (150 ±20MPa) to that of cortical bone. Peelen et al.[18] have reported that the compressive strengths of porous apatite ceramics made by sintering ; ranged from 30 to 170MPa. The maximum strength is almost same as that of the ceramics prepared by sintering at 1050 ℃ after hydrothermal hot-pressing, but the sintering temperature performed by Peelen et al. was quite higher to obtain pure HAp ceramics without dehydration, that was 1250 ℃.

The compressive strength of porous HAp ceramics from the solidified disk prepared by hydrothermal hot-pressing could be varied within wide limits (15 ~ 150±20MPa) by changing the sintering temperature below 1050 ℃. These porous ceramics without dehydration had over 40% porosity with large pore size, therefore the ceramics can be expected to have good bonding-osteogenesis.

4. Summary

The solidified disk of HAp hot-pressed hydrothermally at 300 ℃ under 30MPa for 2h from ultra-fine HAp single crystals had about 51 % relative density. The solidified disk showed homogeneous microstructure, thus sintering at 1050 ℃ for 3h in air brought about densification of the HAp ceramics up to 99% density with few pores and the grain size of about 0.5 μ m.

The solidified disk of HAp hot-pressed hydrothermally at 300 ℃ under 30MPa for 2h from the mixture of Ca(OH)$_2$ and (NH$_4$)$_2$HPO$_4$ showed porous microstructure with about 49% porosity (open pore 40 %). The open pore with the sizes in the order of above 100μ m scarcely decreased by sintering at 1050℃ for 3h in air. This ceramics had comparable compressive strength of 150±20MPa to that of cortical bone.

The microstructure-designed HAp ceramics could be prepared by hydrothermal hot-pressing without dehydration.

References

1) Masaru AKAO, Hideki AOKI and Kazuo KATO, "Mechanical Properties of Sintered Hydroxyapatite for Prosthetic Applications" J. Mater. Sci., 16, 809-812 (1981).

2) Takafumi KANAZAWA, Hideki MONMA, "Rinsan Karusiumu no Kagaku (Chemistry for Calcium phosphate), in Japanese" Kagaku no Ryouiki, 27, [8], 662-672(1973).

3) Hideki MONMA, "Porous Apatite Using Hydraulic Reaction" Ceramics Japan, 23, [8], 745-748(1988).

4) Koji IOKU, Masahiro YOSHIMURA, Shigeyuki SOMIYA, "Post Sintering of Apatite Ceramics from Fine Powders Synthesized under Hydrothermal Conditions" Seramikkusu Ronbunshi (J. Ceram. Soc. Japan), 96, [1], 109-110(1988).

5) Masayuki ASADA, Katsutoshi OUKAMI, Seishiro NAKAMURA and Katsuaki TAKAHASHI, "Microstructure and Mechanical Properties of Non-Stoichiometric Apatite Ceramics and Sinterability of Raw Powder" Seramikkusu Ronbunshi (J. Ceram. Soc. Japan), 96, [5], 595-598(1988).

6) Koji IOKU, Shigeyuki SOMIYA, Masahiro YOSHIMURA, "Dense/Porous Layered Apatite Ceramics Prepared by Post-Sintering" Seramikkusu Ronbunshi (J. Ceram. Soc. Japan), 97, [5], 566-570(1989).

7) Hisashi KUROSAWA, Kazuyuki SHIBUYA, Kenichi MURASE, Shigeru SATO, ATSUSHI Masujima, "(Experimental Study for the Synthetic Porous Apatite as Bone Substitute), in Japanese" Bessatu Seikei Geka, [8], 58-64(1985).

8) Nakamichi YAMASAKI, Kazumichi YANAGISAWA, Mamoru NISHIOKA, "Principle of Hydrothermal Hot-Pressing and its Apparatus" New Ceramics, 2, [10], 81-86, (1989).

9) Hitoshi NISHIZAWA, Haruko TEBIKA, Nakamichi YAMASAKI, "Fabrication of Stabilized Zirconia Compressed Body under Hydrothermal Conditions and Its Sintering" Yogyo-Kyokai-Shi, 92, [7], 420-421(1984).

10) Kazumichi YANAGISAWA, Mamoru NISHIOKA and Nakamichi YAMASAKI, "Solidification of Powders in SiO_2-Fe_2O_3 and SiO_2-ZrO_2 System by Hydrothermal Hot-Pressing" Yogyo-Kyokai-Shi, 94, [11], 1193-1196 (1986).

11) Mamoru NISHIOKA, Kazumichi YANAGISAWA and Nakamichi YAMASAKI, "Solidification of Glass Powders by a Hydrothermal Hot-Pressing Technique" Yogyo-Kyokai-Shi, 94, [11], 1119-1124 (1986).

12) Koji IOKU, Masahiro YOSHIMURA and Shigeyuki SOMIYA, "Hydrothermal Synthesis of Ultrafine Hydroxyapatite Single Crystals" Nippon Kagaku Kaishi, 1988, [9], 1565-1570.

13) JCPDS Card 9-432

14) Takashi YAMAGUCHI, "Characterization Techniques of Ceramics : Properties of Sintered Bodies" Ceramics Japan, 6, [19], 520-529 (1984).

15) Kazushi HIROTA, Yasutoshi T. HASEGAWA and Hideki MONMA, "Densification of Hydroxyapatite by Hot Isostatic Pressing" Yogyo-Kyokai-Shi, 90, [11], 680-682 (1982).

16) Mikiya ONO, "Mukikobunshi-Haiburiddo Porima No Oyo", CMC, Tokyo, 299-315 (1985).

17) Takafumi KANAZAWA, Takao UMEGAKI and Hideki MONMA, "Apatites, New Inorganic Materials" Ceramics Japan, 10, [7], 461-468 (1975).

18) J. G. J. Peelen, B. V. Rejda and K. De Groot, "Preparation and Properties of Sintered Hydroxyapatite" Ceram. Int., 4, [2], 71-74 (1978).

Hydrothermal Treatment of Radioactive Waste:
Solidification of High-Level Radioactive Waste by Hydrothermal Hot-Pressing

Kazumichi Yanagisawa, Mamoru Nishioka, Nakamichi Yamasaki

Research Laboratory of Hydrothermal Chemistry,
Faculty of Science, Kochi University,
Akebono-cho 2-5-1, Kochi-shi 780

Abstract

Simulated high-level radioactive waste was immobilized into a silica matrix by hydrothermal hot-pressing. The optimum conditions to produce a waste form with high mechanical strength and low leachability were determined as follows; starting composition: 21.8wt% waste, 10wt% $Al(OH)_3$, 47.7wt% low-quartz and 20.5wt% amorphous aluminosilicate with the addition of 10N NaOH solution (2.5 cm^3 /20g of starting powder), reaction temperature: 350 ℃, reaction pressure: 66 MPa, reaction time: 6 hours.

The waste form produced under the optimum conditions was mainly composed of low-quartz of the matrix and the waste components (Fe_2O_3, CeO_2, ZrO_2). It was porous, apparent density 2.3 g/cm^3, porosity over 20%, and BET specific surface area 10 m^2/g. It had high mechanical strength, compressive strength 200MPa. The leach rate of the waste form, determined by static leach tests at 90℃ in distilled water for 28 days, was much lower than a concrete waste form and was comparable with glass and ceramic waste forms. The waste form was stable under hydrothermal conditions in comparison with a glass waste form. It had high thermal and thermal shock resistance. Its thermal conductivity was about 0.01 J/cm·sec·K, a value similar to that of a glass waste form.

1. Introduction

In natural circumstances, mineral particles in sediment are transformed into sedimentary rock by a chemical process that reduces the original porosity by compaction and cementation[1]. In laboratories, aggregation of particles by means of dissolution and deposition is usually observed when a large amount of silica powder is hydrothermally treated by a small amount of alkaline solution. In order to enhance the densification to produce hard compacts like sedimentary rock, the solution must be removed from grain boundaries. A hydrothermal hot-pressing technique is intended for artificial transformation of powders into densified bodies under hydrothermal conditions[2]. In this method, an inorganic powder including water or alkaline mineralizer solution is mechamically compressed under hydrothermal conditions from outside an autoclave to expell the medium from grain boundaries, and linkage of grains by the action of hydrothermal solution results in conversion of the powder to a solidified body with high mechanical strength.

The hydrothermal hot-pressing method has two characteristics; continuous compression of starting powder under hydrothermal conditions and space for water retreat. Compression accelerates compaction of the starting powder and prevents development of cracks due to heterogeneous shrinkage. Compaction depends on the rate at which water can be expelled from the starting powder. The space water for retreat, into which water included in the starting powder is released, is essential to the hydrothermal hot-pressing technique. Without the space, water exists in pore space of the starting powder and hinders compaction of the powder.

Under hydrothermal conditions, increase in solubility of many inorganic compounds results in linkage of their particles by means of dissolution and deposition, and the compounds react with each other to produce new compounds with low solubility, which may contribute to cementation of the particles.

Solidified bodies of silica powders such as low-quartz and borosilicate glass[3-6], mixtures of silica powders with metal hydroxides or oxides[7,8], calcium carbonate[9], zirconia[10], and hydroxyapatite[11] have been produced by the hydrothermal hot-pressing technique. Alkaline solutions accelerated the densification of silica powders. Borosilicate glass powders included glass network modifying oxides were densified even by the addition of pure water at low temperatures below 300°C[5]. Shrinkage of the powder started by viscous flow mechanism after soluble components in the glass, such as alkaline metal, were dissolved into water under hydrothermal conditions[6]. Futhermore, this method has applied to immobilization of high-level[12-15] and low-level radioactive wastes[16,17]. This paper deals with the immobilization of high-level radioactive wastes by hydrothermal hot-pressing.

The most generally accepted concept for disposal of high-level radioactive wastes is to incorporate it into a solid waste form that is placed into a deep geologic repository[18]. The waste form is a first barrier in multiple barrier system.

Many kinds of the waste forms have been proposed to immobilize high-level radioactive wastes. Immobilization into borosilicate glass is considered to be the most realizable process. The process is appealing because of its relative simplicity and utilization of conventional glass-making technology. In a quantitative evaluation of candidate waste forms, the high process rating for borosilicate glass and its intermediate product performance score resulted in the overall top-ranking position[19]. It is, however, obvious that glass is thermodynamically unstable relative to a chemically equivalent assemblage of crystalline phases[20]. Under some circumstances, it may devitrify to form the stable crystalline assemblage.

The immobilization of the wastes into crystalline phases has been researched for supercalcine[21-23], synroc[20,24,25], high-alumina tailored ceramics[26-28], and so on. The production process of these ceramic waste forms is rather complicated, though the waste forms produced by the processes have excellent properties in comparison with the glass waste forms.

These processes mentioned above require high temperatures above 1000 °C, which may cause evaporation of volatile radioactive isotopes and corrosion of the process equipment.

Solidification with cement is cost-effective for immobilization of the wastes, because of its ease of processing[29,30]. Concretes formed under elevated temperatures and pressures (FUETAP concretes), however, have higher leach rates than those of other waste forms[19,29].

The hydrothermal hot-pressing technique would be useful to solidification of radioactive wastes, because powders of silica mixed with metal oxides and hydroxides were easily converted to solidified bodies by this technique[7,8]. The technique has the following advantages as a process for solidification of high-level radioactive wastes;

(1) Simplicity of an apparatus and ease of operation.

(2) Low temperature reaction around 300 °C.

(3) Reaction in closed system (solidification in an autoclave), which prevents evaporation of volatile materials.

(4) Low leachability of the solidified bodies, which is supported by the fact

that soluble components in the wastes can be reacted under hydrothermal conditions with each other or a silica matrix to form crystalline phases with low solubility.

(5) Stability of the solidified bodies under repository conditions, because of the products formed under hydrothermal conditions.

In this paper, a simulated high-level radioactive waste was solidified using a silica matrix composed of low-quartz and amorphous aluminosilicate by hydrothermal hot-pressing. The effects of hydrothermal hot-pressing conditions on compressive strength and leachability of the waste forms were investigated to dertermine the optimum conditions to produce a waste form with high mechanical strength and low leachability. The properties of the waste form produced under the optimum conditions were clarified.

2. Experimental

2 — 1 Starting Materials

The silica matrix was a mixture of 70wt% low-quartz powder passed through a 200 mesh (silicastone from Fukushima Prefecture, Japan) and 30wt% amorphous aluminosilicate (siliceous sinter, porous and friable rock deposited by the water of hot springs, from Kagoshima Prefecture, Japan). Their chemical analyses are shown in Table 1. In some cases, $Al(OH)_3$ (Wako Pure Chem. INd., LTD) was added to the mixture. The composition of a simulated high-level radioactive waste from reprocessing of nuclear fuels is shown in Table 2. It was estimated by Japan Atomic Energy Research Institute (JAERI), and Tc, Pm, Ru, Rh, Pd and actinides were simulated by Mn, Nd, Fe, Co, Ni and Ce, respectively[31]. In the waste, Na was not added because NaOH solution was used as a mineralizer to accelerate the solidification by hydrothermal hot-pressing. The waste was prepared by mixing appropriate chemicals in a ball mill.

2 — 2 Apparatus for Hydrothermal Hot-Pressing

An autoclave for hydrothermal hot-pressing used in this study is shown in Fig.1. The cell is made of Hastelloy-C and is composed of an inner and an outer case. The inner case, 1.4cm in inner diameter, is divided into three parts in order to remove a solidified body from the cell without its destruction. The outer case is used to keep the inner case in cylindrical shape. The autoclave is made of carburizing steel with a liner of Hastelloy-C. It has the space for water retreat (about 25cm³) unoccupied by the cell. The starting powder in the cell is uniaxially compressed during hydrothermal treatment, via the push rod of the autoclave with gland packing to prevent leakage.

Table 1. Chemical Analyses of Starting Materials (wt%)

	SiO_2	Al_2O_3	Fe_2O_3	CaO	MgO	Other oxides	Ignition loss
Silicastone [1]	99.5	0.2	0.02	0.02	0.01	0.25	0
Siliceous sinter [2]	71.7	12.9	2.3	2.4	0.4	5.7	4.6

[1] ; from Fukushima Prefecture, Japan.

[2] ; from Kagoshima Prefecture, Japan.

Table 2. Composition of Simulated High-Level
Radioactive Waste (wt%)

Fe_2O_3	39.05	SrO [3]	1.69
MoO_3	8.66	Sm_2O_3	1.64
Nd_2O_3	8.24	MnO_2	1.29
ZrO_2	8.21	TeO_2 [4]	1.12
CeO_2	6.77	CoO	1.00
NiO	4.92	Y_2O_3	1.00
Cs_2O [1]	4.88	Rb_2O [1]	0.61
BaO [3]	3.12	Gd_2O_3	0.19
La_2O_3	2.54	CdO [1]	0.16
Pr_6O_{11}	2.47	Ag_2O [1]	0.15
Cr_2O_3	2.26	Sb_2O_3 [2]	0.13

Added as [1] nitrate, [2] chloride, [3] carbonate, [4] metal.

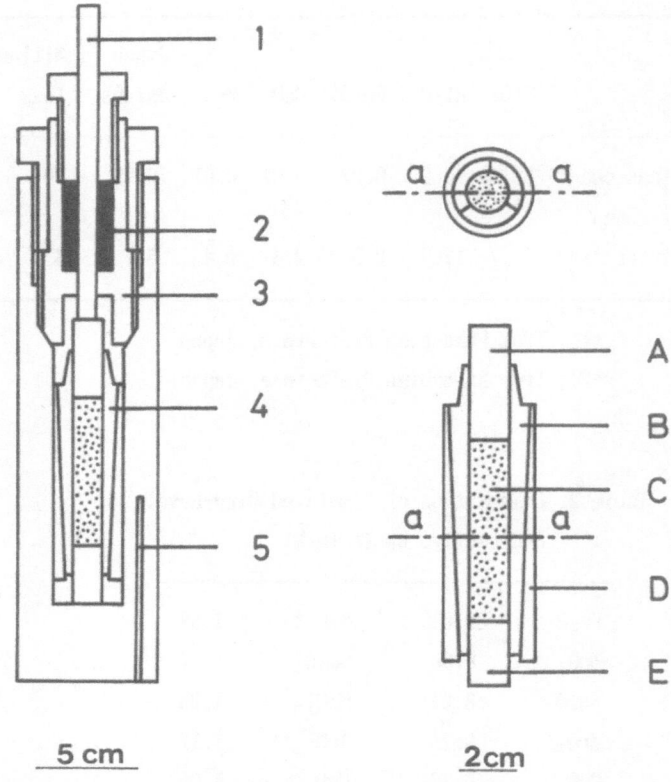

<u>5 cm</u>   <u>2cm</u>

Fig.1 Autoclave and cell for hydrothermal hot-pressing.
       Autoclave(left) made of carburizing steel lined with Hastelloy-C:
            1.push rod, 2.gland packing (copper and cotton yarn),
            3. cone, 4.cell, 5.well for thermocouple.
       Cell(right) made of Hastelloy-C:
            A,E. piston, B. inner case, C. sample, D. outer case,
            The upper figure is a cross-section of aa.

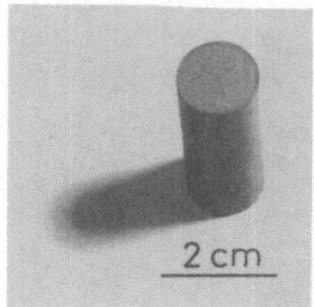

Fig.2 A typical waste form produced by hydrothermal hot-pressing.

2 — 3 Solidification Method

The waste was mixed well in a porcelain motar with the silica matrix to prepare a starting dry powder. The waste loading was expressed as oxides. The dry powder (20g) was kneaded in the motar with a NaOH solution (2.5cm³) to prepare a starting wet powder. The wet powder was transferred into the hot-pressing cell and the cell was put into the autoclave. The autoclave was heated to a desired temperature from 100 to 350℃ at the rate of 30℃/min by an induction heater, and the temperature was kept constant for 0 to 6 hours. During hydrothermal treatment, the powder in the cell was continuously compressed at 17 to 66 MPa. After the treatment, the autoclave was cooled down to room temperature at the rate of about 20℃/min by an electric fan, and a solidified body was taken out of the cell. A typical waste form produced by hydrothermal hot-pressing is shown in Fig.2.

2 — 4 Leach Tests

Three types of leach tests were performed to evaluate the chemical durability of the waste forms; (1) flow leach test by Soxhlet type apparatus, (2) Static leach test under hydrothermal conditions, (3) modified MCC-1 low temperature static leach test.

(1) Flow leach test: Soxhlet type apparatus was constructed of fused silica glass. Cyrindrical specimens for the test (1.4cm in diameter, about 0.8cm in height) were fabricated by diamond-sawing and dry-abrading. The test was performed at 97℃ for 29 days.

(2) Static leach test under hydrothermal conditions: The autoclave for the test, 27.7cm³ in inner volume, was made of carburizing steel and linned by titanium. Buckwalter and Pederson [32] used many kinds of leaching containers for leach tests of a simulated nuclear waste containment glass and reported that relatively minor effects on glass leaching were found in titanium containers in comparison with those measured in Teflon containers. The leachant was distilled water (15cm³). Cubic specimens (5x5x5 mm^3) were fablicated by diamond-sawing and dry-abrading, giving a specimen surface area / leachant volume ratio 1/10 cm⁻¹. The test was performed under hydrothermal conditions up to 350℃ for 1 day. Monolithic glass specimens of JW-A glass waste form (containing the same simulated waste (21.8wt%) that used in this study, prepared by JAERI) were also used for the test.

(3) MCC-1 static leach test: MCC-1 leach test [33,34] is a static low temperature leach test, one of the standard leach tests proposed by the Nuclear Waste Materials Characterization Center established by the U.S.

Department of Energy, to develop standard tests for characterization of the components of the waste package, which include spent fuel, waste forms and so on. The leaching container, 130cm³ in inner volume, was made of Teflon. The leachant was distilled water (40cm³) and each cubic specimen for the test (8.2x8.2x8.2 mm³), fabricated by diamond-sawing and dry-abrading, had surface area of 4cm². The test was performed at 40 and 90 ℃ up to 91 days. The leachant was changed to fresh distilled water when weight loss of a specimen was measured, that is, 3, 7, 14, 28, 56 days in leaching time in this modified MCC-1 leach test.

All leach tests were conducted in duplicate. After the leach tests, each leachant was analyzed by atomic absorption spectrometry. Weight loss was calculated by weight change of a specimen dried at 110 ℃ before and after the leach tests, and normalized by dividing only by geometric surface area of the specimen. Weight loss and elemental mass loss appeared in figures indicated the sum up to a leaching time.

2 − 5 Characterization

The solidified waste forms were cut with a diamond saw to 1.6cm-long sections and dried at 110 ℃ to prepare specimens for strength and density measurement. Uniaxial compression tests were performed at room temperature to measure compressive strength. The compression axis was perpendicular to the base of the cylindrical specimens. The crosshead travel rate of a mechanical machine (Shimazu, RH-100) was 0.02cm/min. The density of the waste form was measured from its weight and apparent volume. The crystalline phases were identified by X-ray powder diffraction (Rigaku 2012). A polished section was examined by a scanning electron microscope (Hitachi S-530). The specific surface area was measured by BET method using N_2 gas as an adsorbent.

The waste forms were heated from 300 to 900℃ for 6 hours, and the weight loss by evapolation was measured. Soxhlet leach tests of the heated specimens were also performed for 1 day. The waste forms heated at 500 ℃ were quenched into water and then their compressive strength was measured. The thermal conductivity of the waste form was calculated from the thermal diffusivity and the specific heat measured at 25 to 900℃ in vacuum (10^{-3}mmHg) by laser flash method (Rigaku PS-7).

3. Results and Discussion

3 − 1 Effect of Hydrothermal Hot-Pressing Conditions on Properties of
       Waste Forms

The weight loss by leach tests at 200℃ for 1 day, the density and the
compressive strength of the waste forms (21.8wt% waste loading) produced by the
use of the silica matrix containing 70wt% low-quartz and 30wt% amorphous
aluminosilicate are shown in Fig.3, vs the hydrothermal hot-pressing condtions;
NaOH concentration, reaction temperature, pressure and time.

The increase in the concentration of NaOH solution from 3 to 10N increased
the compressive strength and the density(Fig.3A). Above 12N, the strength was
reduced and the density was kept constant at 2.7g/cm³. The highest compressive
strength was achieved by the use of 10N NaOH solution. The weight loss was
exponentially increased with the increase in NaOH concentration.

The density was linearly increased along with the increase in the reaction
temperature (Fig.3B). The compressive strength was also increased above 150 ℃
and a peak appeared at 100 ℃, at which the amorphous aluminosilicate in the
silica matrix began to dissolve in a concentrated NaOH solution to form water
glass. The waste form with high compressive strength and low weight loss was
produced at 350℃.

The high reaction pressure increased the compressive strength and the
density, but reduced the weight loss (Fig.3C). The reaction time up to 30
minutes increased the strength and the density (Fig.3D). The long reaction time
(6 hours) reduced the weight loss. It was suggested that the densification of
the starting powder was achieved in a short time up to 30 minutes but the
hydrothermal reaction of the waste with the silica matrix took time to produce
reaction products with low solubility.

Effect of the composition of the starting powder on the leachability, the
density and the compressive strength of the waste forms is shown in Fig.4. The
starting powder including even 70wt% waste could be solidified to a waste form
with high compressive strength over 150MPa (Fig.4A). The waste form with about
20wt% waste loading had high compressive strength and low leachability. Effect
of each waste component on solidification of the silica matrix was examined by
hydrothermal hot-pressing at 300℃ and 28MPa for 3 hours [7]. When Fe_2O_3 was
added to the matrix, a small amount of acmite ($NaFeSi_2O_6$) was formed. Formation
of acmite obstructed the linkage of low-quartz grains and reduced the
compressive strength of the solidified bodies. On the other hand, addition of
ZrO_2 gave no reaction products and accelerated the shrinkage of the starting
powder. The compressive strength was also increased to 600MPa by the addition of

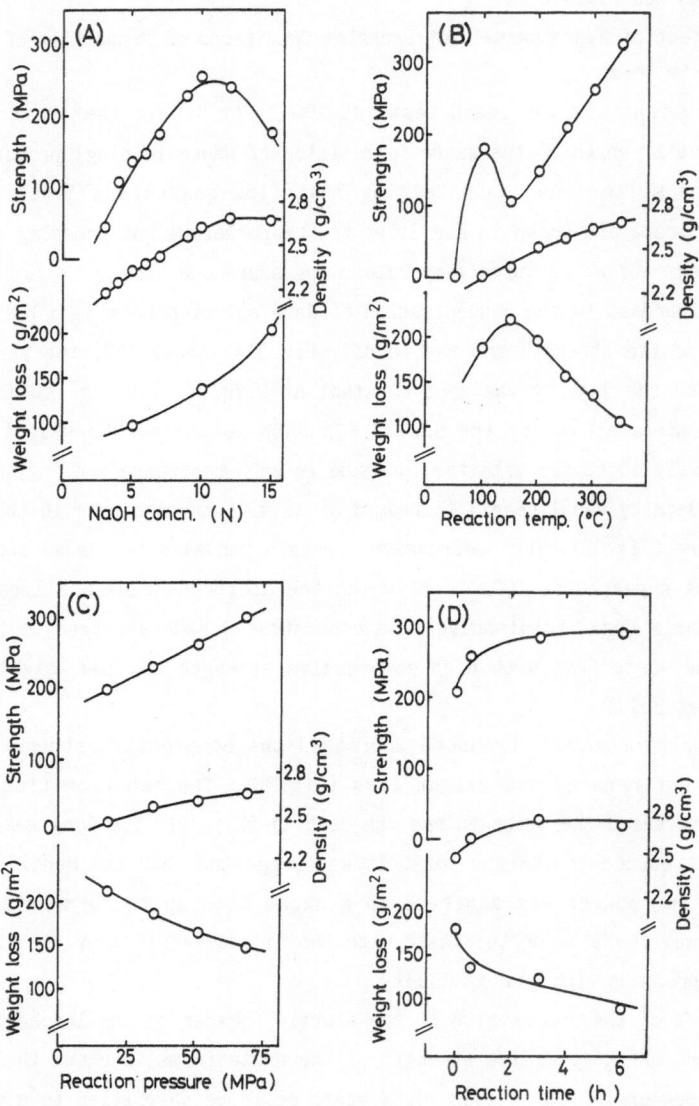

Fig.3 Effect of hydrothermal hot-pressing conditions on properties of waste
forms.

(A) NaOH concentration, (B) Reaction temperature, (C) Reaction pressure,
(D) Reaction time.

Each hydrothermal hot-pressing condition was varied from the standard
condition, i.e., 10N NaOH solution, 300℃, 49MPa, 30 minutes.

Compressive strength, bulk density and weight loss by leach tests at
200℃ for 1 day in distilled water of the produced waste forms, were
measured.

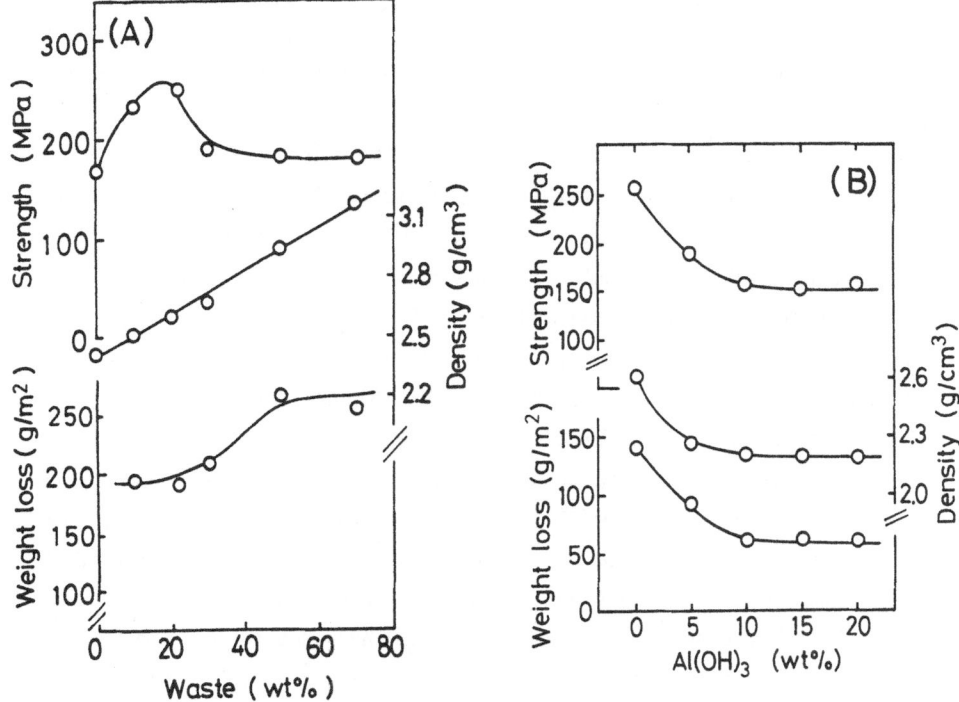

Fig.4 Effect of starting composition on properties of the waste forms.
The waste forms were produced under the standard condition.
(A) Waste loading : Leach tests were performed at 300℃ for 1 day.
(B) Al(OH)₃ addition : Waste loading was 21.8wt%. Leach tests were
performed at 200 ℃ for 1 day.

10wt% ZrO_2. Thus, ZrO_2 in the waste might cause the increase in compressive strength of the waste forms. The density of the waste forms was increased with the increase in waste loading, because the density of the waste was higher than that of the matrix.

The analyses of the leachants showed that the release of Na and Si caused the large weight loss by leach tests. If Na is immobilized into crystalline phases such as aluminosilicates with low solubility, the release of Si from a waste form by leach tests should be also reduced because of the decrease in pH of the leachnts. In order to immobilize Na, $Al(OH)_3$ was added to the starting powder. The addition of $Al(OH)_3$ up to 10wt% reduced the weight loss by the leach test, as well as the compressive strength (Fig.4B); the addition of over 10wt% had no effect on either. The density had the same tendency.

Sodium aluminosilicates such as analcite ($NaAlSi_2O_6$ H_2O) were easily produced from low-quartz and $Al(OH)_3$ within 10 minutes in 5N NaOH solution under hydrothermal conditions at 300 ℃, though the amorphous aluminosilicate of the matrix component was almost completely dissolved into the solution to form water glass and gave no crystalline phase by the same hydrothermal treatment [35]. Thus, Na might be immobilized in aluminosilicates by the reaction with $Al(OH)_3$ and low-quartz during hydrothermal hot-pressing. It was confirmed by separate experiments [36] that Cs could be immobilized into pollucite structure ($CsAlSi_2O_6$), one of the most stable hosts for Cs immobilization, by the same hydrothermal hot-pressing technique using the silica matrix including $Al(OH)_3$.

It was concluded that a waste form with high compressive strength and resistance to leaching could be produced from the mixture of 10N NaOH solution and the starting powder (21.8wt% waste, 10wt% $Al(OH)_3$, 47.7wt% low-quartz, 20.5wt% amorphous aluminosilicate) by hydrothermal hot-pressing at high temperature (350 ℃) and high pressure (66MPa) for a long time (6 hours).

3 − 2 Evaluation of The Waste Form Produced under The Optimum Conditions

Properties of the waste form produced under the optimum conditions mentioned above were clarified to evaluate the waste form produced by hydrothermal hot-pressing.

3 − 2 − 1 Physical Properties

A polished section of the waste form and its Si Kα X-ray image are shown in Fig.5. The X-ray image showed that large grains were low-quartz of the matrix and the waste components existed between low-quartz grains. The waste

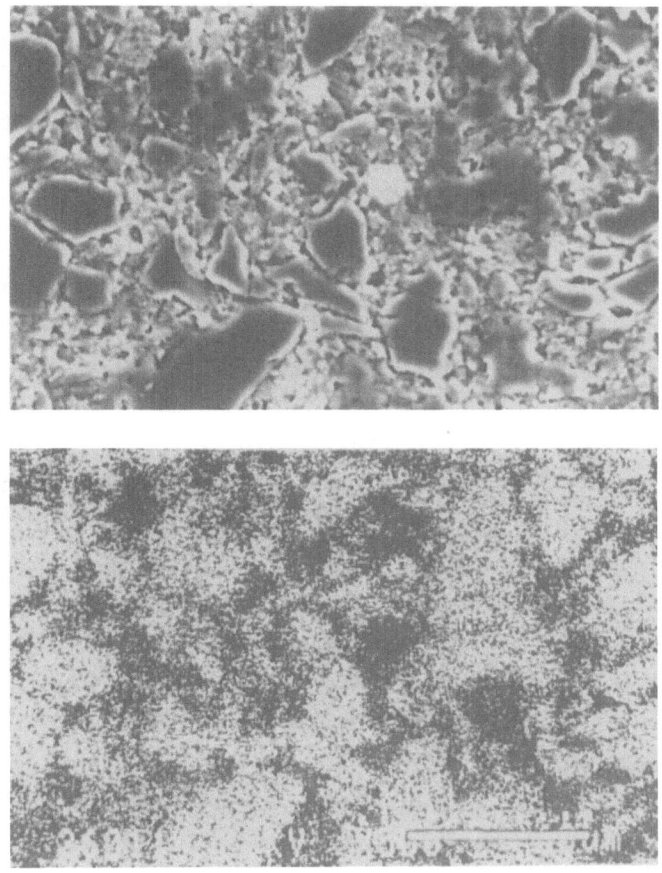

Fig.5 Polished section of the waste form (above) and its Si Kα X-ray image
(below).   (bar:50 μm)

form had many pores and slightly water permeable. Its apparent density was 2.3g/ cm³ and the density measured by the use of a pycnometer was 2.73 g/cm³. The waste form might have closed pores, so that its porosity was estimated to be Over 20%. Its specific surface area was found to be 10 m²/g by BET method.

According to X-ray powder diffraction patterns (Fig.6), the waste form was mainly composed of low-quartz, the main constituent of the silica matrix. Peaks due to α-Fe₂O₃, CeO₂ and ZrO₂ were observed with low intensity. The diffraction patterns showed that MoO₃ and Nd₂O₃ in the waste were disappeared by hydrothermal hot-pressing. A few new peaks with low intensity, probably due to hydrothermal reaction products, were observed, though they were not identified yet. As concerns Sr and Cs, they were immobilized into silicate (SrSiO₃) and aluminosilicate (CsAlSi₂O₆), when each of them was hydrothermally hot-pressed with the silica matrix.

The compressive strength of the waste form (about 200MPa) was higher than that of concrete waste forms (40-100MPa [29]), and correspoded to that of a glass waste form (269MPa [37]). The thermal conductivity of the waste form produced by hydrothermal hot-pressing was reduced with the increase in temperature up to 500℃ and then incresed above 700℃. The value of the thermal conductivity was in the range from 0.008 to 0.012 J/cm·sec·K, similar to that of a glass waste form (0.009 and 0.0125 J/cm·sec·K at room temperature and 400 ℃, respectively [37]).

After the waste form was heated at various temperatures for 6 hours, the weight loss by evaporation and by Soxhlet leach test for 1 day was merasured (Fig.7). The evaporation loss was increased with the increase in heating temperature and reached 2wt% at 900 ℃, probably due to evaporation of hydrated water and anions in the waste such as NO_3^-, CO_3^{2-}, and Cl^-. The waste form heated at 500℃ had hygroscopity. The weight loss by the leach tests was decreased by heating above 300℃. The release of Mo was especially decreased by heating the waste form above 700℃. The waste form was not melted and had no cracks even by heating at 900℃. The heat treatment at 900 ℃ gave no large changes in X-ray powder diffraction patterns (Fig.6). After quenched from 500℃ , the waste form had no cracks and its compressive strength was about 120MPa. On the other hand, many cracks were observed in the quenched JW-A glass waste form.

3 − 2 − 2 Chemical properties

The weight loss of the waste form due to leaching under hydrothermal conditions is shown in Fig.8, together with that of JW-A glass waste form. The

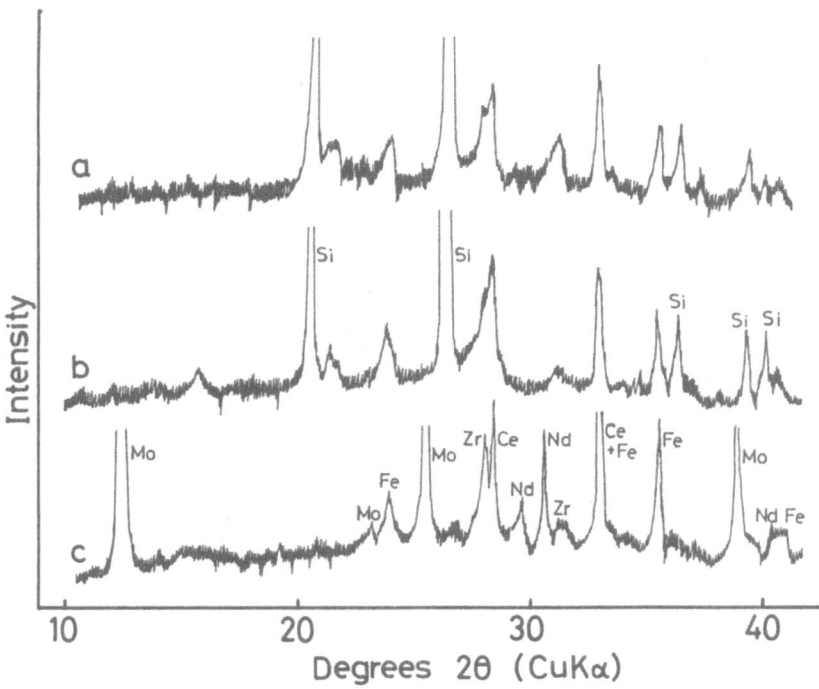

Fig.6 X-ray powder diffraction patterns of the starting simulated waste (a),
the waste form produced by hydrothermal hot-pressing (b) and the waste
form after heated at 900 ℃ for 6 hours (a).

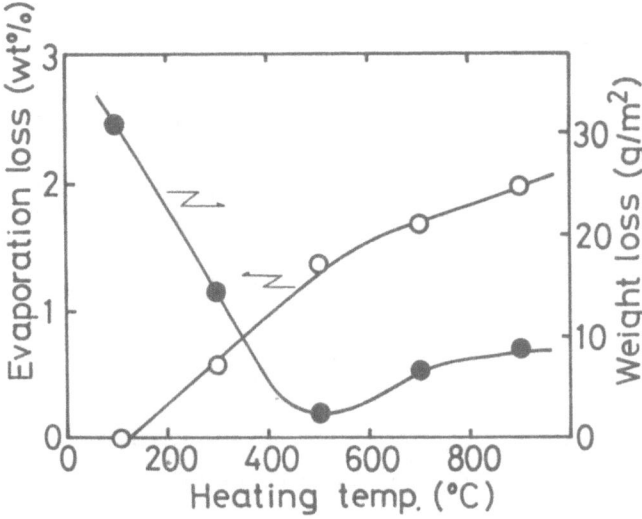

Fig.7 Effect of heat treatment of the waste form on evaporation loss and weight
loss by Soxhlet leach tests for 1 day.

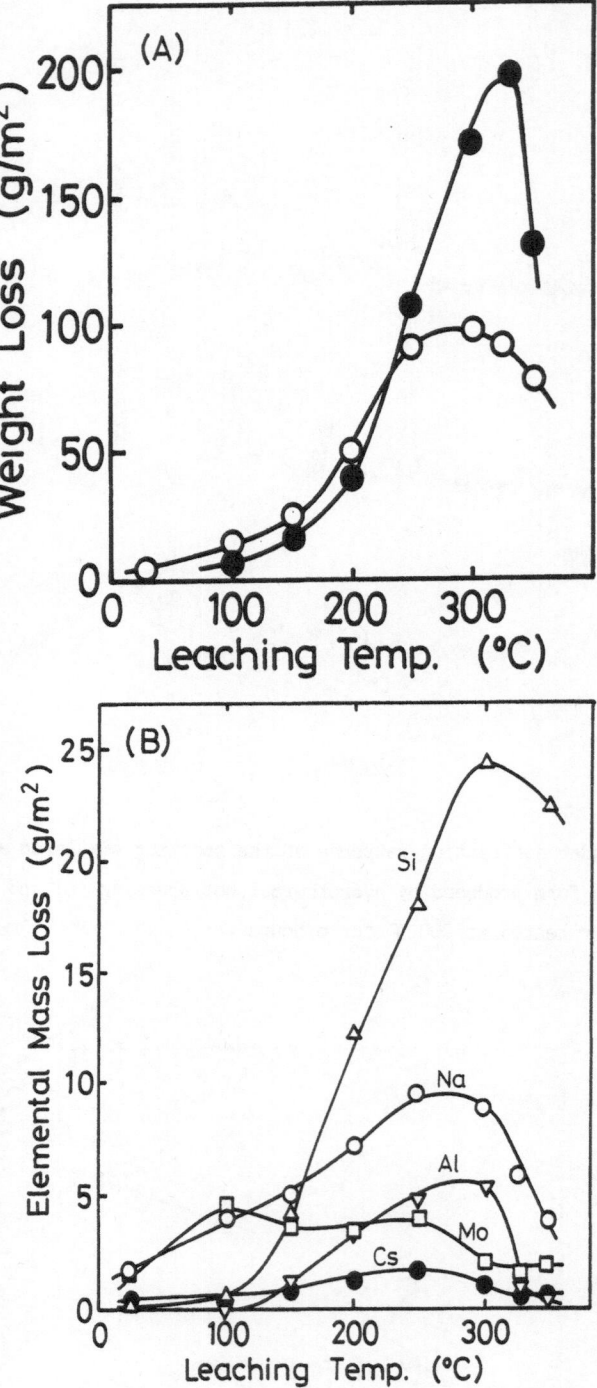

Fig.8 Results of leach tests under hydrothermal conditions for 1 day.
　　　(A) Weight loss of the waste form produced by hydrothermal hot-pressing
　　　　　(○) and of JW-A glass form (●).
　　　(B) Elemental mass loss of the waste form produced by hydrothermal
　　　　　hot-pressing.

weight loss of the waste form was smaller than that of the glass waste form at high temperatures above 250 ℃, but larger below 200℃. The elemental mass loss showed that release of Na and Mo was large at low temperature below 100 ℃ and the release of Si, Na and Al caused the large weight loss at high temperature. When the glass waste form was leached above 150 ℃, a reaction layer was produced on its surface. The thickness of the layer was increased with the increase in leaching temperature. The glass waste form leached above 325℃ was swelled and surrounded by a yellow reaction layer, about 1 mm in thickness (Fig.9). The waste form produced by hydrothermal hot-pressing had no reaction layer and maintained the shape before leaching.

The results of Soxhlet leach test are shown in Fig.10. The leach rate was remarkably reduced with the increase in leaching time, because the weight loss at initial stage of the leach test up to 6 hours was very large. After the leach test for 29 days, the leach rate reached 5.3 g/m^2·d, which was higher than that of a glass waste form (2 g/m^2·d [38]). It may be reduced by further long-time leach tests. In leachants, Sr was not detected. The amount of each element released from the waste form by the leach test up to 6 hours was very large. The release rate of each element was reduced up to 8 days and was constant over 8 days. The amount of Si was largest in released elements by the leach test for 29 days.

The results of MCC-1 leach test are shown in Fig.11. The weight loss by short-time leach tests up to 3 days was very large, as seen in the results by Soxhlet leach test. The leach rate calculated from the weight loss by MCC-1 leach test at 90 ℃ for 3 days was 7 g/m^2·d and was reduced to 0.9 g/m^2.d after the leach tests for 91 days. According to the elemental mass loss, Na and Mo were easily released by the short-time leach tests. In the waste components, MoO_3 might change to sodium molybdate under alkaline hydrothermal conditions. It may be expected that the addition of alkaline-earth metals to the starting powder reduces the leachability because of the formation of alkaline-earth metal molybdate such as $CaMoO_4$ and $SrMoO_4$, which are stable phases with low solubility under repository conditions [39].

The high leach rate of the waste form by the short-time leach tests was explained as follows: The waste forms produced by hydrothermal hot-pressing have many pores, which have been filled with the mineralizer during hydrothermal treatment. The minelralizer may include Na, Si, Al, and soluble elements of the waste components. These elements are not immobilized enough only by drying at 110 ℃ before leach tests and are easily released by an initial washout through porous structure of the waste forms. The same phenomenon was

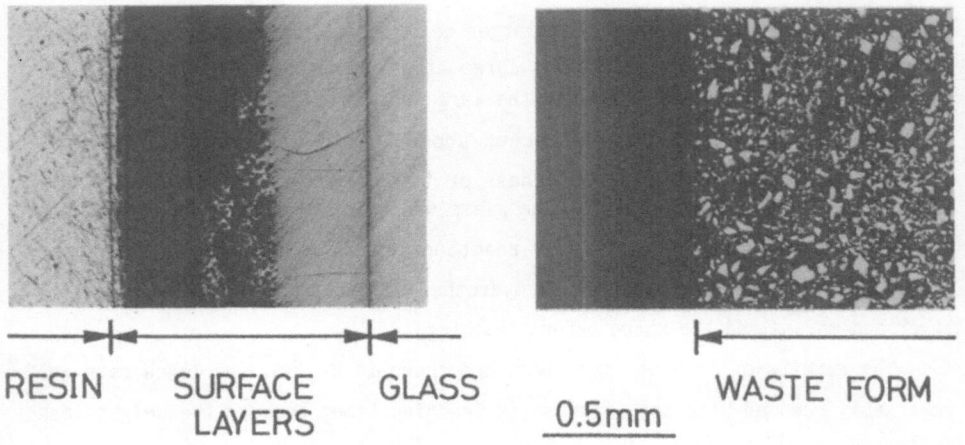

Fig.9 Cross section of specimens leached at 325℃ for 1 day.
JW-A glass form (left) and the waste form produced by hydrothermal
hot-pressing (right).

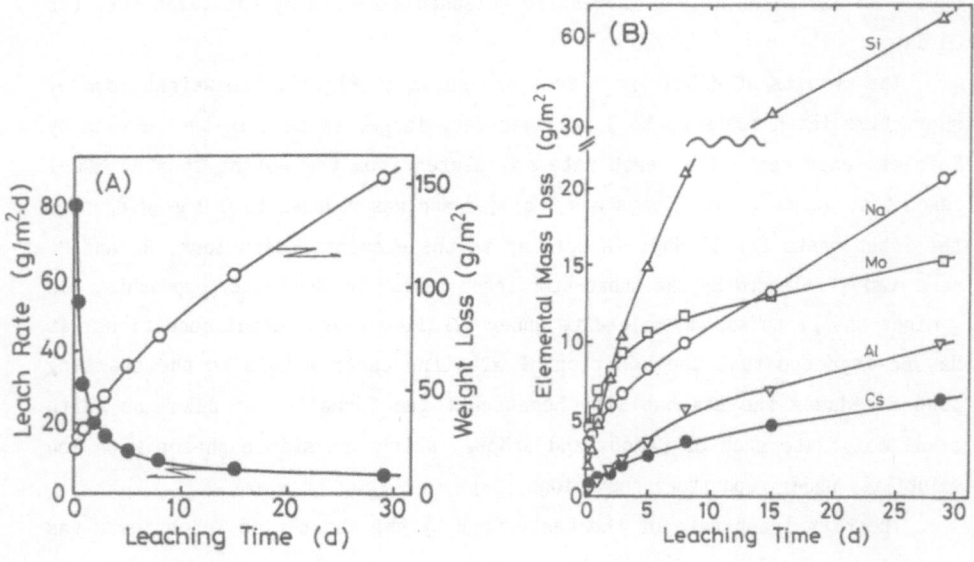

Fig.10 Leach rate and weight loss (A) and elemental mass loss (B) by Soxhlet
leach tests.

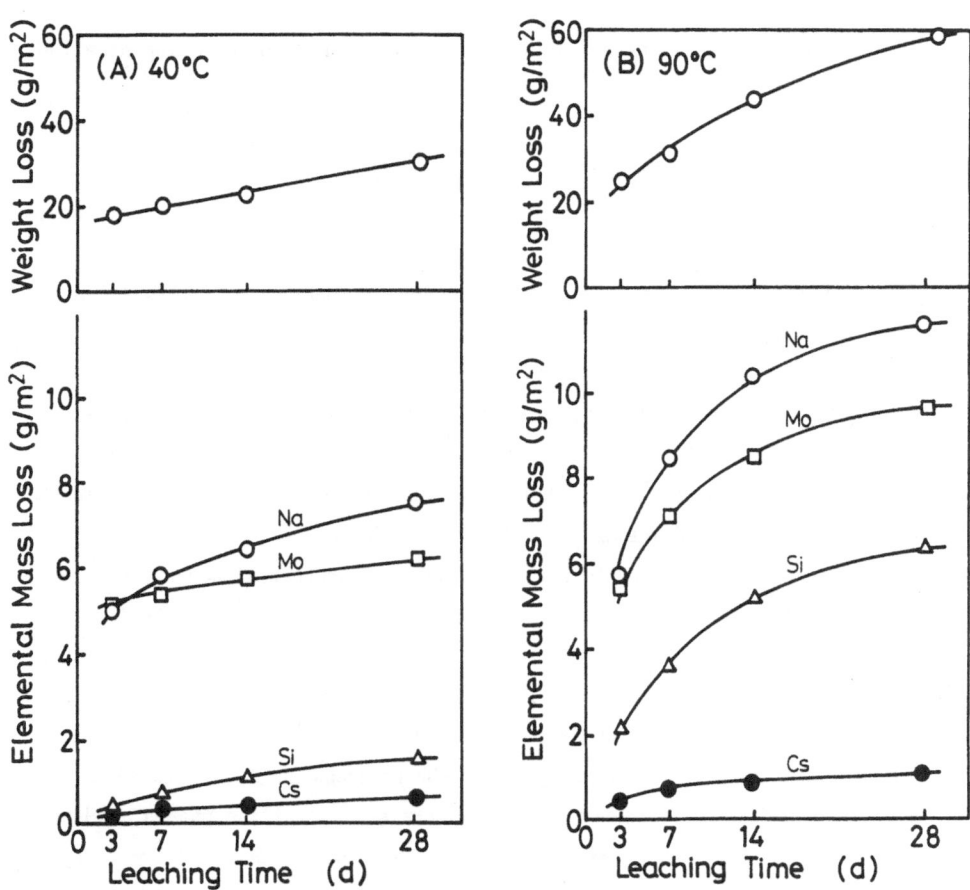

Fig.11 Weight loss and elemental mass loss by modified MCC-1 leach tests at 40 ℃ (A) and 90 ℃ (B).

found when the FUETAP concretes were leached [29]. Furthermore, selective leaching of glass phases was observed in ceramic waste forms [40,41]. In the case of the waste form produced by hydrothermal hot-pressing, a constant leach rate might be achieved after the initial washout followed by the selective leaching of soluble phases. It took 8 days by the Soxhlet leach test to reach a constant leach rate.

Normalized elemental mass loss is usually given by the following equation [34];

$$NL_i = M_i / F_i / SA$$

where NL_i is normalized elemental mass loss (g/ m^2), M_i mass loss of element "i" in leachant (g), F_i fraction of element "i" in unleached waste form (unitless), SA specimen surface area (m^2).

Normalized leach rate of each element (NL_i / leaching time) is shown in Table 3, assuming that a batch of the starting powder gaves a waste form of 20g with 21.8wt% waste loading. The leach rate of Sr was not presented bacause Sr was not detected by any leach tests. The leach rate of Na and Mo was large in all leach tests. The leach rate of Cs, Al and especially Si was increased by the Soxhlet leach test more than that of Na and Mo. By further long-time leach tests, the leach rate of all elements would be reduced.

The leach rate of the waste form produced by hydrothermal hot-pressing was compared with that of other waste forms (Table 4). The leach rate of Cs from the waste form was extremely lower than that of the FUETAP concrete waste form but was higher than that of the borosilicate glass waste form. According to the results of Cs immobilization by hydrothermal hot-pressing using the same silica matrix[36], the waste form including 10wt% CsOH produced at 300 ℃ and 49MPa for 24 hours, consisted of pollucite and low-quartz as crystalline phases, and had low leach rate of Cs, 3.15 g/ m^2·d, determined by Soxhlet leach test for 7 days. It suggested the possibility that a high-level radioactive waste form with lower leachability of Cs would be produced by hydrothermal hot-pressing. The waste form produced by hydrothermal hot-pressing had low leach rate of Sr. When a solidified body was produced by hydrothermal hot-pressing from a borosilicate glass powder containing a simulated high-level radioactive waste, its leach rate of Sr was lower than that of the original glass [4]. It is expected that hydrothermal reaction of Sr with silica may immobilize Sr in more leach resistant structure than glass.

Table 3. Normalized Elemental Leach Rate (g/m² · d)

Element	Soxhlet[1]	MCC-1(90℃)[2]	MCC-1(40℃)[2]
Na	24.57	14.64	9.37
Cs	21.47	3.98	2.21
Al	7.11	1.67	0.57
Si	7.76	0.77	0.19
Mo	38.87	27.42	17.56

[1]: Soxhlet leach tests at 97 ℃ for 29 days.

[2]: Modified MCC-1 leach tests for 28 days in
   distilled water.

Table 4. Comparative Leach Data (g/m² · d; Modified MCC-1
   leach tests at 90 ℃ in distilled water for 28 d.)

Element	HHP	BSG *	SYN *	TC *	FUE *
Cs	3.98	1.12	0.75	0.45	48
Sr	<0.09	<0.001	0.33	0.0011	0.27

HHP: Waste Form Produced by Hydrothermal Hot-Pressing,
BSG: Borosilicate Glass,   SYN: Synrock (Titania Ceramics),
TC: Tailored Ceramics,   FUE: FUETAP Concrete (Formed under
Elevated Temperatures and Pressures)
   (* Bernadzikowski et al. [19])

4. Conclusion

The simulated high-level radioactive waste was immobilized in the silica matrix by hydrothermal hot-pressing technique. The high compressive strength of the waste forms was achieved by the use of 10N NaOH solution. The high reaction temperature and pressure increased the strength and decreased the leach weight loss. The consolidation of the waste forms was achieved in 30 minutes; however, longer reaction times reduced the weight loss. The addition of 10wt% $Al(OH)_3$ to the starting material improved the leachability of the waste forms but also reduced the compressive strength. The waste loading of about 20wt% gave the high compressive strenght. The optimum conditions to produce waste forms with the high compressive strength and the resistance to leaching were as follows within the conditions used in this study; addition of 10wt% $Al(OH)_3$, 21.8wt% waste loading, 10N NaOH solution, 350°C, 66MPa, and 6 hours.

The waste form produced under the optimum conditions was characterized as follows;

(1) Porous structure: The density of the waste form was 2.3 g/cm^3 with porosity over 20% and specific surface area 10 m^2/g.

(2) Silica rich matrix: The waste form was mainly composed of low-quartz, the main consituent of the silica matrix, and included the waste components such as Fe_2O_3, CeO_2 and ZrO_2.

(3) High mechanical strength: The compressive strength was 200MPa, higher than that of a concrete waste form.

(4) Stability at high temperatures: It was not melted even at 900 °C and the heat treatment above 300 °C improved the leachability. The compressive strength of the waste form after quenched from 500°C was 120 MPa.

(5) Medium thermal conductivity: The thermal conductivity of the waste form was about 0.01 J/cm·sec·K, a similar value of that of a glass waste form.

(6) Stability under hydrothermal conditions: The leach weight loss of the waste form was lower than that of JW-A glass waste form above 250 °C and maintained the shape before the leach tests.

(7) High leach rate by short-time leach tests: The weight loss by short-time leach tests was large, but reduced remarkably with the increase in leaching time. The release of Na and Mo was major by low temperature and short-time leach tests.

(8) Low leach rate of Sr and Cs: By any leach tests, Sr was not detected in leachant. The leach rate of Sr and Cs determined by MCC-1 leach tests at 90 °C for 28 days was 0.09 and 3.98 g/ m^2·d, respectively, and much smaller than that of a concrete waste form.

429

References

[1] G.M.Friedman and J.E.Sanders, How sediments Become Sedimentary Rocks, in Principles of Sedimentology, John Wiley & Sons, New York, 1978, pp.143-163.

[2] N.Yamasaki, K.Yanagisawa, M.Nishioka and S.Kanahara, A Hydrothermal Hot-Pressing Method : Apparatus and Application, J.Mater.Sci. Letters, 5 [3], 355-356 (1986)

[3] N.Yamasaki, M.Nishioka, K.Yanagisawa and S.Kanahara, Aggregate Formation of Silica under Hydrothermal Conditions, Yogyo-Kyokai-Shi, 92 [3], 150-152 (1984) (in Japanease)

[4] N.Yamasaki, K.Yanagisawa and M.Nishioka, Hydrothermal Immobilization of Glass Powder Containing Simulated High Level Radioactive Waste, J.At.Energy Soc.Japan, 28 [3], 266-273 (1986) (in Japanease)

[5] M.Nishioka, K.Yanagisawa and N.Yamasaki, Solidification of Glass Powder by A Hydrothermal Hot-Pressing Technique, Yogyo-Kyokai-Shi, 94 [11], 1119-1124 (1986)

[6] K.Yanagisawa, M.Nishioka and N.Yamasaki, Densification Process of Boro-silicate Glass Powders under Hydrothermal Hot-Pressing Conditions, J.Mater.Sci., 24 [11], 4025-4056 (1989)

[7] K.Yanagisawa, M.Nishioka and N.Yamasaki, Solidification of Powders in SiO_2-Fe_2O_3 and SiO_2-ZrO_2 System by Hydrothermal Hot-Pressing, Yogyo-Kkokai-Shi, 94 [11], 1193-1196 (1986) (In Japanese)

[8] N.Yamasaki, K.Yanagisawa and N.Kakiuchi, Production of Hydroxyapatite-Glass Compacts by Hydrothermal Hot-Pressing, J.Mater.Res., in press.

[9] K.Matsuoka, Y.Kuwabara, N.Yamasaki, H.Mitsushio, J.Yamazaki, Hot-Pressing of Calcium Carbonate under Hydrothermal Conditions, Rep.Res.Lab.Hydrothermal Chem., 3 [3], 8-11 (1979) (in Japanease)

[10] H.Nishizawa, H.Tebika and N.Yamasaki, Fabrication of Stabilized Zirconia Compressed Body under Hydrothetmal Conditions and Its Sintering, Yogyo-Kyokai-Shi, 92 [7], 420-421 (1984) (in Japanease)

[11] K.Ioku, T.Kai, M.Nishioka, K.Yanagisawa and N.Yamasaki, Low Temperature Sintering of Hydroxyapatite Ultra-Fine Crystals by Hydrothermal Hot-Pressing, J.Mater.Sci.Letters, to be published.

[12] N.Yamasaki, K.Yanagisawa, S.Kanahara, M.Nishioka, K.Matsuoka and J.Yamazaki, Immobilization of Radioactive Wastes in Hydrothermal Synthetic Rock: Lithification of Silica Powder, J.Nucl.Sci.Technol., 21 [1], 71-73 (1984)

[13] K.Yanagisawa, M.Nishioka and N.Yamasaki, Immobilization of Radioactive Wastes in Hydrothermal Synthetic Rock Ⅲ, Propreties of Waste Form Containing Simulated High-Level Radioactive Waste, J.Nucl.Sci.Technol., 23 [6], 550-558 (1986)

[14] K.Yanagisawa, M.Nishioka and N.Yamasaki, Immobilization of Radioactive Wastes by Hydrothermal Hot-Pressing, Am.Ceram.Soc.Bull., 64 [12], 1563-1567 (1985)

[15] N.Yamasaki, M.Nishioka and K.Yanagisawa, Immobilization of Simulated HLW by Hydrothermal Hot-Pressing of Glass Powder, J.At.Energy Soc.Japan, 30 [9], 815-820 (1988) (in Japanease)

[16] N.Yamasaki, K.Yanagisawa, K.Kinoshita and T.Kashiwai, Solidification of Waste Containing Sodium Borate by Hydrothermal Hot-Pressing, J.At.Energy Soc.Japan, 30 [8], 714-724 (1988) (in Japanease)

[17] K.Yanagisawa, M.Nishioka and N.Yamasaki, Immobilization of Low-Level Radioactive Waste Containing Sodium Sulfate by Hydrothermal Hot-Pressing, J.Nucl.Sci.Technol., 26 [3], 395-397 (1989)

[18] G.L.McVay and C.Q.Buckwalter, Effect of Iron on Waste-Glass Leaching, J.Am.Ceram.Soc., 66 [3], 170-174 (1983)

[19] T.S.Bernadzikowski, J.S.Allender, J.A.Stone, D.E.Gordon, T.H.Gould,Jr, and C.F.Westberry, High-Level Nuclear Waste Performance Evaluation, Am.Ceram.Soc.Bull., 62 [12], 1364-1368,1390 (1983)

[20] A.E.Ringwood, Safe Disposal of High Level Nuclear Reactor Wastes : A New Strategy, Australian Nat.Univ.Press, Camberra, 1978, pp.1-62 .

[21] G.J.McCarthy and M.T.Davidson, Ceramic Nuclear Waste Form Ⅰ, Crystal Chemistry and Phase Formation, Am.Ceram.Soc.Bull., 54 [9], 782-786 (1975)

[22] G.J.McCarthy and M.T.Davidson, Ceramic Nuclear Waste Form Ⅱ, A Ceramic-Waste Composite Prepared by Hot Pressing, Am.Ceram.Soc.Bull., 55 [2], 190-194 (1976)

[23] G.J.McCarthy, High-Level Waste Ceramics : Materials Considerations, Process Simulation, and Product Characterization, Nucl.Technol., 32 [1], 92-105 (1977)

[24] A.E.Ringwood, S.E.Kesson, N.G.Ware, W.Hibberson and A.Major, Immobilization of High Level Nuclear Reactor Wastes in SYNROC, Nature, 278, 219-223 (1979)

[25] W.J.Buyk, D.J.Cassidy, C.E.Webb and J.L.Woolfrey, Fabrication Studies on Perovskite, Zirconolite, Barium Aluminum Titanate, and Synroc-B, Am.Ceram.Soc.Bull., 60 [12], 1284-1288 (1981)

[26] A.B.Harker, C.M.Jantzen, P.E.Morgan and D.R.Clarke, Tailored Ceramic
Nuclear Waste Forms : Preparation and Characterization, in Scientific
Basis for Nuclear Waste Management Ⅲ, ed. J.G.Moore, Plenum, 1981,
pp.139-146.

[27] P.E.Morgan, D.R.Clarke, C.M.Jantzen and A.B.Harker, High-Alumina
Tailored Nuclear Waste Ceramics, J.Am.Ceram.Soc.,64 [5], 249-258 (1981)

[28] D.R.Clarke, C.M.Jantzen and A.B.Harker, Dissolution of Tailored Ceramic
Nuclear Waste Form, Nucl.Chem.Waste Manage., 3 [1], 59-66 (1982)

[29] L.R.Dole, J.G.Moore, G.C.Rogers, G.A.West, H.E.Devaney, M.T.Morgan,
E.M.McDaniel and J.H.Kessler, Cementitious Radioactive Waste Hosts
Formed under Elevated Temperatures and Pressures (FUETAP Concretes),
in Scientific Basis for Nuclear Waste Management Ⅳ, ed. S.V.Topp,
North-Holland, 1982, pp585-593.

[30] B.E.Scheetz, D.M.Roy, C.Tanner, M.W.Barnes, M.W.Grutzeck and S.D.Atkins,
Properties of Cement-Solidified Radioactive Waste Forms with High Levels
of Loading, Am.Ceram.Soc.Bull.,64 [6], 687-690 (1985)

[31] T.Banba, H.Kimura, M.Senou, H.Mitamura, U.Kiriyama, T.Furuya, S.Tashiro,
K.Araki and H.Amano, Study of Actual and Aimulated HLW Composition,
JAERI-memo 8982 (1980) (in Japanese)

[32] C.Q.Buckwalter and L.R.Pederson, Inhibition of Nuclear Waste Glass
Leaching by Chemisorption, J.Am.Ceram.Soc., 65 [9], 431-436 (1982)

[33] J.E.Mendel, R.P.Tarcotte and J.H.Wastsik,Jr., Leaching Test Methods for
Waste Forms, Trans.Am.Nucl.Soc., 34 193-194 (1980)

[34] D.M.Strachan, B.O.Barnes and R.P.Tarcotte, Standard Leach Tests for
Nuclear Waste Materials, in Scientific Basis for Nuclear Waste
Management Ⅲ, ed. J.G.Moore, Plenum, 1981, pp347-354.

[35] K.Yanagisawa, M.Nishioka and N.Yamasaki, Immobilization of Radioactive
Wastes in Hydrothermal Synthetic Rock Ⅱ, Hydrothermal Synthesis of
Pollucite, J.Nucl.Sci.Technol., 21 [7], 558-560 (1984)

[36] K.Yanagisawa, M.Nishioka and N.Yamasaki, Immobilization of Cesium into
Pollucite Structure by Hydrothermal Hot-Pressing, J.Nucl.Sci.Technol.,
26 [3], 395-397 (1989)

[37] F.Kamei, H.Igarashi, N.Sasaki and H.Nagaki, Preprint 1983 Annu. Mtg. At.
Energy Soc.Japan, F34, (1983) (in Japanease)

[38] K.Ishiguro, N.Kawanishi, N.Sasaki, H.Nagaki and M.Yamamoto, Growth of
Surface Layer during The Leaching of The Simulated Waste Glass and Its
Barrier Effects on The Leaching, in Scientific Basis for Nuclear Waste
Management Ⅵ, ed. D.G.Brookins, Plenum, 1983, pp135-142.

[39] S.Komarneni, R.Roy and D.M.Roy, Evaluation of Strontium Molybdate (SrMoO$_4$) in Repository Simulating Tests, Nucl.Technol., <u>62</u>, 71-74 (1983)

[40] D.R.Clarke, Preferencial Dissolution of an Intergranular Amorphous Phase in A Nuclear Waste Ceramics, J.Am.Ceram.Soc., <u>64</u> [6], C89-90 (1981)

[41] F.J.Ryerson, F.Bazan, and J.H.Campbell, Dissolution of a Nuclear Waste Ceramics : An Experimental and Modeling Study, J.Am.Ceram.Soc., <u>66</u> [6], 462-470 (1983)

GROWTH AND CHARACTERIZATION OF SOME NEW SUPERIONIC PHOSPHATES

K. Byrappa
The Mineralogical Institute
University of Mysore, Manasagangotri
Mysore 570 006 India

A B S T R A C T

The search for new sodium superionic conductors is becoming very popular with
the development of NASICON whose ionic conductivity is equivalent to that of
Na β - alumina. NASICON poses a challenge to Materials Scientists in understanding
its structure and conduction mechanism due to the lack of single crystals, non-stoi-
chiometry in the composition, zirconium deficiency, etc. It is a solid solution
between $NaZr_2P_3O_{12}$ - $Na_4Zr_2Si_3O_{12}$. Then came several new triorthophosphates,
which became popular as NASICON analogues with their simple structures and stoichiometric composition. However, all the compounds whether NASICON or NASICON analogues always had only the triorthophosphate end members and their structures were
directly related to $Na_3Sc_2P_3O_{12}$. Here, the author reports the ionic conductivity
in condensed phosphates, particularly pyrophosphates, for the first time, showing
high ionic conductivity. These pyrophosphates have been grown by hydrothermal method.
The author has reported some of these pyrophosphates in brief as the perspective
superionics very recently. These condensed phosphates are much easier to obtain in
the form of single crystals with stoichiometric composition. The structures vary
from the regular NASICON type. Although, cations form the usual octahedra, Na^+ atoms
lying in the cavities, but the framework linking differs. The conductivity is attributed to the diffusion of Na^+ through a network of tunnels in a rigid structure
made up of pyrophosphate anions sharing corners with ZrO_6 / MO_6 octahedra. This
has opened a new chapter in the search for new Na^+ superionic conductors even among
the condensed phosphates and silicates. The pyrophosphates considered here have a
wide range of cationic groups. The author has discussed in the present work, the
growth, structure and impedence spectroscopy of these new pyrophosphate superionics.

1. INTRODUCTION

In recent years, there has been a continuous search for new superionic conductors for various device applications like high temperature solidstate batteries, fuel cells, specific ion and gas sensors, electro-chromic displays, ion-exchanged membranes and so on. The superionic conductors exhibit ionic conductivities comparable to those of molten salts of liquid electrolytes while still being in solid phase. These materials include the following types:

Fast ionic conductors
Fast proton conductors
Fast anionic conductors

There is no precise definition for superionic materials so far. According to Suresh Chandra[1], a superionic solid should show the following characteristics:

1. ionic bonding
2. high electrical conductivity (10^{-4} to 10^{-1} $Ohm^{-1} Cm^{-1}$)
3. principal charge carriers are ions
4. very low electronic conductivity
5. low activation energy (of the order 0.1 eV to 1.0 eV)

Ionic conductivity in solids was first reported by Faraday.[2] Subsequently Warburg[3] described the migration of Na^+ ions through glass and its precipitation on the surface of glass when a direct current flowed through the glass. Nernst[4] reported in 1899, high ionic conductivity in mixed oxides at high temperature. In 1914 Tubandt and Lawrence[5] detected the extraordinary high silver conductivity of the α - phase of AgI which exists above 147° C. Its conductivity ranges between 1.2 and 2.6 $(\Omega\ Cm)^{-1}$ which is comparable to the best conducting liquid electrolytes.

Subsequently a lot of other superionic conductors showing uni-dimensional, two-dimensional, and three-dimensional ionic conductivity were reported. Some of the most famous superionics are Naβ - alumina, Naβ'' - alumina, titanates, boracites, silicates, phospho-silicates, phosphates (NASICON and its analogues), sulphates, etc. NASICON became the most popular superionic conductor with a three-dimensional ionic conductivity in contrast to the Na β'' - alumina and all other earlier reported superionics which exhibited one or two - dimensional ionic conductivity.[6] But today, NASICON poses a challenge to solidstate physicists with reference to its structure, stoichiometry, synthesis, zirconia deficiency, conduction mechanism and a host of other problems.[7]

The study of Na^+ ion conductors has become very popular, particularly after the development of the NASICON group of superionic conductors. In fact, NASICON ($Na_{1+x} Zr_2 Si_x P_{3-x} O_{12}$; $0 < x < 3$) is a solid solution in the system $NaZr_2P_3O_{12}$ - $Na_4Zr_2Si_3O_{12}$ and is being studied extensively by a large group of workers throughout the world, because of its applications as a solid electrolyte in high temperature Na/S batteries.[2] The high ionic conductivity, which is higher than Na β - and Na β'' - alumina is mainly due to the high mobility of Na^+ ions through tunnels in a rigid structure of PO_4/SiO_4 tetrahedra sharing corners with ZrO_2/MeO_6 octahedra. Many variations have been investigated by appropriate substitutions in the NASICON system. The basic structure of most of these derivatives remains that of $Na_3Sc_2P_3O_{12}$. NASICON has two structure types viz. NASICON and anti-NASICON. When Hong[8] reported the structure of NASICON for the first time (fig.1), only one site for Na^+ was proposed. Recently Boilet et al[10] have identified five different structural sites for Na^+ , out of these only Na_1^+ is mobile. Therefore, the material scientists continue to strive to develop new superionic conductors with simple structures, and phosphates are found to be the most suitable ones.

II. SUPERIONIC PHOSPHATES

The study of superionic behaviour in phosphates is at least one decade old. Some of the phosphates, particularly condensed phosphates were reported as superionic conductors on par with silicates.[11-13]

The ionic conductivity in these phosphates is only in the order of 10^{-5} to

436

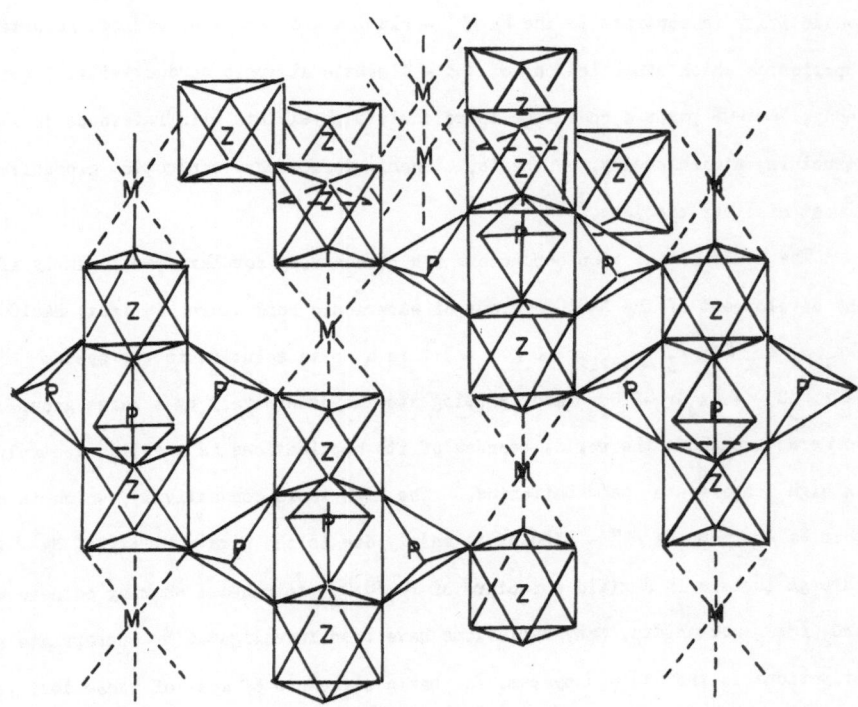

Fig.1. Structure of NASICON.[8]

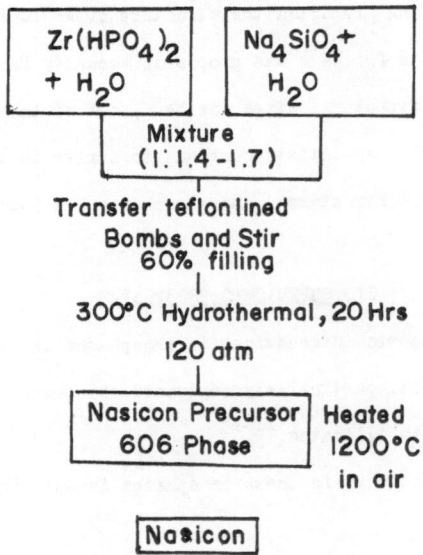

Fig.2. Flow diagram showing the hydrothermal procedure to
develop NASICON.[21]

10^{-7} (Ω $Cm)^{-1}$. The interest in the development of pure superionic phosphates appeared only recently, when there was a search for new Na^+ superionic conductors they become popular as NASICON analogues. The structures of these phosphates consist of a three dimensional framework of atoms. Based on these a series of new NASICON analogues like $Na_3M^{+2}(PO_4)_3$, $M = Sr,Mg,Fe,Mn$ [14] ; $Na_2(R,M^{+3})M^{+4}(PO_4)_3$ [15]; $A_3M_2^{+3}(PO_4)_3$, $A = Li,Na,Ag,K;$ $M = Cr,Fe$ [16,17], etc. were synthesized only with an intention to develop simple compounds with stoichiometric composition and relatively simple structures. However, all these new compounds did contain only the ortho-group of radicals irrespective of a wide variety of them. In fact, it was found that these structures are directly related to the basic NASICON structure reported by Hong.[8] The author has recently reported for the first time pyrophosphates showing high ionic conductivity. Hitherto none of the phosphates having condensed radicals and showing a high ionic conductivity has not been reported. The crystal chemical characteristics of these new superionic phosphates will be discussed later.

III. CRYSTAL GROWTH

In general the NASICON group of ionic crystals are being obtained by two techniques like flux growth and solid state reactions. Melt technique has not been attempted not only because, of the higher melting points, but also due to the multiple phase transitions. As it was earlier stated that NASICON is a solid solution in the system $NaZr_2P_3O_{12}$ - $Na_4Zr_2Si_3O_{12}$, this has resulted in the synthesis of only twinned crystals. The single crystal growth has not been attempted by these two techniques, and this led to the lack of data on the conduction mechanism. Although, several refinement techniques were adopted on NASICON, the results varied from one group to another. Also they found the existence of zirconium deficiency associated with the presence of condensed radicals like $[P_2O_7]$, $[P_3O_9]$, $[Si_3O_8]$, etc. in the final products. If one can overcome all these problems, it would be easy to understand NASICON clearly. Hence, in the last few years there is a strong move to look for other Na^+ superionics. The group led by the author has successfully obtained a new series of Na^+ superionics belonging to NASICON family, but the pure end members of phosphates. The group has obtained crystals of $Na_2(R,M^{+3})M^{+4}P_3O_1$ where M^{+3}= Al,Fe,Cr,Bi; R = rare earths, M^{+4} = Zr,Ti and a group of pyrophosphate by various techniques like solid phase synthesis, flux growth and hydrothermal technique. The latter to be the most suitable not only to grow crystals but also to understand the phase equilibria within the given system, which is highly essential to develop a growth technology for any compound. Unfortunately the literature available on the growth of superionics, particularly belonging to NASICON family is highly limited and only a few scanty reports and in some cases only a hydrothermal processing of the materials obtained by solid state reactions are adopted.[21,22] For example, the authors[21] have adopted the hydrothermal procedure to develop NASICON and it is shown in the form of a flow chart (fig.2). It was found that by hydrothermal method NASICON indicates that insufficient silicon is incorporated into the resultant products. Also these authors indicated that it is possible to develop a stoichiometric NASICON compound through hydrothermal reactions. The problem only arises when silicon is added into the system. Keeping this as a prelude, the present author has obtained pure phosphate phases by hydrothermal technique which helped in several ways. The presence of condensed radicals like $[P_2O_7]$, $[P_3O_9]$, $[P_4O_{12}]$ and $[P_5O_{14}]$ in the in the pure triorthophosphates could be avoided, because these condensed radicals normally form the bottle necks for the conducting media. Similarly the formations of twinning was reduced and it is permitted in many cases to obtain pure single crystals of a particular phase. Here the author discusses in brief the development of the hydrothermal growth of phosphates having condensed radicals. The synthesis of phosphates by hydrothermal method in general started during 1950s. Both rare earth orthophosphates and aluminium-orthophosphates were obtained at this time. The higher volatility of the P_2O_5 creates a major problem in the synthesis of rare earth phosphates under hydrothermal conditions. The formation of orthophosphates from HCl solutions at 105 - 300° C and at a pressure of 90 atms has been described by earlier workers.[23] Monocrystals of Monozite $CePO_4$ were synthesized from aqueous solutions of 85 % H_3PO_4 at 300° C using teflon liners.[24] The $GdPO_4$ crystals were obtained using aqueous solutions of HCl and $NH_4H_2PO_4$ at temperature 150 - 160°C and pressure 1.5 kbars.[25] Similarly Berlinite, $AlPO_4$ was obtained.[26,27] However, the first systematic attempt to grow condensed phosphates under hydrothermal conditions was made by Yoshimura et al.[28] In general the hydrothermal technique is quite complicated, particularly for the growth of phosphates. This may be due to the following reasons:

1. Lack of growth technology
2. High volatility of phosphorus at higher temperature
3. High reaction susceptibility of phosphorus
4. Corrosive nature of phosphorus at higher temperature

Of course, one can overcome all these problems through a systematic and careful study of phosphates using the modern instrumentation technique. High volatility of phosphates can be controlled using a special outlet to the crystallization chamber. The other important problem that phosphorus corrodes even the metals like Pt, Ir, Nb, Au and Ni can be solved by replacing them by vitreous carbon glass and teflon liners. The main drawback in using the vitreous carbon glass liners is sealing. If one can seal these vitreous carbon glass liners by some means, the growth of pure phosphates can be carried out even at higher temperature by hydrothermal method.

In the present work the author reports the hydrothermal growth of some superionic phosphates, particularly the pyrophosphates $Na_2MZr(P_2O_7)_2$, (M = Ni,Co) and $(Na_{0.66}Zr_{0.33})P_2O_7$.

The synthesis of phosphates by hydrothermal technique normally results in the crystallization of orthophosphate, because of the higher P_2O_5 and H_2O pressures. It has been earlier studied in detail with reference to the crystallization of condensed radicals by hydrothermal techniques which insists normally lower water vapour pressure.[28-30] The author[31] has made some conclusions based on the water vapour pressure to explain the possible reasons for the absence of the condensed phosphates in nature. Keeping this in mind the author has carried out the present hydrothermal synthesis experiments at relatively low pressure and temperature i.e. at 220° C and 50 to 60 atms. The alkaline component was taken in the form of a solution (NaOH) with a known molarity and this solution acts as a mineraliser. The starting materials such as the oxides of zirconium and the desired transitional metal were taken in required amounts in a teflon liner, later 85 % H_3PO_4 was poured into it. The superionic phosphates under consideration were obtained under the following molar ratio:

$$Na_2O : M^{+2}O \quad M_2^{+3}O_3 : M^{+4}O_2 : P_2O_5 \;=\; 4.5 : 1 : 0.8 : 10$$
$$Na_2O : ZrO_2 : P_2O_5 \;=\; 4 : 1 : 8$$

The growth of pyrophosphate crystals was carried out in Morey type autoclaves (length 15 cm, internal diameter 3.5 cm) with teflon liners of capacity 25 ml. The cross section of the autoclave used in the synthesis is shown in fig.3. The crystallization was carried out by spontaneous nucleation which was controlled by a systematic rate of heating. The starting materials such as $M^{+4}O_2$ was taken in the required amount in a teflon liner. The liner was filled with 85 % H_3PO_4 . The alkaline component (Na_2O) was taken in the form of a solution (NaOH) with a known molarity and this solution acts as a mineralizer. The author could synthesize pyrophosphate superionics under the following conditions: temperature 250° C, pressure 150 atms, mineralizer 1.5 M NaOH.

In the growth of double pyrophosphate superionics, the addition of surplus P_2O_5 did not change the resultant product which can be explained from the influence of H_2O vapour pressure in the system. Surplus P_2O_5 should actually result in the formation of simple or compound orthophosphates. But owing to the lower water vapour pressure maintained within the system, the reproducibility of pyrophosphates could be achieved. The following reactions would explain the synthesis of the present compounds:

(1) $NaOH + M(NO_3)_2 + Zr(NO_3)_4 + H_3PO_4 \rightleftharpoons Na_2MZr(P_2O_7)_2 + (NO_2 + H_2O)\uparrow$

(2) $Na_2O + MO + ZrO_2 + P_2O_5 \rightleftharpoons Na_2MZr(P_2O_7)_2$

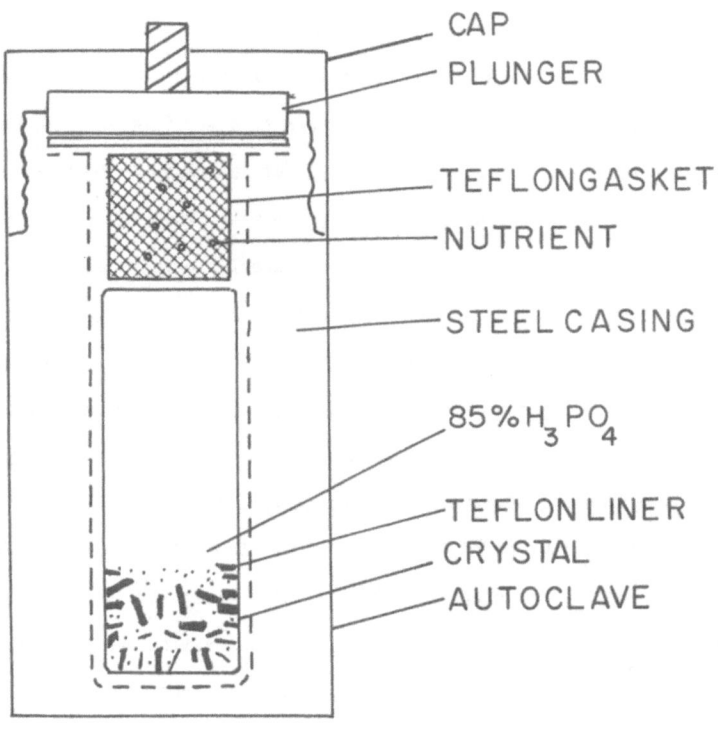

CAP
PLUNGER

TEFLONGASKET
NUTRIENT

STEEL CASING

$85\% H_3PO_4$

TEFLON LINER
CRYSTAL
AUTOCLAVE

Fig.3. Hydrothermal apparatus.

The crystals obtained were superior in size from 0.5 to 2 mm. Twinning was almost absent and crystals have well developed morphology with vitreous lustre. The characteristic photographs of these crystals are shown in figs.4a,b,c. In the growth of pyrophosphates $(Na_{0.66} Zr_{0.33})_2 P_2O_7$ the experiments were carried out exactly under similar conditions as in the previous case. However, the difference is in the ratio of the starting materials and the type of trivalent metals added into the nutrient. Both Bi and Ce were added into the nutrient in the form of oxides and nitrates respectively in relatively larger quantity. But surprisingly both Bi and Ce have not entered into the composition of this pyrophosphates. This has led to think about role of cations in the crystallization of these superionic phosphates. The results are in good agreement with the earlier reports that the divalent metals enter easily into the structure than except a few trivalent metals.[14] Hence, in the present work phosphates containing divalent metals like Ni, Co, Mn, Cu have been obtained in the form of well developed single crystals with perfect morphology. The authors have carried out a large number of experiments using trivalent metals like lanthanides, transitional metals like Cr,Fe,Bi,Al,Co. But it was found that only Al enters the composition more easily than other trivalent metals. Similarly the lanthanides, inspite of surplus concentration, did not enter the composition and therefore, only a limited number of superionic phosphates containing trivalent metals were obtained. The best ones are $Na_2(R,Al)ZrP_3O_{12}$: R = La,Ce,Nd. The presence or absence of rare earth was confirmed through ICP and XRF. Even in the earlier report where $Na_3M_2(PO_4)_3$ M = Sc,Cr,Fe was obtained,[16] the crystals were not of good quality and showed multiple phase transitions, which varied among Sc - Cr - Fe. This is mainly connected with the framework structure of these phosphates. The important aspect of this report is the growth of condensed phosphates by hydrothermal technique at relatively lower temperature. In the earlier reports, the growth of RP_5O_{14} [28,29] and $MNdP_4O_{12}$ [30] was carried out at elevated temperatures in order to get rid off the possibilities of the entry of OH^- molecules in the grown crystals. Further

443

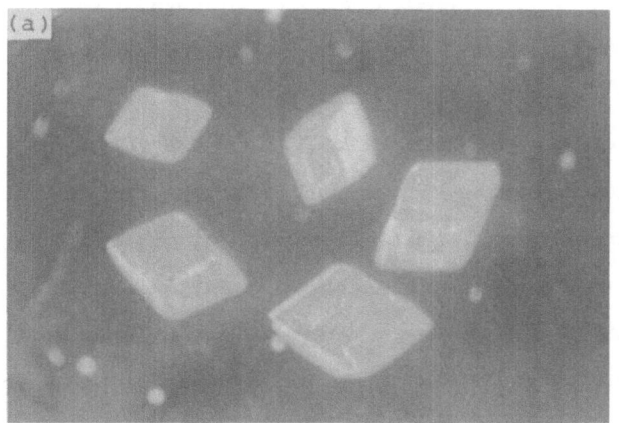

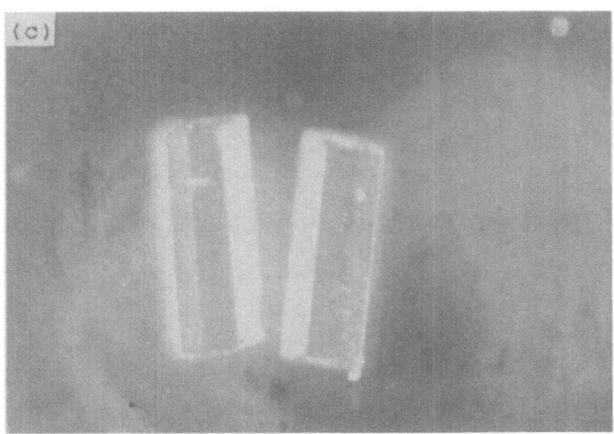

Fig.4. Characteristic photographs of $Na_2CoZr(P_2O_7)_2$ crystals.

the size of the RP_5O_{14} crystals obtained by hydrothermal technique was only 0.1 to 0.3 mm at 650° C.[28] Here the author reports the growth of very well developed crystals with a size of 1 - 4 mm at relatively very low temperature. Thus concluding the role of cations and volatiles like P_2O_5 and H_2O in the crystallization of these superionic phosphates. The study of infrared spectra of these phosphates showed the absence of mixed radicals in the resultant products. This is the most important pre-requisite for any superionic compound. The addition of trivalent metal into the nutrient yielded several metastable phases also in the final products. Thus the hydrothermal growth of superionic phosphates has greater scope in order to obtain good single crystals which in turn helped to understand the conduction mechanism.

The hydrothermal growth elucidation of mixed double pyrophosphates has been carried out in greater detail, particularly with reference to the cobalt bearing pyrophosphates. The morphology of $Na_2CoZr(P_2O_7)_2$ depends mainly upon the concentration of P_2O_5 in the system. The fig.5 shows the schematic diagram, how the morphology varies with the concentration of P_2O_5 . It is clearly observed that the prism faces dominate over the pyramidal faces as the P_2O_5 concentration increases. Similarly, an attempt has been made to study the growth rate of $Na_2CoZr(P_2O_7)_2$ crystals agains the concentration of P_2O_5 and H_2O fig.6. These figures show a critical point beyon which the growth rate falls sharply. This helps to fix the optimum conditions for th growth of $Na_2CoZr(P_2O_7)_2$. In order to study the growth rate and other morphological features a good and well developed spontaneously grown crystal was used as a seed which was kept at the bottom of the autoclave. Although, the growth elucidation for $Na_2CoZr(P_2O_7)_2$ is given in brief, it emphasises the importance of such studies for the successful growth of any superionic phosphate in general. However, a further detailed research is highly essential to understand the growth technology for these superionic phosphates.

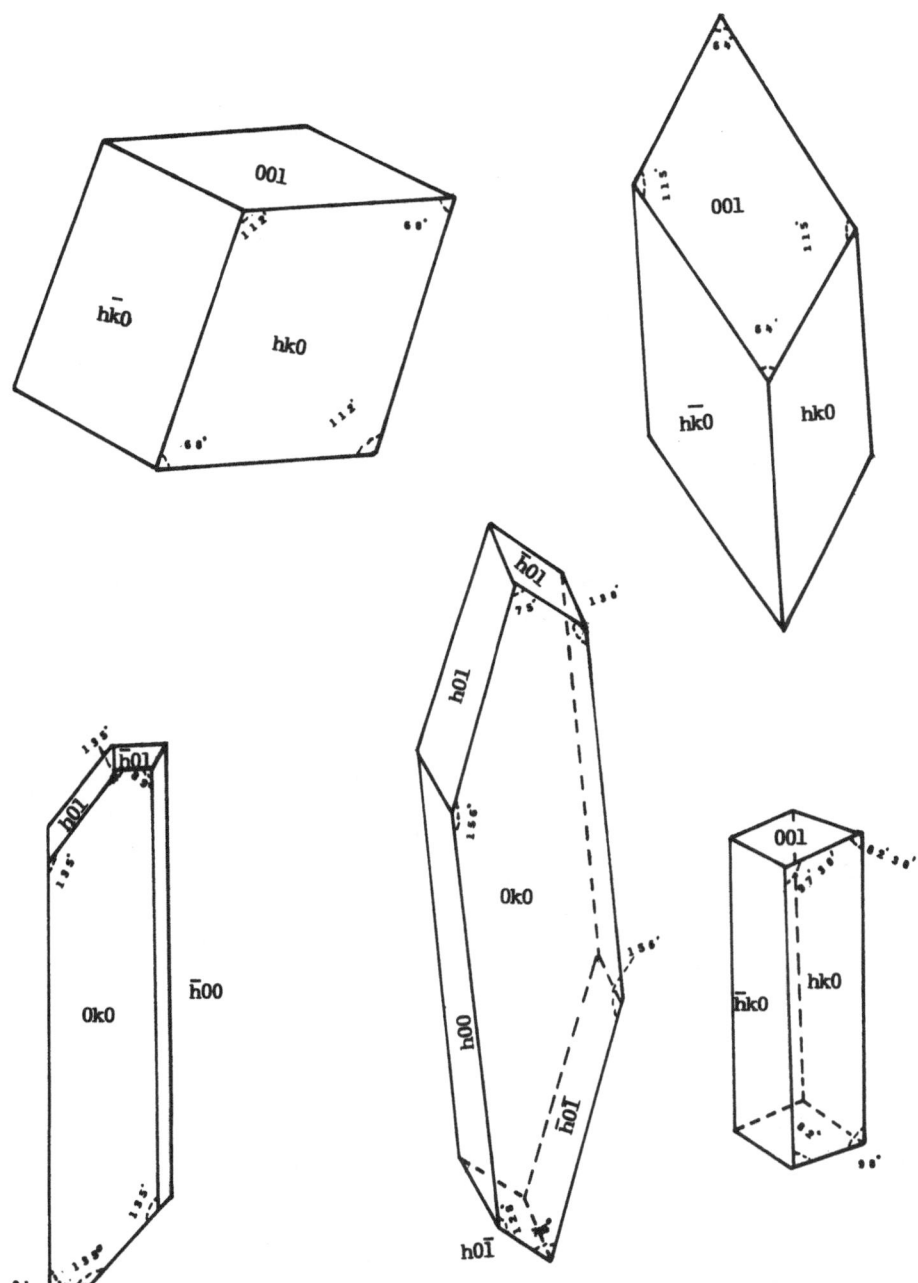

Fig.5. Schematic diagram showing morphological variation of
$Na_2CoZr(P_2O_7)_2$ with P_2O_5 concentration.

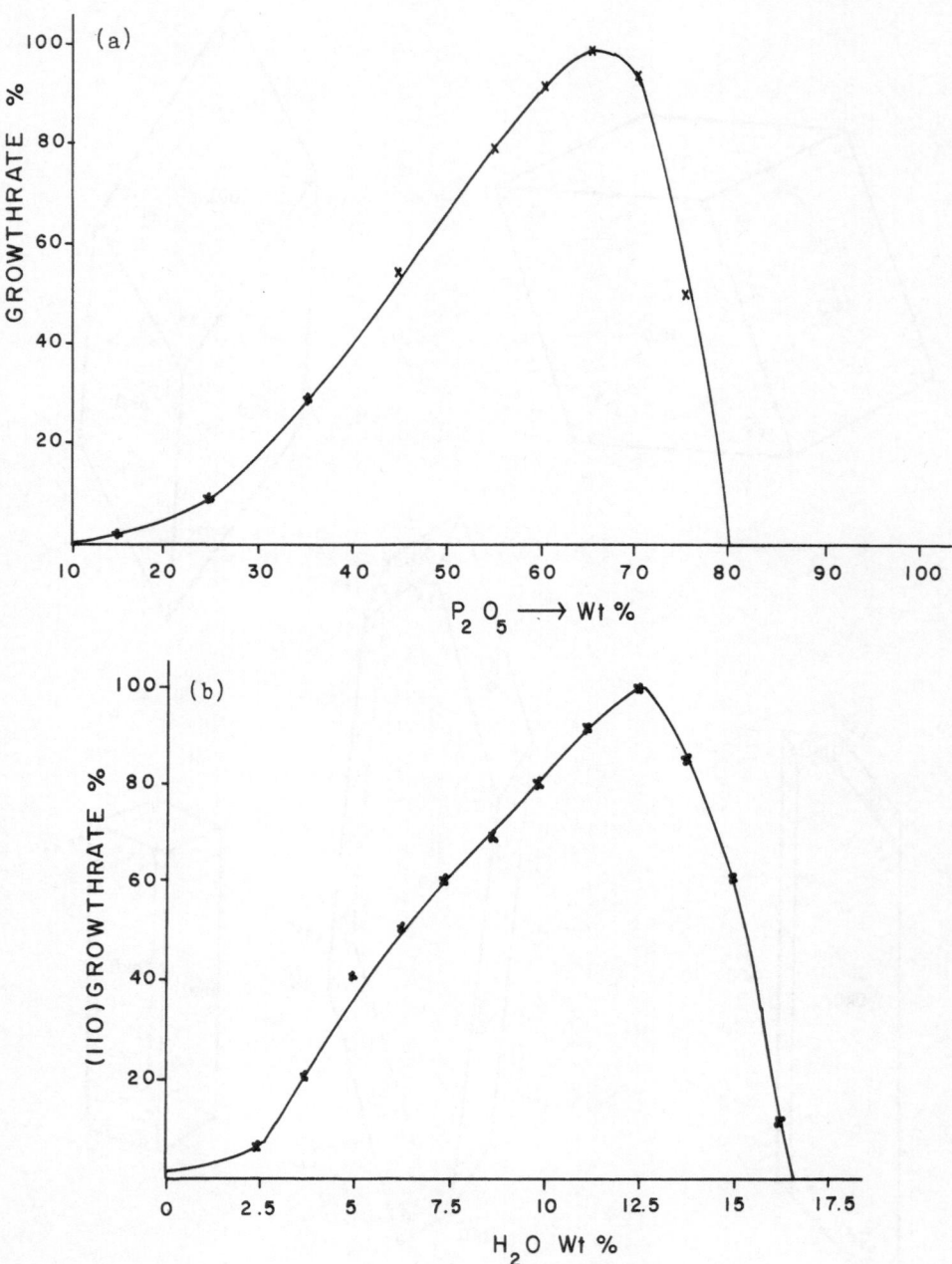

Fig.6. Dependence of the growth rate of $Na_2CoZr(P_2O_7)_2$ crystals against the concentration of P_2O_5 (a) and H_2O (b).

IV. CHARACTERIZATION

IV.a) Crystal Structure: The crystal structure of NASICON and some NASICON analogues are given earlier in section 1. The crystal structure of these compounds and bond types directly control the single crystal growth of these superionic phosphates and the possible cationic entry into the final composition and also the application or superionic behaviour. Hence, the study of crystal structure is an important aspect in the growth and characterization of superionics. With this aim a single crystal structural refinement work has been carried out for some of the representative compounds in each group:

$$Na_2MZR(P_2O_7)_2 \qquad M = Co,Ni$$
$$(Na_{0.66}Zr_{0.33})_2 P_2O_7$$

Although, the cations form usual octahedra and Na atoms lying in the cavities but framework linking differs in each type. The cell parameters for these compounds are given in table 1.

In the structure of $Na_2MZr(P_2O_7)_2$, two different dipyrophosphate anions connect Zr distorted octahedra signallizing the main bond chains of the framework. Each Zr atom situated conventionally in the origin sharing an edge with the second octahedron in the structure occupied by Ni or Co. Two independent Na^+ atoms are located in irregular polyhedra. The projection of the structure of $Na_2NiZr(P_2O_7)_2$ is given in fig.7. Close examination shows that two independent Na^+ atoms are located in irregular polyhedra and most apparent diffusion path of Na to be along (001) direction. Average distances of both $Na(1) - O = 2.628$ Å, and $Na(2) - O = 2.564$ Å are comparable to the values reported in other Na ionic conductors.

In the structure of $(Na_{0.66}Zr_{0.33})_2P_2O_7$, Na^+ and Zr atoms are located in the same general position surrounded by a distorted octahedra in each diphosphate Oxygen atoms. The PO_4 tetrahedra in each diphosphate anion are related by a binary axis possessing through the oxygen atom linking both P - atoms. Diphosphate anions are arranged in a sort of trigonal packing slices parallel to (001) with thickness

TABLE 1. Cell parameters of some superionic pyrophosphates

Compound	Crystal system	Space group	Axial lengths, Å			Axial angles, °			V, gcm^{-3}
			a	b	c	α	β	γ	
$Na_2NiZr(P_2O_7)_2$	Triclinic	P1	6.461(3)	7.257(4)	6.501(3)	123.24	91.95	93.79	253.5(1)
$Na_2CoZr(P_2O_7)_2$	Triclinic	P1	6.535(3)	7.266(4)	6.496(3)	122.96	92.28	93.75	257.2(1)
$(Na_{.66}Zr_{.33})_2P_2O_7$	Orthorhombic	-	6.867(5)	12.345(4)	27.527(5)	-	-	-	2333(2)

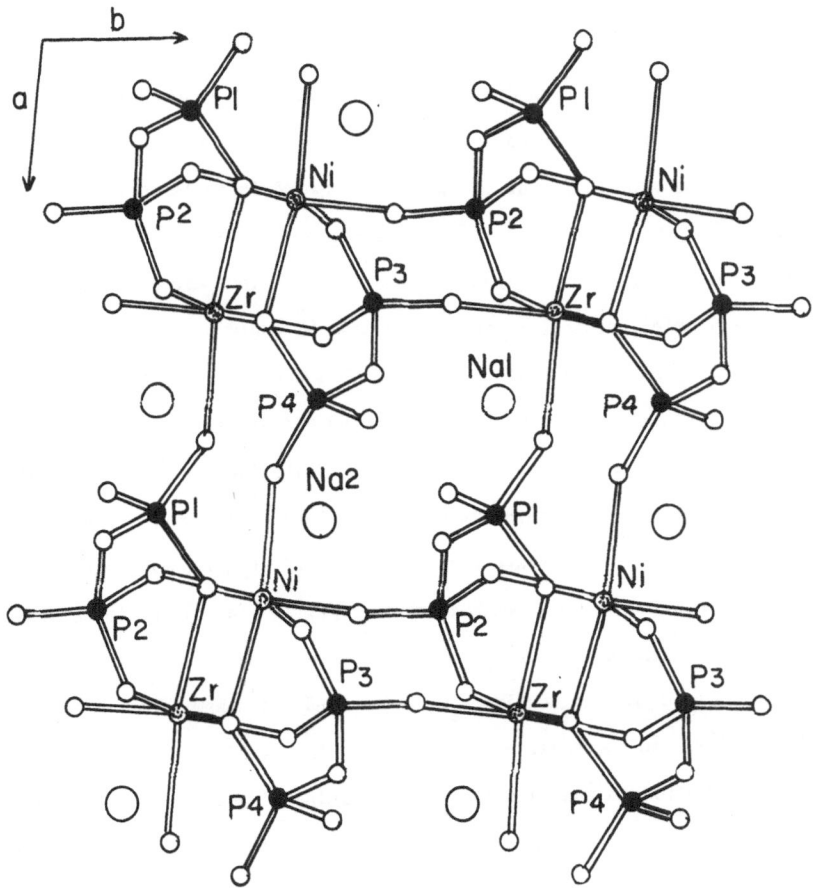

Fig.7. Projection of the structure of $Na_2NiZr(P_2O_7)_2$.

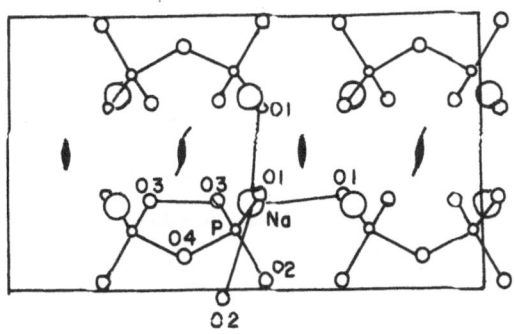

Fig.8. Projection of the structure of $(Na_{.66}Zr_{.33})_2P_2O_7$

C/8 . In these slices, (Na,Zr) octahedra from strips parallel to (110) by sharing edges which in turn connect different anions.

(Na + Zr) - O bonding distances are intermediate between typical Na - O distances in similar ionic conductors (2.6 Å) and those encounter in ZrO_6 octahedra (2.15 Å). The composition of the present compound is to be compared with the so called NASICON (sodium, zirconium, scandium phosphoro-silicates) ratio Na/(Zr,Sc) is 3/2, but in the present structure, Na^+ and Zr atoms are occupying the same crystallographic position. It is thought that Si could be introduced in this struc- ture partially substituting for P, making this compound very similar in composition to non-stoichiometric NASICON described by Rudolf et al.[32] Indeed $(Na_{0.66}Zr_{0.33})_2 P_2O_7$ could be orthorhombic, structure type synthesized by the above authors as a precursor of the non-stoichiometric. Figure 8 shows a projection of the structure down the a-axis. The structure consists of a kind of trigonal arrangement of anions in slices parallel to (001) with thickness C/8. In each slice, the anions are linked together by chains of $(Na,Zr)O_6$ octahedra parallel to either (110) or (-110). Each octahedron is connected to five P_2O_7 anions and 3 of them in the slice through atoms O(1) and O(2). In consecutive slices related by Z fold axes, octahedra chains in directions (110) and (-110) share edges defining a distorted tetrahedron formed by O(1) atoms around the empty 8a Wyckoff position with symmetry 222 (fig.9). The blocks founded by 2 of such slices are connected through O(2) atoms.

Thus the above structural study shows clearly, even the condensed phosphates can go as superionic conductors and hence it has opened a new chapter in solid state science in the search for new superionic conductors even among condensed phosphates and silicates.

IV. b) Differential Thermal Analysis: Differential thermal analysis was carried out for representative samples using Stanton Redcroft apparatus with a temperature programmer. The internal standard used was pure Al_2O_3 powder. The DTA curves of $Na_2(La,Al)ZrP_3O_{12}$ and $Na_2M^{+2}Zr(P_2O_7)_2$ where M = Co,Ni, crystals exhibit phase

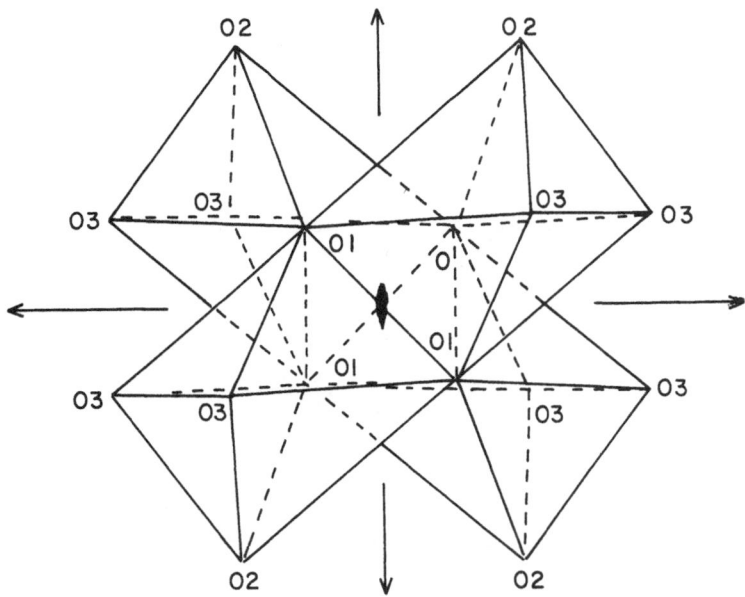

Fig.9. Octahedral chains in directions   110   and   -100
share edges defining a distorted tetrahedron formed by
O(1) atoms around the empty 8a Wyckoff position with
symmetry 222.

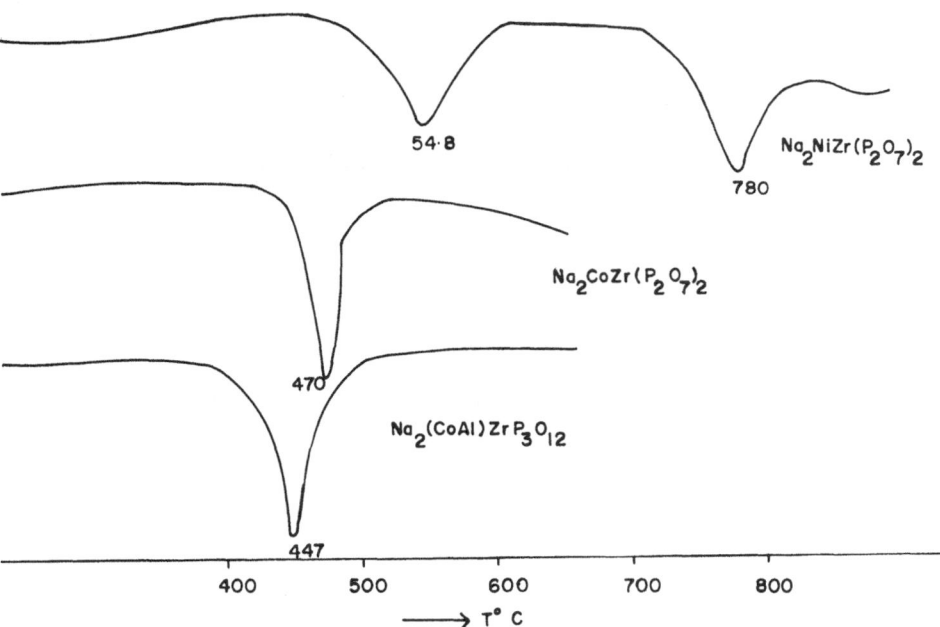

Fig.10. DTA curves for some representative superionics.

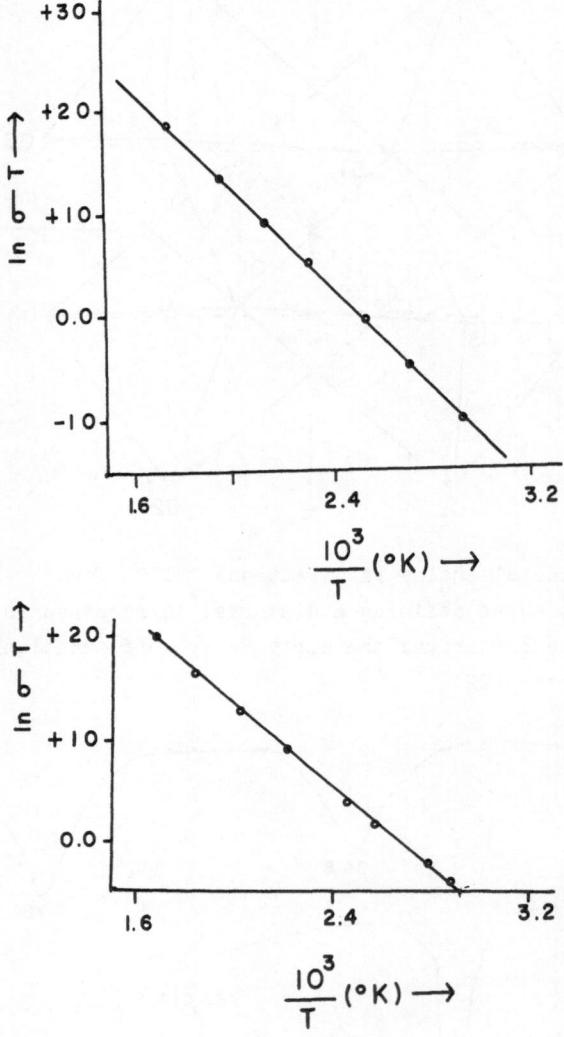

Fig.11. Arrhenius plots of $\ln \sigma T$ Vs. $10^3/T$.

transitions at various temperatures (fig.10).

$Na_2(La,Al)ZrP_3O_{12}$ and $Na_2CoZr(P_2O_7)_2$ have only one endothermic phase transition at $447^{\circ}C$ respectively. Whereas Ni bearing crystals show two endothermic phase transitions at 548° and 780° C. Hence, these materials are very good prospective superionic materials.

IV. c) Impedance Spectroscopy: The impedance measurements were carried out according to the technique adopted by earlier workers.[33] The crystals of pyrophosphates by applying 5 tn^{-2} pressures. The pellets (diameter 10 mm, thickness 3 mm, compactness 85 - 90 %) were provided with blocking silver electrodes on either side. The pellets were sintered in a vacuum chamber (10^{-2} torr) at 300° C. An impedance analyser (HP 4192A, USA) was used. The pellets were placed inside a cylinder which was evacuated (10^{-2} torr) and kept inside a tubular furnace (with a three terminal arrangement).

The ionic conductivity measurements were carried out from room temperature up to 300° C within a wide range of frequencies (dc to MHz). The Arrhenius plots of $\ln \sigma$ T against $10^3/T$ are shown in fig.11. The activation energy values were calculated and are given in the corresponding Arrhenius plots. The ionic conductivity values in general vary from 10^{-2} to 10^{-3} $(Ohm,Cm)^{-1}$ at about 300° C.

The values of ionic conductivity and the congenial structure of these pyrophosphates make them prospective superionic materials.

V. CONCLUSIONS

A high ionic conductivity in these new pyrophosphates and the congenial structures have led to the search for new soidum superionics not only among ortho-group of phosphates, but also among other condensed phosphates and even silicates, which together form a greater prospective to the field of superionics.

454

REFERENCES

1. S.Chandra, Superionic Solids, Elsevier Science Publishers, p.17 (1981).

2. M.Faraday, Experimental Researches in Electricity, Taylor and Francis, London, p.120 (1839).

3. E.Warburg,"Sodium ions migration through glass," Wiedemann Ann. Phys., **21** 622-24 (1884).

4. W.Nernst, "Ionic Conductivity in Mixed Oxides," Z.Electrochem., **6** 41-44 (1899).

5. C.Tubandt and E.Lawrence, "Silver Conductivity in Alpha-AgI," Z.Physik Chem., **87** 513-43 (1914).

6. J.B.Goodenough, H.Y.-P.Hong and J.A.Kafalas, "Fast Na^+ Ion Transport in Skeleton Structures," Mat. Res. Bull., **11** 203-20 (1976).

7. K.Byrappa and G.S.Gopalakrishna, "A Critical Survey on the Study of Alkaline Rare Earth Phosphate and NASICON Systems with a Special Reference to the Hydrothermal Method," Prog. Crystal Growth Charact., **7** 89-105 (1985).

8. H.Y.-P.Hong, "Structure of $Na_3Zr_2PSi_2O_{12}$," Mat. Res. Bull., **11** 173-76 (1976).

9. V.A.Efremov and V.B.Kalinin,"Determination of Crystal Structure of $Na_3Sc_2(PO_4)_3$, Kristallografia **23** 703-708 (1978).

10. J.P.Boilot, G.Collin and Ph.Colomban, "NASICON and Related Compounds: A Review," Ed. T.W.Wheat, Progress in Solid Electrolytes, Energy, Mines, Resources, Ottawa, 91-122 (1983).

11. J.B.Goodenough, "Fast Ionic Conductors," International School on Advanced Materials,of Erice, Italy (1980).

12. K.Byrappa, Ph.D. Thesis, Moscow State University (1981).

13. L.O.Atomyan, O.S.Filipenko, V.I.Panomarev, L.S.Leonova and E.A.Ukshe, "Crystal Structure and Ionic Conductivity of Solid Electrolytes $M_5RESi_4O_{12}$ where M = Na, Ag, RE = Sm $\rightarrow$ Lu," Solid State Ionics **14** 137-42 (1984).

14. A.Feltz and S.Barth, "Preparation and Conductivity Behaviour of $Na_3M''Zr(PO_4)_3$ M'' = Mn, Mg, Zn," Solid State Ionics **9 & 10** 817-22 (1983).

15. K.Byrappa, G.S.Gopalakrishna, A.B.Kulkarni and V.Venkatachalapathy, "Synthesis and Characterization of $Na_2(R,Co)Zr(PO_4)_3$ Crystals," J. Less Common Metals **110** 441-44 (1985).

16. M. de la Rochere, F.d'Yuoire, G.Collin, R.Come's and J.P.Boilot, "NASICON type Materials - $Na_3M_2(PO_4)_3$ (M = Sc,Cr,Fe): Na^+ - Na^+ Correlations and Phase Transitions," Solid State Ionics **9 & 10** 825-28 (1983).

17. F.d'Yvoire, M.Pintard-Screpel, E.Bretcy and M.de la Rochere, "Phase Transition and Ionic Conduction in 3d Skeleton Phosphates $A_3M_2(PO_4)_3$: A = Li,Na,Ag,K; M = Cr,Fe," Solid State Ionics **9 & 10** 851-58 (1983).

18. K.Byrappa, G.S.Gopalakrishna and S.Gali, "Synthesis and Characterization of a New Superionic Polyphosphate," Indian J. Phys., **63A** 321-25 (1989).

19. S.Gali, K.Byrappa and G.S.Gopalakrishna, "Structure of $Na_2Mzr(P_2O_7)_2$ (M = Ni, Co)," Acta Cryst. (in press).

20. S.Gali and K.Byrappa, "Structure of $(Na_{0.66}Zr_{0.33})_2 P_2O_7$," Acta Cryst. (in press).

21. A.Clearfield, S.Subramanian, W.Wong and P.Serus, "The Use of Hydrothermal Procedures to Synthesize NASICON and Some Comments on the Stoichiometry of NASICON Phases," Solid State Ionics **9 & 10** 895-902 (1983).

22. F.Genet and M.Barj, "Hydrothermal Synthesis and Recrystallization of Compounds belonging to the NASICON Family: Synthesis and Crystallization of Sodium Zirconium Silicon Oxide - $Na_4Zr_2Si_3O_{12}$," Solid State Ionics **9 & 10** 891-93 (1983).

23. M.Carron, M.Morse and K.Murata, "Relation of Ionic Radius to Structure of Rare Earth Phosphates, Arsenates and Vanadates," Amer. Min., **43** 985-86 (1958).

24. J.W.Anthony, "Hydrothermal Synthesis of Monazite," Amer. Min., **42** 904-6 (1957).

25. H.M.Kurbamov, V.Yu.Kara-Ushnov and B.C.Halikov, "Orthophosphates of Rare Earth Elements," Danish Publishers, Dushanbe, p.128 (1981) (in Russian)

26. V.M.Jahn and Kordes, "Hydrothermal Synthesis of Large Aluminium Phosphate Crystals," Chem. Earth **70** 75-78 (1953).

27. J.M.Stanley, "Hydrothermal Growth of $AlPO_4$," Ind. Eng. Chem., **46** 1684-89 (1954).

28. M.Yoshimura, K.Fuji and S.Somiya, "Phase Equilibria in the System $Nd_2O_3 - P_2O_5$ - H_2O at 500° C under 100 MPa and Synthesis of NdP_5O_{14} Crystals," Mater. Res. Bull., **16** 327-31 (1981).

29. L.Yonghua, M.Ninghai, Z.Qinglian, S.E.Endong, W.Mingyi, L.Shuzhen and W.Shixue, "Crystal Structure of Erbium Pentaphosphate," Kexue Tongbao **27** 1126-30 (1982) (in Chinese)

30. K.Byrappa and B.N.Litvin, "Hydrothermal Synthesis of Mixed Phosphates of Neodymium and Alkaline Metals $(Me_2O \cdot Nd_2O_3 \cdot 4 P_2O_5)$," J. Mater. Sci., 18 703-08 (1983).

31. K.Byrappa, "A Possible Reasons for the Absence of Condensed Phosphates in Nature. Phys. Chem. Minerals 10 94-95 (1983).

32. P.R.Rudolf, M.A.Subramanian, and A.Clearfield, "The Crystal Structure of a non-stoichiometric NASICON," Mater. Res. Bull., 20 643-51 (1985).

33. J.M.Hodge, M.D.Ingram and A.R.West, J. Electroanal. Chem., 74 125-29 (1976).

457

INDEX OF CONTRIBUTORS